PHYSICAL METALLURGY

Physical Metallurgy

SECOND EDITION

PETER HAASEN

TRANSLATED BY JANET MORDIKE

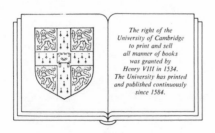

*The right of the
University of Cambridge
to print and sell
all manner of books
was granted by
Henry VIII in 1534.
The University has printed
and published continuously
since 1584.*

CAMBRIDGE UNIVERSITY PRESS

Cambridge

London   New York   New Rochelle

Melbourne   Sydney

PHYSICS

73006749

Published by the Press Syndicate of the University of Cambridge
The Pitt Building, Trumpington Street, Cambridge CB2 1EP
32 East 57th Street, New York, NY 10022, USA
10 Stamford Road, Oakleigh, Melbourne 3166, Australia

Originally published in German as *Physikalische Metallkunde*
by Springer-Verlag, 1984, and © Springer-Verlag, Berlin Heidelberg 1974

First published in English by Cambridge University Press 1978 as
Physical Metallurgy
Second edition 1986
English translation © Cambridge University Press 1978, 1986

Printed in Great Britain at the University Press, Cambridge

British Library Cataloguing in publication data
Haasen, P.
  Physical metallurgy. – 2nd ed.
  1. Physical metallurgy   2. Metallography
  I. Title   II. Physikalische Metallkunde.
  English
  003.695′1   TN690

Library of Congress Cataloguing-in-publication data
Haasen, P. (Peter)
  Physical metallurgy.

  Translation of: Physikalische Metallkunde.
  Includes bibliographical references and index.
  1. Physical metallurgy.   I. Title.
TN690.H123 1986   669′.9   86-6059

ISBN 0 521 32489 0 hard covers
ISBN 0 521 31037 X paperback

(First edition ISBN 0 521 21548 X hard covers
            ISBN 0 521 29183 6 paperback)

CONTENTS

PREFACE TO
THE ENGLISH EDITION

A number of English and American colleagues have expressed a desire for an English translation of this book, which they feel would be useful in final year B.Sc. Honours Metallurgy or master's degree courses. I am very pleased that two of my friends and former co-workers, Janet and Barry Mordike, have undertaken this work since they are familiar with the Göttingen as well as the British lecture courses. I would like to thank them for their careful and sensitive translation.

The English edition has provided the opportunity to correct a number of errors in the German version which have been pointed out to me by students and colleagues, especially Professor A. W. Sleeswyk, Drs H. H. Homann and V. Schlett. SI units are now used wherever it seemed advisable. Drs L. Schultz and R. Wagner have helped in proof-reading.

Caen, July 1976   Peter Haasen

Preface to the Second English Edition

The second edition of this book follows that of the German original as it was revised and enlarged in 1984. I am pleased that many English and American colleagues use the text in their courses and give me the benefit of their comments. Drs H. G. Brion and F. Wöhler helped in proof-reading; Dr F. Faupel corrected an error.

Göttingen, Summer 1985   Peter Haasen

PREFACE

Since 1959 the Institute of Metal Physics, Göttingen University, has offered a third year course in Physical Metallurgy for undergraduates who have successfully completed the basic Physics course. The same need is felt at many German universities because of the still relatively good career prospects for physical metallurgists in research, development and industrial production of metallic materials. Metals technology itself is extraordinarily diverse in character and inviting as a field of research. In addition, other industries involved in the production and processing of solid materials (ceramics, semiconductors, plastics) are making increasing use of metallurgical techniques. Formerly metallurgists generally began as chemists or mechanical engineers. Today a degree in physics can be an ideal starting qualification for a career in this field. The basic physics course includes not only a rigorous grounding in mathematics but also lectures and practical classes in experimental physics and a course in theoretical physics including an introduction to quantum theory. In the third year, introductory lectures on solid state physics lead into physical metallurgy. In Germany these lectures are often based on the textbook by C. Kittel [1.1].

The present volume continues to build on the material contained in textbooks of solid state physics such as Kittel. It was therefore felt unnecessary to include the fundamental concepts of crystallography and the electron theory of metals. In taking the basic principles of solid state physics for granted, this book differs from other textbooks of physical metallurgy which endeavour to cover this background material but at a more elementary level. The latter are intended primarily for students following the straightforward metallurgy courses offered by certain German universities from the first year onwards. Cottrell's book [1.2], for example, which was used originally in Göttingen, contains an introduction to quantum mechanics, the periodic system, theory of bonding in crystals, crystal structure, electron theory of metals, thermodynamics and statistical mechanics, etc., in the first seven of its fifteen chapters. The present textbook, however, assumes that the reader possesses a knowledge of these subjects at the level presented in Kittel [1.1] and goes on to apply them to metallurgical problems. In addition to the physics students mentioned earlier, who are just at the threshold of their metallurgical career, this book is also intended for metallurgists who have learned solid state physics in the course of their on-going education and now seek to deepen their knowledge of the physical foundation of their own subject. I feel that up to now no suitable book has been available, at least not in German (see [1.7]). The book by Paul Shewmon [1.3] which sprang from a series of guest lectures given in Göttingen in 1963–64 provides a good introduction to some branches of metallurgy. This book and subsequent discussions with Erhard Hornbogen during his period at the Göttingen Institute between 1965–69 provided considerable stimulus for the present volume.

The physics-oriented Göttingen metallurgy lectures are accompanied by a practical course. Before embarking on this, the students are required to study a specially written script, which is elucidated further in a series of tutorials. An introduction to the most important experimental methods for the investigation of metals also exists in book form [2.3]. Although these texts give a good description of the actual experimental procedure it was felt necessary in the present volume to include the principles of certain techniques used in metallurgy. Those given in chapter 2 are either relatively unfamiliar to physicists and solid state physicists or they are referred to frequently in subsequent chapters. There is no need for the reader to study the whole of chapter 2 before going on to later chapters. Instead he will find it expedient to refer back when he needs more detailed information about a particular experimental method.

Further information can be found in the literature. I have taken pains to cite recent publications which give a good review of the subject and treat the material in a critical fashion in the hope that the reader will thus find his way more easily to the copious and often bewildering original literature. I have therefore been unable to do justice to the contributions of all the authors involved, individually and in chronological order. I hope that I can be forgiven for this, bearing in mind that the present volume is intended as a textbook rather than a collection of review articles. Indeed its aim is to introduce the reader to the fundamental aspects of metallurgy by means of quantitatively formulated physical arguments, and, with the aid of appropriate literature references, to give him an up-to-date physical interpretation of metallurgical phenomena. A more detailed description of the field is given in the new edition of the book *Physical Metallurgy*, edited by R. W. Cahn and P. Haasen, North Holland, Amsterdam, 1983.

In bringing this plan to fruition I have been assisted by several generations of Göttingen University students and in particular by a group of close co-workers. J. Dönch, H. Steinhardt, W. Schröter and R. Wagner have undertaken a critical reading of the text and offered numerous suggestions for improvement. In addition, friends and colleagues have read and constructively criticized the chapters dealing with their specialist fields of research. I would especially like to thank U. Gonser, G. Kostorz, B. L. Mordike and H. Wever for suggestions on the second edition. The colleagues at Stanford University provided a stimulating atmosphere in 1984 by awarding me the W. Schottky Visiting Professorship. The second edition tries to take into account the progress of our field and to correct errors that have become apparent during use and the translation into other languages. I have decided against incorporating a chapter on 'gases in metals' as its outline tended to become a book of its own, and there are such books already, e.g. [1.4], [1.5], [1.6].

Stanford, California, Summer 1984   Peter Haasen

MEANING OF FREQUENTLY
USED SYMBOLS

A amplitude of wave, of dislocation stress at unit distance

a lattice parameter, shear strain

$\dot{a}$ strain rate

a_i activity of component i

B dislocation friction coefficient

$\mathbf{b}$ Burgers vector

c height of tetragonal cell (c/a axial ratio), velocity of light

c_i volume concentration of component i

c_j number of jogs per unit length

c_V fraction of vacant lattice sites

c_I fraction of interstitial atoms

c_D fraction of double vacancies

c_t transverse sound velocity

c_v specific heat at constant volume

D diffusion coefficient

$\tilde{D}$ concentration dependent, inter-diffusion coefficient

D_L lattice diffusion coefficient

D_G grain boundary diffusion coefficient

D_S surface diffusion coefficient

d (atomic) distance, grain size

$\mathbf{ds}$ line element

E internal energy

E_F Fermi energy

$\bar{E}$ energy per unit area

E_L line energy of dislocation

E Young's modulus

$\mathscr{E}$ electric field strength

EN   electronegativity

e strain energy density

e/a number of electrons per atom

F free energy

f scattering amplitude of an atom, free energy per unit volume, number of degrees of freedom

$\not{f}$ correlation factor

$\mathbf{g}$ reciprocal lattice vector

H enthalpy, magnetic field strength

h distance, Miller index

$\hbar$ Planck's constant divided by 2π

I intensity, current

Im   imaginary part of a complex quantity

J flux

j flux density

K force, reaction rate constant

$\mathbf{k}$ wave vector

k Boltzmann constant, Miller index

k_0 distribution coefficient

k_y Petch parameter

k_z oxidation rate parameter

L latent heat, domain size, dislocation mean free path

l length, Miller index

M mass, Taylor factor, period of superlattice

M_{ij} Onsager mobility parameter

m atomic mass, slope of liquidus

m' strain rate sensitivity

m^* effective mass of electron

m_s Schmid factor

N number of atoms, dislocation density group number in the periodic system, number of cycles

$\mathbf{n}$ normal to a plane

n number of nearest neighbours, number of components

n_i number of atoms of component i

P^{AB} probability for AB pair

$\mathbf{p}$ phonon wave vector

p pressure

Q activation energy of grain boundary motion

Q^* heat of transport

G Gibbs free energy, shear modulus
Re   real part of a complex quantity
r position, distance, number of phases
r_0 atomic radius
S entropy, lamellar spacing
s distance (from **g**) in reciprocal space
     degree of long-range order
T absolute temperature
T_m melting temperature
t time, thickness
U voltage, activation energy of dislocation motion
u displacement, vibrational energy
V volume, (electrostatic) potential
v velocity
W work
w width of dissociation of a dislocation
X volume fraction
x coordinate
y coordinate
Z partition function, ionic charge
z coordinate
α angle
α_m degree of short range order in mth shell
β angle, spinodal wave number
Γ jump frequency
γ stacking fault energy
γ_i activity coefficient of component i
Δ angle
δ size misfit, grain boundary thickness

q cross-section
R rate
ε strain, pair interchange energy
η (shear) modulus defect, grain boundary (diffusion) factor
ζ width of dislocation core
Θ angle, Debye temperature, work hardening coefficient
κ wave vector difference
κ coefficient of gradient energy
Λ Slip line length parameter
λ wave length, angle
μ linear absorption coefficient
μ_i chemical potential of component i
ν frequency, Poisson's ratio
ν_i molar fraction of component i
ν_0 attack frequency
ν_V vacancy jump frequency
ξ extinction distance
ρ density of dislocation line elements
ρ_{el} electrical resistivity
ρ_m mass density
σ stress, electrical conductivity
τ relaxation time, shear stress
Φ complex scattering amplitude
φ angle, statistical scattering amplitude
χ angle
ψ wave function
Ω atomic volume
ω angular frequency, number of states of energy E

CONVERSION OF UNITS

1 joule = 1 newton metre = 1 watt second = 0.24 calories = 10^7 ergs

1 eV/atom = 23 kilocalories/mole = 96 kilojoules/mole = R (11 600 kelvin), R ≈ 2 calories/mole kelvin

1 kgf/mm² ≈ 10 meganewtons/metre² = 10^8 dynes/centimetre² ≈ 7×10^2 psi = 10^7 pascal

1
Introduction

Since metals constitute a class of solids it might well be asked what aspects of Physical Metallurgy do not already fall within the scope of solid state physics. If Kittel's book is taken as a guide there is at least one basic metallurgical concept which is unfamiliar to solid state physicists. Metallurgy relates the properties of metals and metallic 'mixtures' or *alloys* to their *microstructure*. Whereas solid state physics is based on the *crystal structure* of a single crystal in which all the atoms occupy sites in a three-dimensional lattice, metallurgy takes into account that the perfect regularity of the arrangement is often restricted to microscopic regions and differs from that in neighbouring regions. In other words, superimposed on a macroscopic piece of metal there is another pattern known as the *microstructure*, much coarser than the crystal structure which forms the foundation of solid state physics.

It happens that many of the properties of metals, especially those that are technologically important, are determined by the microstructure. The most important property from the point of view of mechanical engineering is the strength. This is strongly influenced by the microstructure and is thus one, but by no means the only, *microstructure sensitive property*. In order to define 'strength' and give it some physical meaning (chapters 12 and 14) we must first (chapters 3, 4, 11, 13, 15, etc.) examine the microstructure of metals and describe it quantitatively.

A second difference between solid state physics and metallurgy is that the former is concerned with simple, pure materials and the latter with alloys. The physicist is inclined to consider the dependence of a property on the composition of the material as a chemical phenomenon. It is, however, by no means true that a given composition defines the properties of an alloy. For example, the 'hardness' HV (see section 2.6.3) of a steel consisting of iron with $\frac{1}{3}\%$ carbon can have values of between 10^9 and $7 \times 10^9 \, \text{N/m}^2$ depending on how the material has been heat-treated. This treatment changes the distribution of the carbon in the iron, hence the microstructure and hence the mechanical properties. The distribution which arises is determined by the thermodynamics and kinetics of the alloy system and thus by a few parameters describing the energy, entropy and geometry. These will be discussed in chapters 5, 8 and 10 before examining typical atomic configurations in alloys (e.g. in chapters 7 and 9). The thermodynamics and kinetics of atomic distributions really fall within the domain of physical chemistry, and it was from physical chemistry that Gustav Tammann developed physical metallurgy in Göttingen in the period between 1903 and 1938. The physicist is inclined to enquire into the atomistic origin, in the case of metals the electron–theoretical explanation of the energy terms determining the thermodynamics and kinetics of metallic systems. Some theories along these lines are described in chapter 6 by considering the present state of the theory of multi-particle, multi-component systems, thermodynamics itself is much better able to yield information and be tested experimentally and is therefore discussed in relative detail in chapter 5.

This book is designed to guide the reader from the simpler to the more complex metallurgical phenomena and to give them some physical meaning, starting from the central concept of microstructure in chapter 3. A physicist using this book should be motivated to master what may first appear to be tiresome subjects such as chemical thermodynamics or dislocation theory. The logical conclusion to the book is a discussion of the principal object of metallurgy, namely the mechanical hardening of metals (and the reverse process on recrystallization, chapter 15). In view of our long standing interests in Göttingen, it would have been natural to add two further chapters parallel to mechanical hardening, namely hardening of ferromagnetic materials and hardening of superconductors, see [14.11]. These processes produce additional important properties in both these classes of material, which are for example exploited in permanent magnets and superconductors able to carry a high current in a magnetic field. These two chapters would, however, have required so many new concepts and so much basic knowledge that they exceeded the framework of a book designed for an introductory lecture course. I trust, however, that the growing general interest in new metallurgical developments in the field of ferromagnetic materials and superconductors will justify a discussion of this subject in a second volume in the foreseeable future.

2

Experimental methods for the physical examination of metals

Metallurgists employ a number of experimental methods not normally encountered by the solid state physicist. Several of these will be described and critically assessed in the following chapter because the data they can provide will be drawn upon in subsequent chapters. Naturally many experimental techniques are standard for both solid state physicists and metallurgists, for example X-ray methods for the determination of crystal structure, lattice parameters and crystallographic orientation [2.1], [2.2], based on the Bragg relationship for constructive interference of X-rays scattered by the atoms in the lattice. Similarly, both physicists and metallurgists make use of measurements of electrical conductivity, Hall voltage, macroscopic density and its variation with temperature (thermal expansion), although the metallurgist usually measures the latter unidimensionally in a dilatometer. Nor do measurements of elastic moduli, specific heat or magnetic susceptibility introduce the physicist to any new methods. Results of these and related investigations will be referred to without any detailed description of the procedure. Relevant techniques are summarized in [2.3], [2.4], [2.5].

2.1   Microscopy of surfaces

Apart from their crystal structure, metals possess a *microstructure*, in that they consist of differently oriented grains or differently constituted phases. The observation and if possible quantitative description of this microstructure is the aim of *metallography*, which employs optical, electron, ion and X-ray microscopical methods some of which are described in sections 2.2 and 2.4.

Often the geometrical dimensions of the microstructure can be resolved in the light microscope. Because of their opacity, metals must be observed in reflection in the vertical microscope. For this purpose the surface is made optically plane by grinding, followed by mechanical, chemical or electrolytic *polishing*. As a rule the polished metal surface appears smooth

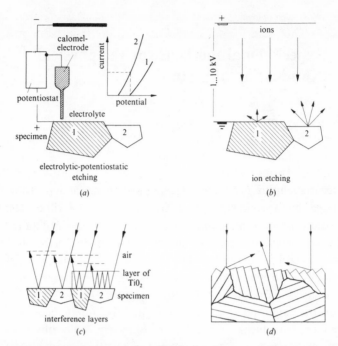

Fig. 2.1. Methods of revealing microstructural constituents ((*a*) to (*c*)) after E. Hornbogen and G. Petzow, *Z. Metallkunde*, **61** (1970), 81).

and uniformly coloured under the microscope. In order to distinguish the individual microstructural components, different techniques are used to produce or accentuate contrast. One way of producing specific changes in the specimen surface is by *etching*. In addition to empirical chemical techniques, reproducible physical methods (several of which are illustrated in fig. 2.1) are becoming increasingly important. Whereas in figs. 2.1(*a*) and (*b*) the structural components are revealed by the removal of different amounts of material, fig. 2.1(*c*) illustrates the exploitation of the different reflectivities of light at the interface to a vapour deposited layer. Chemical etching reagents often expose particular crystallographic planes as a series of steps, as shown in fig. 2.1(*d*). Differently oriented grains reflect the light in different directions and thus appear of different brightness in the vertical microscope. If the reflecting lattice plane is known, the orientation of the grains (or crystallites) in the specimen can be determined. 'Orientation' signifies the position of the crystal lattice in relation to the external features of the specimen (surface, longitudinal axis).

On the other hand, the contrast of a microstructure can be enhanced by the use of a suitable optical arrangement, e.g. by observation in polarized light, using phase contrast, interference contrast or in 'dark field'. In the latter, the contours of a surface relief are revealed in glancing incidence (more strongly than in vertical incidence in 'bright field'). The surface relief can be measured quantitatively using an *interference microscope* which has a vertical resolution of a fraction (about $\frac{1}{20}$) of the wavelength of light used. (In the case of multiple beam interference, structures with depth variations of 3 nm can be resolved.) The resolution in the plane of observation is, however, that of the optical microscope, about 300 nm.

A better resolution of the microstructure observed directly in reflection can be obtained using electron waves. (Observations in transmission will be discussed in section 2.2.) In *reflection electron microscopy* a glancing beam of electrons impinges on the surface. The electrons are scattered mainly in the direction symmetrical to the normal to the surface and can then be used to form a magnified image. The resolving power is several tens of nanometers but the surface appears very distorted because of the oblique angle of observation. The *emission electron microscope* avoids this disadvantage by forming the image with secondary electrons released directly from the surface at right angles. UV light is one possible primary exciting radiation unless thermal emission is used, in which case this electron microscope is particularly suitable for the high resolution observation of high temperature processes.

In contrast to the microscopes described above the *scanning electron microscope* does not use electron optics for magnification but scans the specimen surface with a very finely focussed electron beam (diameter $\geqslant$ 20 nm). The secondary electrons are amplified and control the brightness of a television tube, the scanning system of which is synchronized with that of the primary electron beam (fig. 2.2). Because of the asymmetrical ray path, the brightness, i.e. the intensity of the secondary electrons emitted from a particular point, depends on the local surface relief as well as on the material of the specimen. This together with the extraordinarily good depth of focus (e.g. 35 μm at a magnification of 1000) produces a three-dimensional image.

The primary electron beam can also excite the characteristic X-radiation of the target atoms. If the spectrum of this fluorescent radiation is analysed using a standard crystal, a localized chemical analysis of the specimen is obtained. In addition to the topographical image of the surface due to the secondary electrons (fig. 2.3) the scanning technique can be used to obtain other pictures using the various characteristic X-radiations which then show the distribution of particular chemical elements (fig. 2.4). This analysis can be undertaken quantitatively by X-ray emission in the electron beam

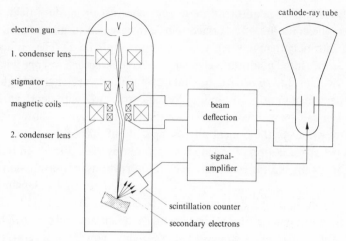

Fig. 2.2. Schematic representation of the scanning electron microscope, after [2.8].

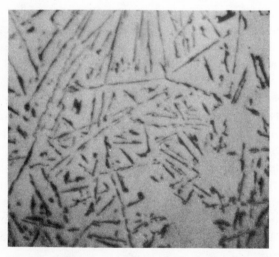

Fig. 2.3. Solidified Al–11.7 wt.% Si alloy (185 ×).

microprobe. Concentrations $\geqslant 10^{-4}$ in surface regions of about $1\,\mu m$ diameter (equivalent to 10^{-11} g of the element) can be detected.

Optical microscopy and general metallographic techniques are described in [2.3], [2.5], [2.6a], [2.6b], [2.6c], [2.6d] and electron microscopical techniques in [2.7], [2.8], [2.9]. Using stereometric methods and automatic instruments, it is now possible to give a quantitative description of the

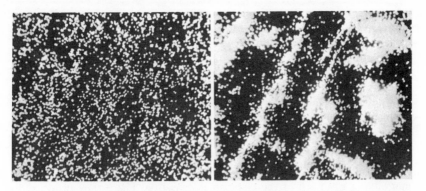

Fig. 2.4. Distribution of the components in the alloy in fig. 2.3; (*a*) Al
X-ray fluorescence, (*b*) Si X-ray fluorescence of the same area in the
microprobe. (G. Horn, Doduco, Pforzheim, W. Germany.)

microstructure [2.6*d*] (see section 3.3). Whereas the metallographic
methods described up to now have made use of radiation reflected from a
polished surface, the following technique requires thin films transparent to
electrons.

2.2   Transmission electron microscopy (TEM)

At the present time, TEM represents one of the most important
investigatory techniques in metallurgy. A thin film of metal (about 100 nm)
is produced from a macroscopic specimen and then viewed in transmission
in a normal electronmicroscope (using accelerating voltages ≤ 120 kV).
Electron microscopes using even higher accelerating voltages (1 MeV) are
being used to an increasing extent and these of course permit thicker foils to
be observed. A well established variation is the production of a *replica* of the
metal surface for examination at high resolution in the electron microscope.
In this technique the specimen is 'shadowed' by oblique deposition of a
heavy metal to accentuate the contrast and then coated, usually with
carbon, by deposition from the vapour phase. The film is peeled off the
surface and placed on the specimen grid. The thickness variations in the
shadowed film correspond to the surface relief of the specimen and the
replica yields a three-dimensional image of the surface topography
(extinction contrast, see section 2.2.1.1). Surface details greater than 2 nm
can be rendered visible in this manner.

In the case of so-called *extraction replicas* the aim is to remove inclusions
or small particles of a second hard phase with the replica, having previously
loosened them by suitably etching the primary material (matrix). The

particles together with the replica can then be observed in the transmission electron microscope with a view to determining their structure, size distribution, etc.

Much richer in contrast and more characteristic of the interior of the specimen is a thin film of the specimen itself. Thinning is carried out by a special technique, usually electrolytic polishing [2.7], [2.8]. At the tapered edges of a freshly formed perforation the metal film is thinner than 150 nm and thus transparent to electrons at normal accelerating voltages ($\leqslant 120$ kV). Depending upon the microstructural features in their path, a proportion of these electrons reach the display screen by way of the magnifying ray path of the electron microscope. There they produce an intensity profile ('contrast'). The interpretation of this contrast in terms of the microstructure which gave rise to it is a science in its own right, which must be described in somewhat more detail in the following. In addition a normal *electron diffraction pattern* according to Bragg's law can be obtained from the specimen from which its crystal structure, orientation, etc., can be derived.

2.2.1   Contrast theory
2.2.1.1   Microscopically homogeneous specimen

The simplest case of contrast in the electron microscope arises on an amorphous layer which might consist of regions of different density (A, B, fig. 2.5). The electrons passing through B are more strongly scattered than those passing through A. Most of the scattered electrons are caught by the objective aperture. The intensity from A on the fluorescent screen of the electron microscope is thus greater than that from B. In the case of crystalline layers the scattering of the electrons is strongly anisotropic in accordance with Bragg's Law ('diffraction'). The aperture angle of the electron microscope is so small ($\approx 10^{-2}$ radians at 100 kV) that in general none of the reflections (except those of the zeroth order) pass through the aperture (bright field case, fig. 2.6(a)). It is, however, possible to displace the aperture or tilt the primary beam I_0 so that only a particular reflection passes through the aperture (dark field case, fig. 2.6(b)). Only by using special imaging techniques can both the zero-order beam and the diffracted beam be made to pass together through the aperture and interfere to yield a true Abbé image (fig. 2.6(c)). Otherwise a 'contrast' is obtained with I or I_1 alone by virtue of the inhomogeneity of the specimen. This can be calculated making the following assumptions [2.31].

(a) Only elastic scattering according to Bragg's Law is taken into account.

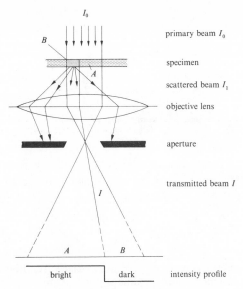

Fig. 2.5. Production of contrast in amorphous objects with regions A, B of different density, after [2.8].

(*b*) Only one transmitted and one diffracted beam are taken into account: $I_0 = I + I_1$ (two beam case).

(*c*) Interaction between I and I_1 is neglected. This is justifiable for $I \gg I_1$; in this way the Bragg reflection itself is excluded from the discussion (kinematical theory).

(*d*) The 'image' is composed of intensities from all regions of the specimen which is therefore considered to be divided into columns $\Delta x \times \Delta y \times t$ (t = layer thickness, see figs. 2.9 and 2.10). The intensities of all columns with longitudinal axes in the z-direction are calculated independently.

I_1 is calculated by summing the amplitudes in direction $\mathbf{k}$ of all secondary waves of the atoms (n) of one column (position vector $\mathbf{r}_n$)

$$A_1(\mathbf{k}) = \sum_n f_n \exp[-2\pi i((\mathbf{k} - \mathbf{k}_0)\mathbf{r}_n)] \tag{2-1}$$

where f_n is the scattering amplitude of the individual atom proportional to its atomic form factor, $2\pi \mathbf{k}_0$ is the wave vector of the incident wave. In accordance with assumption (*c*) we must keep a finite distance $\mathbf{s}$ (corresponding to an angular difference $\Delta\theta > 0$) from the Bragg reflection itself, i.e. $\mathbf{k} - \mathbf{k}_0 = \mathbf{g} + \mathbf{s}$, where $\mathbf{g}$ is the reciprocal lattice vector of the lattice plane (*hkl*). If all the atoms are alike, i.e. $f_n = f$, and replacing the sum by an

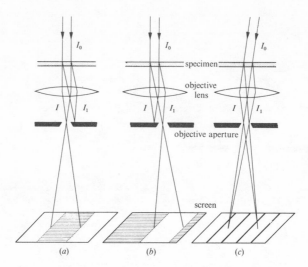

Fig. 2.6. Production of contrast in crystalline objects: (*a*) bright field contrast; (*b*) dark field contrast; (*c*) image produced by interference.

integral, then for lattice constant a we have

$$A_1(\mathbf{s}) = \left(\frac{f}{a}\right) \int_{-t/2}^{t/2} \exp[-2\pi i s_z \cdot z] \, dz = f \frac{\sin(\pi t s_z)}{\pi t s_z} \frac{t}{a}. \tag{2-2}$$

The scattered intensity $I_1 = A_1^2$ and the intensity of the zero-order beam complementary to it oscillate as a function of s_z and t. The s_z variation is realized in the case of a continuously bent layer; intensity bands are obtained as a function of position, so-called *bending contours*. If, however, the orientation, i.e. $\mathbf{s}$, is held constant and t is varied, e.g. in a tapered layer such as is present after electrolytic thinning, so-called *thickness contours* are obtained for I_1, fig. 2.7. The periodicity of the intensity due to the path length t defines a periodicity length $t_g = 1/s_z$. The kinematical theory breaks down for $s_z = \mathbf{s} \to 0$. According to the then valid *dynamical theory* t_g is finite for $s \to 0$ and must be replaced by

$$t_g^{\text{eff}} = \frac{1}{\sqrt{(s^2 + \xi_g^{-2})}} \tag{2-3}$$

in which the extinction distance ξ_g is invsersely proportional to the structure factor of the reflection $\mathbf{g}$ and hence a material parameter (for Al (111): $\xi_g = 55.6$ nm, (222) 137.7 nm; Au (111) 15.9 nm, (222) 30.7 nm). The layer thickness can be determined by counting the thickness contours from the edge.

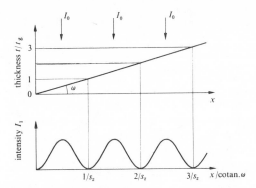

Fig. 2.7. Distribution of the intensity scattered by a wedge-shaped layer.

Recently the *high resolution electron microscopy* of very thin and well oriented crystals has become very important ($t \approx 10$ nm). It tries to 'image' the two-dimensional arrangement of atomic columns in the true sense of Abbé by interference of the diffracted beams (see fig. 2.6(c)). The method relies on a well-corrected electron microscope the point-to-point resolution of which is determined by its spherical aberration C_s as $r_{min} = B\lambda^{3/4}C_s^{1/4} \approx$ 0.3 nm (using the electron wave length λ corresponding to an acceleration voltage of 200 kV). The thin crystalline foil produces only a weak phase contrast from column to column which by defocussing the specimen is transformed into amplitude contrast on the screen. It is necessary to compare the images so obtained with images simulated on the computer in order to avoid artefacts especially with the atomic resolution of crystal defects [2.24], [2.25].

A graphical method in the complex number plane, the *amplitude phase diagram*, is often used to obtain the resultant amplitude A_1 at the exit of the column from the superposition of the secondary waves of the column atoms. Their amplitudes are equal but they are out of phase by an amount equal to the path difference ($\Delta = 2\pi s_z z$). If the path difference $z = 1/s_z$, the two waves are in phase (Bragg reflection), $\Delta = 2\pi$. For other z and Δ, the elementary waves can be added by means of a vector polygon as shown in fig. 2.8 for the wedge. At the first intensity zero a full circle has been completed with a diameter in space (in units of f) of $1/\pi s_z$.

2.2.1.2 Specimen with lattice defects

In this case instead of occupying their ideal lattice sites $\mathbf{r}_n$ the atoms

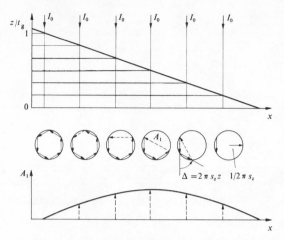

Fig. 2.8. Amplitude phase diagrams (APD) for diffraction at a wedge.

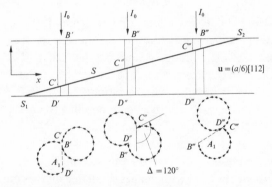

Fig. 2.9. Diagonal stacking fault S in a layer and APD for three columns: the resulting amplitudes $B'D'$ and $B'''D'''$ are finite and equal but $B''D''$ is zero.

in the column are displaced by $\mathbf{u}_n$. Then

$$A_1 \approx \frac{f}{a} \int_{-t/2}^{t/2} \exp[-2\pi i s_z \cdot z] \exp[-2\pi i \mathbf{g}\mathbf{u}] \, \mathrm{d}z. \tag{2-4}$$

Special cases [2.10]

(a) Let a *stacking fault S* in the fcc lattice (see section 6.2.1) cut the column as shown in fig. 2.9. The atoms below S are then displaced by $\mathbf{u} = (a/6)$ [112]. If this is observed for the (200) reflection with $\mathbf{g} = (2/a)$ [100] an additional phase angle Δ_+ is introduced where $\Delta_+/2\pi = \mathbf{g}\mathbf{u} = \frac{1}{3}$ and a phase discontinuity of 120° appears at point C in the amplitude phase diagram.

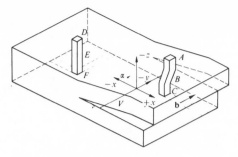

Fig. 2.10. Distortion of two columns ABC, DEF in the vicinity of a screw dislocation (V).

The diagram then continues on a second circle. If point C is shifted from S_1 to S_2 the length of the vector $A_1 = \overline{BD}$ varies periodically and the stacking fault appears as a system of light and dark bands. This is a special case of the thickness contours derived above which can also be used to reveal grain boundaries and twin boundaries.

(b) *Screw dislocation*, fig. 2.10 (see section 11.2.1).

$$\mathbf{u} = \mathbf{b}\,\frac{\alpha}{2\pi} = \frac{\mathbf{b}}{2\pi}\tan^{-1}\frac{z}{x}$$

and   (2-5)

$$\Delta_+ = (\mathbf{gb})\tan^{-1}\frac{z}{x}$$

where $(\mathbf{gb})$ is a whole number n or zero depending on the reflecting lattice plane. If $\mathbf{b}$ lies in this plane the dislocation does not give rise to any contrast: thus if $\mathbf{g}$ is known the direction of the Burgers vector $\mathbf{b}$ can be determined.

According to (2-5), the additional phase angle Δ_+ has a different sign on the left and on the right of the dislocation ($x \lessgtr 0$). On one side therefore the phase angle $\Delta = 2\pi s_z z$ resulting from the path difference is increased and the circle in the amplitude phase diagram tightened and on the other side it is diminished and the circle enlarged, as shown in fig. 2.11. Thus in bright field (in the light of I) the dislocation appears as a black line accompanying the dislocation on one side, see fig. 2.12. The width Δx of the contrast $|2\pi s_z\, \Delta x| \approx 2$ (for $n = 2$) depends on the tilt s of the reflecting lattice plane and hence on the angle of incidence of the beam on the specimen. It is necessary therefore to be able to tilt the specimen in the microscope.

For a 'mean' value of $s = 3 \cdot 10^{-2}\,\text{nm}^{-1}$ (according to dynamical theory) $\Delta x \approx 10\,\text{nm}$ is the typical contrast width of the dislocation.

A much narrower contrast is obtained in dark field using a reflection

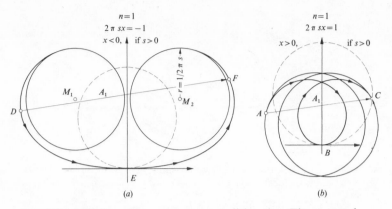

Fig. 2.11. APD for columns near a screw dislocation. The scattered amplitude $\overline{AC}$ for a column ABC at $x>0$ (b) is smaller than that $\overline{DF}$ of the column DEF at $x<0$ (a). The dotted circles apply to the perfect crystal. After [2.8].

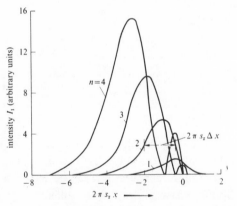

Fig. 2.12. Intensity losses due to scattering at a screw dislocation at $x=0$ for different values of $n=$**gb**. After P. B. Hirsch and [2.8].

which is only *weakly excited* in the perfect lattice (with $|s\xi_g| \gg 1$, e.g. for Si (666), one finds I_1 (perfect) $\ll 1$). The strongly rotated lattice planes in the dislocation core (fig. 2.13) may, however, reflect a strongly excited beam (e.g. Si (111)) through the aperture onto the screen. The corresponding dislocation contrast then is only 1 nm wide and permits the observation of dissociated dislocations, see sections 11.3.3 and [2.24], [2.26].

(c) *Edge dislocation* (see section 11.1.1). The amplitude phase diagram is shown in fig. 2.13 without derivation. Here again reinforced reflection is

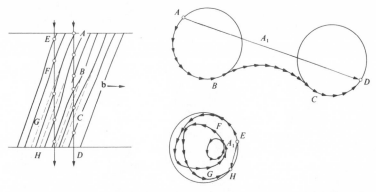

Fig. 2.13. APD for two columns near an edge dislocation. After [2.10].

found on one side of the dislocation and diminished reflection on the other as long as $\mathbf{gb} \neq 0$.

(*d*) *Coherent precipitate with distortions* (see chapter 9). There are various ways in which a precipitated particle of a second phase can produce a contrast in TEM. Either its composition, and hence its extinction, can differ from that of the matrix, or electrons can experience a phase discontinuity at the interface, or finally, and this will be the only case considered here, the particle can distort the surrounding matrix. A sphere of radius $r_0(1 + \delta)$ is fitted into a hole of radius r_0. This produces a radial displacement field with a strength at a distance r

$$\mathbf{u} = \frac{\delta r_0^3}{r^3} \mathbf{r}. \tag{2-6}$$

The additional phase angle $\Delta_+ = 2\pi(\mathbf{gr})(\delta r_0^3/r^3)$ vanishes in the direction of the intersection between the reflecting plane (hkl) and the particle with the result that the contrast has a 'coffee-bean' profile (fig. 2.14) consisting of two dark half-moons, rather than the symmetry of the strain field.

2.2.2   *X-ray topography*

X-rays can be used in a similar way to electrons to reveal lattice distortions, although the (primary) magnification is only about 1:1 on a photographic plate placed just in front of or behind the specimen if the observation is made in reflection/transmission. Various mechanisms based on dynamical theory are used to produce contrast. On one hand the so-called 'primary extinction' is reduced in distorted crystalline regions which therefore reflect more strongly (at the Bragg angle) than perfect crystalline regions. In thin crystals ($\mu t < 1$, $\mu =$ linear absorption coefficient) distorted

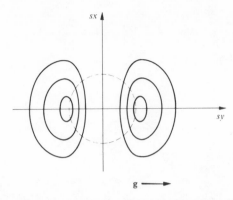

Fig. 2.14. Contrast profile of a coherently distorting, spherical precipitate (dotted) for reflection at planes $\mathbf{g} = (hkl)$.

crystalline regions transmit more X-radiation. On the other hand in thick, perfect crystals ($\mu t > 10$) the phenomenon of 'anomalous transmission' of X-radiation (Borrmann effect) is exploited. Anomalous transmission is not possible in distorted crystalline regions which appear dark. It is possible to obtain complementary contrasts from the same dislocations in one and the same crystal (Ge, 0.5 mm thick) using W K_α radiation ($\mu t < 1$) and Ag K_α radiation ($\mu t = 10$). The contrast widths are 5 μm and 15 μm respectively. The local resolution of lattice defects is thus considerably less than in the TEM but the distortions can be analysed very accurately. The reader is referred to [2.1] and [2.11] for a more detailed description of this and related X-ray topographical techniques.

2.3   Diffuse X-ray scattering

Deviations from the ideal crystal structure, i.e. lattice distortions on an atomic scale and non-random atomic distributions in alloys, cause 'diffuse' X-ray scattering outside the ideal Bragg reflections. The study of this diffuse scattering is of the utmost importance for the accurate determination of the atomic structure of alloys. Detailed accounts are given in [2.12] and [2.13].

2.3.1   *Diffuse scattering by lattice distortions*

Let the nth atom in the crystal be displaced $\mathbf{u}_n$ from its ideal position $\mathbf{r}_n$. (Averaging over all n, let $\bar{\mathbf{u}}_n = 0$.) The scattered X-ray amplitude for a special angle of deflection, i.e. location $\boldsymbol{\kappa} = \mathbf{k} - \mathbf{k}_0$ in the reciprocal lattice, is then

$$A_1(\boldsymbol{\kappa}) = \sum_n f_n \exp[-2\pi i \boldsymbol{\kappa}(\mathbf{r}_n + \mathbf{u}_n)]. \tag{2-7}$$

With a complex scattering amplitude $\Phi_n \equiv f_n \exp[-2\pi i \boldsymbol{\kappa} \mathbf{u}_n]$ and a distance vector $\mathbf{r}_m \equiv \mathbf{r}_{n+m} - \mathbf{r}_n$ the intensity becomes

$$I_1 = A_1 A_1^* = \sum_{m,n} \Phi_n \Phi_{n+m}^* \exp[2\pi i \boldsymbol{\kappa} \mathbf{r}_m]. \tag{2-8}$$

Φ_n is now split up into the amplitude $\bar{\Phi}$ of the undistorted lattice and φ_n due to the distortion

$$\Phi_n = \bar{\Phi} + \varphi_n \quad \text{with} \quad \bar{\varphi}_n = 0 \quad \text{and} \quad |\varphi_n| \ll |\bar{\Phi}|.$$

Thus for N atoms

$$I_1 = N|\bar{\Phi}|^2 \delta(\boldsymbol{\kappa} - \mathbf{g}) + \sum_{m,n} \varphi_n \varphi_{n+m}^* \exp[2\pi i \boldsymbol{\kappa} \mathbf{r}_m]$$

$$\equiv I_g + I_{\mathrm{diff}}. \tag{2-9}$$

The first term gives a sharp reflection at the (hkl) plane with the reciprocal lattice vector ('relvector') $\mathbf{g}$ but its intensity is reduced as compared with that from the undistorted lattice, since for $\boldsymbol{\kappa} = \mathbf{g}$

$$|\bar{\Phi}|^2 = |\overline{f_n \exp[-2\pi i \mathbf{g} \mathbf{u}_n]}|^2 \approx |\overline{f_n(1 - 2\pi^2 (\mathbf{g} \mathbf{u}_n)^2)}|^2$$

$$\approx |\overline{f_n \exp[-2\pi^2 (\mathbf{g} \mathbf{u}_n)^2]}|^2. \tag{2-10}$$

The exponential factor describing the attenuation is known as the Debye–Waller factor $\exp[-M]$. If the displacement $\mathbf{u}_n$ results from *thermal vibrations* of the lattice, then after averaging over the angle we obtain

$$|\bar{\Phi}|^2 = \bar{f}^2 \exp\left[-\frac{2\pi^2}{3} g^2 \overline{u_T^2}\right] \equiv \bar{f}^2 \exp[-2M] \tag{2-11}$$

where $\overline{u_T^2}$ is the mean square of the thermal displacement of an atom in the lattice.

A similar reduction in the intensity of the ideal reflection is obtained in the case of random lattice distortions due to solute atoms. There is a corresponding increase in the second term of (2-9), i.e. in the diffuse scattering outside the main reflections. Again for the thermal scattering by a lattice of like atoms we have

$$\varphi_n \varphi_{n+m}^* \equiv \{\Phi_n - \bar{\Phi}\}\{\Phi_{n+m}^* - \bar{\Phi}^*\}$$

$$\approx \bar{f}^2 \{\exp[-2\pi i \boldsymbol{\kappa} \mathbf{u}_n] - 1\}\{\exp[2\pi i \boldsymbol{\kappa} \mathbf{u}_{n+m}] - 1\}$$

$$\approx \bar{f}^2 4\pi^2 (\boldsymbol{\kappa} \mathbf{u}_n)(\boldsymbol{\kappa} \mathbf{u}_{n+m}). \tag{2-12}$$

Expanding the phonon spectrum in terms of plane waves of wave vector $\mathbf{p}$

$$\mathbf{u}_n = \sum_p \mathbf{u}_p \cos(2\pi \mathbf{p}\mathbf{r}_n)$$

we obtain for a special $\mathbf{p}$ with the addition formulas for the cos

$$\varphi_n \varphi_{n+m}^* = \bar{f}^2 2\pi^2 (\kappa \mathbf{u}_p)^2 (\cos 2\pi \mathbf{p}(2\mathbf{r}_n + \mathbf{r}_m) + \cos 2\pi \mathbf{p}\mathbf{r}_m). \qquad (2\text{-}13)$$

On summing over all values of n, the first term vanishes leaving

$$I_{\text{diff}} = \bar{f}^2 \pi^2 (\kappa \mathbf{u}_p)^2 \sum_m (\exp[2\pi i(\kappa + \mathbf{p})\mathbf{r}_m] +$$

$$+ \exp[2\pi i(\kappa - \mathbf{p})\mathbf{r}_m]). \qquad (2\text{-}14)$$

This corresponds to scattering at points $\pm\mathbf{p}$ on either side of a reflection $\mathbf{g}$, i.e. to satellites of the Bragg reflection. For a spectrum of $\mathbf{p}$ the reflections are accompanied by diffuse intensity, the distribution of which permits measurement of the thermal wave spectrum. The distortion field of an impurity atom in a corresponding Fourier series can give rise to a similar secondary intensity and can thus be analysed using the diffuse scattering ('Huang scattering'). In the special case of the dilation centre of equation (2-6), $|u_p| \approx \delta r_0^3 / p\Omega$, which corresponds to a relatively broad, diffuse 'halo' round the reflection. Overall, the distortion scattering increases with $\mathbf{g}^2$ for high index reflections. In contrast to Compton scattering I_{diff} is coherent and therefore dependent on the atomic configuration.

2.3.2　Diffuse solid solution scattering

If the scattering amplitudes f_A and f_B of the alloy partners A and B are different, diffuse so-called 'Laue-scattering' again occurs. At a composition v_A, v_B (atomic fractions) and in the absence of distortion we have for an A atom on site n (with $v_A + v_B = 1$)

$$f_n - \bar{f} = f_A - v_A f_A - v_B f_B = v_B(f_A - f_B)$$

and for a B atom

$$f_n - \bar{f} = v_A(f_B - f_A). \qquad (2\text{-}15)$$

If P_m^{AB} is the fraction of sites in the mth shell round an A atom occupied by B atoms, the fraction of AB pairs in a distance r_m is $v_A P_m^{AB}$ which make a contribution of

$$(-v_A^2 v_B(f_A - f_B)^2 P_m^{AB}) \quad \text{to} \quad \varphi_n \cdot \varphi_{n+m} = (f_n - \bar{f})(f_{n+m} - \bar{f}).$$

Treating AA, BA and BB pairs in a similar way since $P_m^{AA} + P_m^{AB} = 1$ we obtain:

$$\sum_n \varphi_n \varphi_{n+m} = v_A v_B (f_A - f_B)^2 \{ -v_A P_m^{AB} + v_B (1 - P_m^{AB}) - v_B P_m^{BA} +$$

$$+ v_A (1 - P_m^{BA}) \} = v_A v_B (f_A - f_B)^2 \{ 1 - P_m^{AB} - P_m^{BA} \}.$$

$$(2\text{-}16)$$

Using $N v_A P_m^{AB} = N v_B P_m^{BA}$ for the total number of AB pairs this becomes

$$\sum_n \varphi_n \varphi_{n+m} = v_A v_B (f_A - f_B)^2 \frac{P_m^{AA} - v_A}{1 - v_A} \equiv v_A v_B (f_A - f_B)^2 \alpha_m \qquad (2\text{-}17)$$

where

$$\alpha_m \equiv \frac{P_m^{AA} - v_A}{1 - v_A} \qquad (2\text{-}18)$$

defines the short-range order coefficient for the mth shell, which will be discussed in detail in chapter 7. In the special case of nearest neighbours, $m = 1$, the sign of α_1 determines various limiting cases of the atomic configuration

$$\alpha_1 = 0, \qquad P_1^{AA} = v_A \quad \textit{statistical arrangement of nearest neighbours}$$

$$\alpha_1 = +1, \qquad P_1^{AA} = 1 \quad \text{complete 'clustering'}$$

$$\alpha_1 < 0, \qquad P_1^{AA} = 0 \quad \text{perfect 'short-range order'}$$

$$(\textit{special case } \alpha_1 = -1 \textit{ for } v_A = \tfrac{1}{2}).$$

Thus the intensity of the diffuse solid solution scattering is

$$I_{\text{diff}}^{\text{Laue}} = v_A v_B (f_A - f_B)^2 \sum_m \alpha_m \exp[2\pi i \kappa \mathbf{r}_m] \qquad (2\text{-}19)$$

whereby according to the definition of equation (2-8) m passes through all the neighbours of an atom. α_m is, however, constant for all neighbours in one shell so that here m can represent the index of the shell. Furthermore because $P_0^{AA} = 1$ at $r_0 = 0$, $\alpha_0 = 1$. It is thus possible to derive the α_m values from $I_{\text{diff}}(\kappa)$ by Fourier analysis. Usually α_1 is the dominating factor and its sign can be ascertained from the intensity curve in the vicinity of the transmitted beam ($\kappa = 0$) or of a reflection. ($I_{\text{diff}}^{\text{Laue}}$ is in fact invariant to substitution of $(\kappa + \mathbf{g})$ for κ.) For *clustering* the curve shows a relative maximum and for *short-range order a relative minimum above a homogeneous diffuse background* $v_A v_B (f_A - f_B)^2$. If $\alpha_m = 0$ and $m > 0$, only the background is present. In the case of short-range order the diffuse intensity accumulates between the normal reflections in the reciprocal lattice. These

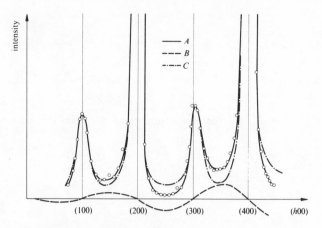

Fig. 2.15. Measurements of the diffuse X-ray scattering along $[h00]$ in Cu_3Au quenched from 500 °C, with CuK_α radiation (curve A). Curve B is the coupling term $I_{\text{diff}}^{\text{C}}$ determined by the different sizes of the atoms. The difference $C = A - B$ indicates short range order (in the case of (100), (300)). After B. E. Warren, *Trans. AIME*, **233** (1965), 1802.

are the points at which the superlattice reflections occur in the case of long-range order. The α_m values are periodic in m for long-range order.

2.3.3 *Solid solution scattering in the presence of distortions*

In this case we have a coupling term $I_{\text{diff}}^{\text{C}}$ in addition to $I_{\text{diff}}^{\text{Huang}}$ and $I_{\text{diff}}^{\text{Laue}}$ which remains linear in the product $(f_A - f_B) \cdot (\kappa \mathbf{u}_{PA})$ when distorting A atoms are incorporated in a B-lattice ($\mathbf{u}_{PA}$ is a Fourier transform of this distortion $\mathbf{u}_A = \delta_A r_0^3 \mathbf{r}/r^3$). It is interesting that $I_{\text{diff}}^{\text{C}}$ is asymmetrical on either side of the reflection ($\pm \kappa$), the sign of the asymmetry being governed by that of the product $\delta_A(f_A - f_B)$, which can therefore be determined from it. Very detailed information concerning the atomic configuration in solid solutions can thus be obtained from the diffuse X-ray scattering. Fig. 2.15 shows an example.

2.3.4   *Small angle scattering [2.29], [2.30]*

The diffuse X-ray (or neutron) scattering near the primary beam (at $\mathbf{g} = 0$) is caused by inhomogeneities which are larger than the lattice parameter. In particular, clusters of B atoms in an A-rich matrix formed early in precipitation will scatter according to equation (2-19) with $\alpha_m = 1$

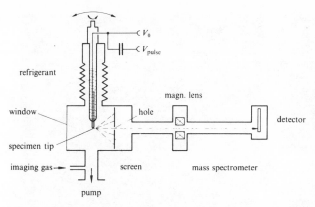

Fig. 2.16. Field ion microscope with moveable tip and mass spectrometer attachment [2.15].

for all m within a certain radius, $\alpha_m = 0$ outside, the intensity

$$I_{SAS} = N_B V_B (f_A - f_B)^2 \int_{V_B} \exp\left(-2\pi i(\kappa r)\right) \frac{dV}{V_B}. \qquad (2\text{-}14a)$$

Here V_B and N_B are volume and number density of the B-rich clusters, assumed to be small. For small scattering angles $(\kappa r) \ll 1$ ('SAS') one obtains by expansion (as in the case of equation (2-10)) the Guinier approximation [2.29]

$$I_{SAS} \approx N_B V_B (f_A - f_B)^2 \exp(-4\pi^2 \kappa^2 R_G^2 / 3). \qquad (2\text{-}14b)$$

This allows to determine the cluster radius $R^2 = \frac{5}{3} R_G^2$.

2.4 Field ion microscopy (FIM)

2.4.1 Investigation of lattice defects

With the field ion microscope it is possible to form an image of the arrangement of individual atoms on the surface of a pointed metal tip and, in conjunction with a mass spectrometer, to identify them chemically [2.14]. A recent experimental arrangement for this purpose is shown schematically in fig. 2.16 [2.15]. The pointed tip of the material to be examined has a radius ≈ 100 nm and is located about 10 cm from a fluorescent screen. The intermediate space is first evacuated to 10^{-8} Pa and then filled with 'imaging gas' (He, Ne or another inert gas) to 10^{-2} Pa. A positive potential V_0 of about 10 kV is then applied to the pointed tip relative to the screen. This creates a field of several 10 V/nm at the surface of the specimen. Atoms of the imaging gas are attracted to the pointed tip by

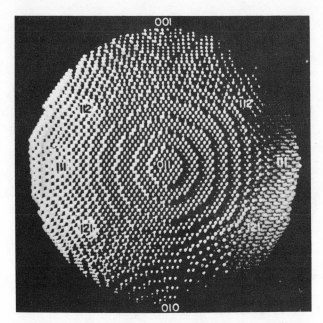

Fig. 2.17. Hard sphere model of a metal tip with the [011] orientation.

polarizing forces, thermalized and ionized by the tunnel effect especially at surface irregularities. The positively charged gas ion is then accelerated radially from the point of ionization to the screen where it excites scintillations. The intensity on the screen represents a reasonably accurate image of the steplike geometry of the tip. The steps ionize most strongly and therefore appear brightest as is shown by a comparison of a model of spheres (fig. 2.17) with the field ion micrograph of a pointed tungsten tip (fig. 2.18(*a*)). The resolution is about 1 nm. If the material of the tip consists of two phases the ionization rate of the imaging gas at them may differ; thus small particles of a second phase in the matrix become visible (fig. 2.18(*b*)). Lattice defects in the pointed tip such as dislocations, grain boundaries, stacking faults and antiphase boundaries in ordered structures are revealed as are vacancies or interstitial atoms arising for example from bombardment with high energy particles (irradiation damage). Caution must of course be exercised in the interpretation because these are defects in a surface layer which is further distorted by the high electric field strength.

2.4.2   *The atom probe*

An investigation into the bulk of the specimen is possible by removal of successive atomic layers from the pointed tip. This can be

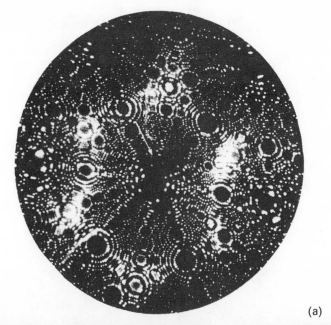

(a)

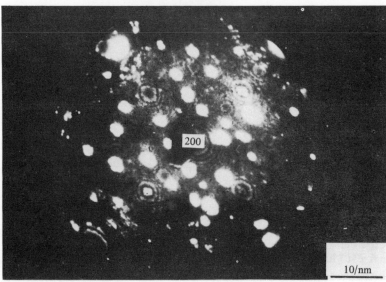

(b)

Fig. 2.18. (*a*) Field ion micrograph of a tungsten tip [2.15]. (*b*) FIM image of copper with cobalt precipitates. After H. Wendt.

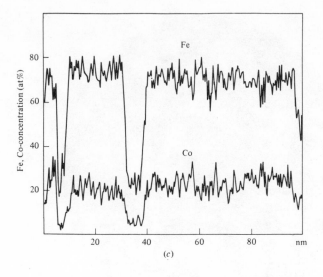

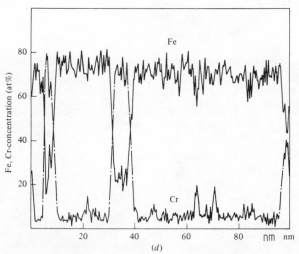

Fig. 2.18. (*c*) and (*d*) Concentration profiles in depth of a heterogeneous Fe 28% Cr 15% Co alloy [2.28].

achieved by short voltage pulses lasting several 10^{-8} sec (field evaporation) using the apparatus shown in fig. 2.16. The metal ions removed from the tip are also accelerated towards the fluorescent screen which contains a hole about 1 mm in diameter through which ions from a particular location on the tip pass into a mass spectrometer. It is thus possible to obtain a chemical analysis of a surface region with atomic dimensions using the time of flight principle at a mass resolution of 1:250 (at $T < 80$ K). By micromanipulation of the tip it is possible to focus each point resolved on the screen on to the input of the mass spectrometer and hence to analyse particles of a second phase morphologically, structurally and chemically *in statu nascendi*. Figs 2.18(c) and (d) show, for example, concentration profiles for iron, chromium and cobalt in a decomposed permanent magnet Fe 28%, Cr 15% Co [2.28]. Evidently, chromium-rich particles (68% Cr) have formed in a matrix of only 6% Cr at 22% Co. This really provides metallurgy with an instrument of the highest precision and versatility the potential of which becomes increasingly exploited [2.27].

2.5   **Thermal analysis** [2.3], [2.16]

Using the methods discussed in the preceding sections, it has become possible, although sometimes at the cost of some effort or complex apparatus, to describe the structural and microstructural state of a metal or alloy with ever increasing accuracy. In general a change in this state is accompanied by a reduction in the free energy of the system. Measurements of the corresponding gain or loss of heat can therefore provide direct information about the driving forces responsible. Thermal analysis has always been an important investigatory technique in metallurgy. The physicist meets it only in the form of a measurement of the specific heat, in other words of the amount of heat necessary *to raise the temperature* by one degree. The specific heat shows an anomaly during a phase change (in the case of melting, a singularity). In metallurgy, it is usual to plot a *cooling curve* because it is easier to achieve uniform cooling (e.g. by stirring the melt until it solidifies) than uniform heating. The anomalies in the temperature–time curve at a change of state are then correspondingly more distinct.

The normal cooling curve of a specimen heated to a temperature T_1 and placed at $t = 0$ in an insulated container at temperature T_0 obeys the time law for the difference between specimen temperature T and T_0

$$(T - T_0) = (T_1 - T_0) \exp(-t/t_0) \tag{2-20}$$

in which t_0 is proportional to the mass and specific heat of the specimen and also depends on the geometry and thermal contact with the environment. The rate of cooling should be as low as possible. Temperature gradients in the specimen are to be avoided. The mass of the specimen should be large compared with that of the crucible and thermocouple. If a change of state occurs in the specimen, e.g. a first-order transformation, which releases heat at a fixed temperature, a stationary period is observed in the cooling process, fig. 2.19(a). The stationary temperature identifies the transformation and the length of the stationary period gives the quantity of

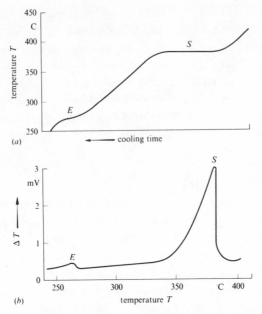

Fig. 2.19. Thermal analysis of a Zn–11.3 at.% Al alloy. S solidification; E eutectoid decomposition; (a) cooling curve; (b) differential thermal analysis.

material transformed and the heat of transformation. More general transformations, rather than the first-order transformation considered above, also change the normal course of the curve.

The often weak thermal effects associated with transformations in the solid state can be displayed in a magnified form by using a differential plot of the cooling curves for a dummy specimen and a transforming specimen which are in thermal contact (see fig. 2.19(b), *differential thermal analysis*). Well thought out apparatus has been designed for this which in particular permits a calibration of the heat absorption by simulation of the $\Delta T(t)$ curve with supplementary electrical heating. Using this method the heat of transformation can be measured to an accuracy of 5% in addition to which it is possible to detect the course of transformations as a function of temperature with a high degree of sensitivity. Naturally it is possible to follow changes in other physical properties in order to assess changes in state but calibration is often more difficult than in the case of the caloric effects, which are in fact responsible for initiating the transformation.

2.6 Mechanical testing methods

Since technological interest in metals centres mainly on their

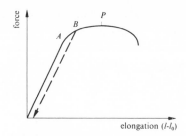

Fig. 2.20. The force necessary to deform a tensile specimen at a given
rate. Up to *A* the extension is (approximately) *elastic* (Hooke's law). If
the specimen is unloaded at point *B*, a fraction of the extension
remains and is thus *plastic*.

mechanical properties, the corresponding investigatory methods are of
considerable significance in metallurgy. It is not the elastic properties which
are being considered here but rather the phenomenon of irreversible or
plastic deformation. The characteristic parameters of plastic deformation
will be defined independently of the mechanism on the basis of various
mechanical testing procedures, the limits of which will thereby become
obvious.

2.6.1   *The tensile test* [2.17]

A rod-shaped specimen of a rectangular or circular cross-section is
deformed axially at a constant rate of extension (dynamic experiment). The
force K necessary for further deformation at a given time is recorded as a
function of time or of extension $(l - l_0)$. Fig. 2.20 illustrates a typical curve.
Only at small forces is the extension elastic, that is it returns to zero when
the load is removed. Greater extensions do not return to zero when the load
is removed and are therefore of a plastic nature. The greater the plastic
deformation, the greater is the force required for further deformation, in
other words the specimen work hardens. When the rate of hardening
becomes zero, the specimen fractures.

In order to be able to describe this phenomenon quantitatively,
experimental parameters are introduced which are independent of the
initial length (l_0) and cross-section (q_0) of the specimen. These are the
(engineering) stress $\sigma_n = K/q_0$ and the nominal (or engineering) strain $\varepsilon_n =
(l - l_0)/l_0$. Since the volume generally remains constant on plastic
deformation, the cross-section q decreases with strain according to
$q \cdot l = q_0 \cdot l_0$. Thus the true stress $\sigma = K/q$ can easily be calculated from σ_n
and ε_n. It is approximately equal to the resistance to deformation (flow
stress) of the material at every point on the $\sigma(\varepsilon_n)$ or work hardening curve

although, it being an externally applied stress, it is obvious that a clear distinction must be made between them. Similarly the extension of the specimen should be related to the actual length rather than to the initial length. This defines the so-called logarithmic strain

$$\varepsilon = \int_{l_0}^{l} \frac{dl'}{l'} = \ln l/l_0 = \ln(1 + \varepsilon_n) \tag{2-21}$$

which in the limiting case $\varepsilon_n \ll 1$ coincides with the engineering strain ε_n.

All these definitions assume that the specimen deforms uniquely along its length. This presupposes firstly an initial length $(l_0 \gg \sqrt{q_0})$ clearly defined by suitable specimen holders. Secondly, in the case of small variations in cross-section δq along the length of the specimen, the deformation must not become unstable. In other words, the rate of work hardening must be sufficient to stop further deformation at these points. This can be expressed by the *condition for plastic stability* $\delta \dot{q}/\delta q < 0$. The δ's refer to changes along the length of the specimen. The condition means that at random localized constrictions $(\delta q < 0)$ the rate of necking $(-\dot{q})$ must be smaller than elsewhere in the specimen $(\delta \dot{q} > 0)$ in order that the deformation at these points does not proceed in an unstable manner. This quotient can be derived from the material parameters of the specimen as follows. The following relationships are valid

$$\delta K = 0 = \sigma \, \delta q + q \, \delta \sigma \tag{2-22}$$

i.e. the same load acts on every specimen element;

$$\delta \sigma = \frac{\partial \sigma}{\partial \varepsilon}\bigg|_{\dot{\varepsilon}} \delta \varepsilon + \frac{\partial \sigma}{\partial \dot{\varepsilon}}\bigg|_{\varepsilon} \delta \dot{\varepsilon} \tag{2-23}$$

i.e. the flow stress of the material depends on the local deformation and rate of deformation, whereby

$$\delta \varepsilon = -\frac{\delta q}{q}; \quad \delta \dot{\varepsilon} = -\frac{\delta \dot{q}}{q} + \frac{\dot{q} \, \delta q}{q^2} \quad \left(\dot{\varepsilon} = -\frac{\dot{q}}{q} \right). \tag{2-24}$$

If a relative work hardening coefficient is defined

$$\Theta \equiv \frac{1}{\sigma} \frac{\partial \sigma}{\partial \varepsilon}\bigg|_{\dot{\varepsilon}}$$

and a strain rate sensitivity

$$m' \equiv \frac{1}{\sigma} \frac{\partial \sigma}{\partial \ln \dot{\varepsilon}}\bigg|_{\varepsilon}$$

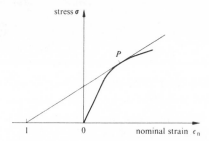

Fig. 2.21. The Considère construction of the tangent to the
stress–strain curve which passes through the point $\varepsilon_n = -1$ gives the
range of stable tensile deformation (up to point P).

which can naturally both still depend on σ, ε and $\dot{\varepsilon}$, the last three equations
can be summarized as

$$\frac{\delta\dot{q}}{\delta q}\cdot\frac{q}{\dot{q}}=\frac{1}{m'}\,(-1+\Theta+m').\tag{2-25}$$

The factor $q/\dot{q}$ is negative in the tensile test. Plastic stability in the tensile test
therefore requires (at $m' \geqslant 0$) that

$$\Theta+m'\geqslant 1.\tag{2-26}$$

There are two limiting cases: (1) $m' \ll \Theta$. In this case the tensile specimen
undergoes stable deformation in the region of the *Considère construction*
(fig. 2.21) up to the point P on the stress–strain curve, characterized by the
condition

$$\frac{\mathrm{d}\sigma}{\mathrm{d}\varepsilon}=\sigma\quad\text{or}\quad\frac{\mathrm{d}\sigma}{\mathrm{d}\varepsilon_n}=\frac{\sigma}{1+\varepsilon_n}\quad\text{or}\quad\frac{\mathrm{d}\sigma_n}{\mathrm{d}\varepsilon_n}=\frac{\mathrm{d}K}{\mathrm{d}l}=0.\tag{2-27}$$

Further strain occurs in the necked-down region in which the specimen
ultimately fractures. (2) $\Theta \approx 0$, in high temperature deformation. In this case
stable strain is just possible for $m' \approx 1$, i.e. $\sigma \propto \dot{\varepsilon}$, viscous flow. This case of
very high homogeneous strain is realized in 'superplastic' alloys (see section
12.6.2).

In the *compressive test* the analogue to the instability described above is
Euler instability, or the buckling of long specimens $(l_0 \geqslant 4\sqrt{q})$ [2.18]. It is
clear that in the compressive test very heavy deformation can only be
described meaningfully in terms of ε and not of ε_n. There appears to be a
limiting value $\varepsilon_n^{\text{compression}} \Rightarrow (-1)$ corresponding to $\varepsilon_n^{\text{tension}} \Rightarrow \infty$, whereas ε can
become infinite in both cases.

2.6.2 *The creep test*

In the so-called static experiment (creep test) the load, or preferably the stress σ, on the specimen is held constant and the strain is recorded as a function of time (or $\dot{\varepsilon}$ as a function of ε). If an equation of state $f(\dot{\varepsilon}, \varepsilon, \sigma) = 0$ holds good, the same information can be obtained from creep curves (for various σ) as from stress–strain curves (for various $\dot{\varepsilon}$). The increase in flow stress with strain for $\dot{\varepsilon} = \text{const.}$ corresponds to a decrease in $\dot{\varepsilon}$ with strain at $\sigma = \text{const.}$ in the creep test. The plastic instability of the tensile test has its analogue in 'tertiary' creep, during which $\dot{\varepsilon}$ again increases.

A particular type of creep under decreasing stress is observed in the *stress relaxation experiment*. A specimen is deformed at constant $\dot{\varepsilon}$ to a stress σ_0. The deformation machine is then halted and the specimen allowed to creep under the initial stress σ_0. Each extension $d\varepsilon$ unloads the specimen by $(-d\sigma) = \hat{E}\, d\varepsilon$ whereupon the creep rate again decreases. $\hat{E}$ is an elastic modulus of the machine/specimen system. The behaviour of the material $\dot{\varepsilon}(\sigma) = \dot{\sigma}(t)/\hat{E}$ is obtained from the measured time dependence of the relaxation $\sigma(t)$. Work hardening of the material during relaxation can be neglected because $\hat{E}$ is usually so large. The material parameter m' defined above can be obtained directly from the time dependence of the stress relaxation $m' = \partial \ln \sigma / \partial \ln \dot{\sigma}$.

2.6.3 *The hardness test*

The mechanical investigations described above cause considerable damage to the specimen. This is not the case in the 'hardness' test, which measures the resistance of a material to penetration by a harder test body. This can be a steel ball (Brinell hardness HB) or a diamond pyramid on a square base with an apex angle of 136° (Vickers hardness HV). The hardness is defined as the quotient of the test load and the area of the indentation it produces (measured in N/m^2). The state of stress under the test body is very complex, especially under the sphere, and so correspondingly is the plastic flow of material induced by the test body. A generally valid relationship between the hardness and the flow stress in a uniaxial tensile test is therefore not to be expected even when the diameter of the indentation relative to that of the test body is held constant to within certain limits (0.2 to 0.7) and the duration of loading is specified. Often $HB \approx HV \approx 3\sigma_0$, where σ_0 is a suitably defined stress at the start of the dynamic tensile test. Usually special hardness testing machines are used. If not, as in the case of microhardness testing, the diamond indenter is attached directly to the front lens of a metallurgical microscope. Hardness measurements are of great practical importance in metallurgy. A discussion of the mechanical principles is given in [2.19].

2.7   Investigation of anelasticity

According to Zener [2.20] reversible mechanical behaviour of a solid which nevertheless depends on the loading time is described as 'anelastic'. It is determined by atomic movements in the interior of the solid and is thus intermediate between elastic and plastic deformation as defined above. Anelastic phenomena are usually investigated under an alternating load in which case the anelasticity or internal friction manifests itself as a loss in energy per cycle, in other words as damping. The 'time-dependent linear standard solid' can be described generally by the relationship

$$\sigma + \tau_1 \dot{\sigma} = \hat{E}_R(\varepsilon + \tau_2 \dot{\varepsilon}). \tag{2-28}$$

The significance of the three constants in this equation is illustrated by the following experiments:

(a) At time $t = 0$ the specimen is instantaneously strained by ε_0, after which $\dot{\varepsilon} = 0$. The stress relaxes in a relaxation time τ_1 from an initial value zero to an equilibrium value $\hat{E}_R \varepsilon_0$. The solution of (2-28) is

$$\sigma(t) = \hat{E}_R \varepsilon_0 (1 - \exp[-t/\tau_1]). \tag{2-29}$$

$\hat{E}_R$ is the 'relaxed elastic modulus'.

(b) At time $t = 0$ a stress σ_0 is suddenly applied to the unstrained specimen. τ_2 is the relaxation time for the strain to reach its ultimate value of $\sigma_0/\hat{E}_R$ according to

$$\varepsilon(t) = \frac{\sigma_0}{\hat{E}_R} (1 - \exp[-t/\tau_2]). \tag{2-30}$$

(c) σ_0 is increased by $\Delta\sigma$ in the infinitesimally short time δt. Integration of (2-28) with respect to time for $\delta t \to 0$ yields

$$\tau_1 \Delta\sigma = \hat{E}_R \tau_2 \Delta\varepsilon. \tag{2-31}$$

We define an unrelaxed modulus $\Delta\sigma/\Delta\varepsilon \equiv \hat{E}_U$ given by

$$\hat{E}_U = \hat{E}_R \frac{\tau_2}{\tau_1}. \tag{2-32}$$

Normally $\hat{E}_U > \hat{E}_R$, that is $\tau_2 > \tau_1$.

For cyclic stressing $\sigma = \sigma_0 \exp[i\omega t]$ we expect $\varepsilon = \varepsilon_0 \exp[i(\omega t - \varphi)]$ and

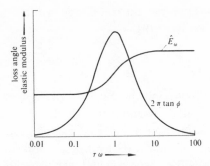

Fig. 2.22. Elastic modulus $\hat{E}_\omega$ and energy loss per cycle $\Delta u/u \approx 2\pi \tan \varphi$ as a function of the vibrational frequency ω of the standard solid.

obtain the phase angle φ from (2-28) as

$$\tan \varphi = \frac{\omega(\tau_2 - \tau_1)}{1 + \omega^2 \tau_1 \tau_2}. \tag{2-33}$$

We then define a modulus $\hat{E}_\omega = \sigma_0/\mathrm{Re}\ \varepsilon(t=0)$ as the ratio of the stress to that part of the strain which is in phase with the stress. From (2-28) we obtain

$$\hat{E}_\omega = \hat{E}_R \frac{1 + \omega^2 \tau_2^2}{1 + \omega^2 \tau_1 \tau_2} \tag{2-34}$$

substituting the definitions $\tau \equiv \sqrt{(\tau_1 \tau_2)}$ and $\hat{E} \equiv \sqrt{(\hat{E}_U \hat{E}_R)}$ equations (2-33) and (2-34) can be rewritten

$$\tan \varphi = \frac{\hat{E}_U - \hat{E}_R}{\hat{E}} \frac{\omega\tau}{1 + \omega^2 \tau^2} \equiv \frac{\Delta\hat{E}}{\hat{E}} \frac{\omega\tau}{1 + \omega^2 \tau^2} \tag{2-35}$$

$$\hat{E}_\omega = \hat{E}_U - \frac{\hat{E}_U - \hat{E}_R}{1 + \omega^2 \tau^2} \equiv \hat{E}_U - \frac{\Delta\hat{E}}{1 + \omega^2 \tau^2}. \tag{2-36}$$

In fig. 2.22 these functions are plotted against $(\omega \cdot \tau)$ on a logarithmic scale. At $\omega\tau = 1$, $(\tan \varphi)_{max}$ has a maximum value $(\frac{1}{2})\Delta\hat{E}/\hat{E}$ and $\hat{E}_{1/\tau} = (\hat{E}_U + \hat{E}_R)/2$ has decreased by exactly a half.

The relative energy loss per cycle in the case of forced vibrations is $\Delta u/u = 2\pi \sin \varphi$ and is thus identical with $2\pi \tan \varphi$ at small ϕ. In the case of freely decaying vibrations of frequency ω, the logarithmic decrement is $\pi \tan \varphi$. Both the mean relaxation time τ and the relaxation strength $\Delta\hat{E}/\hat{E}$, which are characteristic of the mechanism of internal friction can be derived from such measurements and from measurements of the variation of the $\hat{E}$ modulus with frequency. Three frequency ranges are accessible to

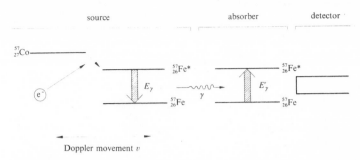

Fig. 2.23. Nuclear reactions in the source and absorber in resonance absorption in the case of iron.

measurement. The natural vibration of rods a few mm in diameter and several cm in length lies in the kHz range, signals in the MHz range can be transmitted through solid bodies with quartz oscillators, and finally frequencies in the Hz range can be studied in an experiment in which a wire specimen is coupled to a mass with a large rotational inertia and made to vibrate as a torsion pendulum. Not all the energy losses in solids can be described as relaxation phenomena by means of the time-dependent linear standard body, which, for example, does not permit the internal friction to depend on amplitude. Higher derivatives with respect to time are not taken into account either, which means that inertia effects are excluded (see section 11.4.2).

2.8   Mössbauer effect [2.21]–[2.23]

In 1958, R. L. Mössbauer discovered recoilless nuclear resonance absorption of γ-radiation which has since found increasing application in metallurgy as elsewhere. A source is required containing the original isotope which decays into an excited state of a Mössbauer isotope (fig. 2.23). The excited state of this nucleus decays to the ground state under emission of a γ-quantum. The accompanying recoil energy must be absorbed by the atom bound in a crystal. The probability that the recoil does not excite lattice vibrations but is absorbed by the crystal as a whole is described by the Debye–Waller factor, (2-11). In this case the Mössbauer spectral line associated with the transition is extremely sharp (relative line width $\approx 10^{-13}$). It can be absorbed by a second Mössbauer isotope in an absorber with only a small energy shift $\Delta E_\gamma \approx (v/c)E_\gamma$, which can be compensated by the Doppler effect if the source is moving at a velocity v relative to the absorber ($c =$ velocity of light). For metallurgical purposes, ^{57}Fe, ^{119}Sn and several rare earths have proved most suitable as Mössbauer isotopes. In the

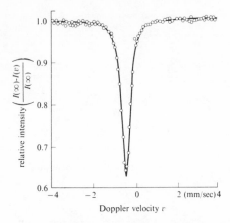

Fig. 2.24. Mössbauer spectrum of a ^{57}Co source at 80 K with an austenitic steel absorber (300 K), line displacement 0.438 mm/sec, line width 0.471 mm/sec. After U. Gonser [2.21].

case of ^{57}Fe, the energy of the γ-quantum is $E_\gamma = 14.4$ keV, the line width $\approx 10^{-8}$ eV and the necessary Doppler velocity of the order $v \approx 0.1$ mm/sec.

Fig. 2.24 shows a typical absorption spectrum of ^{57}Fe nuclei which in the absorber are located in an austenitic steel and in the source in copper. The changes in the position of the line (from $v \propto \Delta E_\gamma = 0$) and in the line width (from the natural line width) are determined by the different environments of the nuclei in source and absorber. These conditions can now be investigated with the aid of the Mössbauer effect.

Interactions of the Mössbauer nucleus with its environment
(a) Electric monopole interaction (isomer shift)

A Coulomb interaction exists between the nucleus, which has a mean square radius of $\langle R^2 \rangle$, and the electronic shell which has an s-electron density $|\psi(0)|^2$ at the nucleus. If the radii of the ground and excited states differ by $\Delta R/R$ and if the electron density at the nucleus differs in source and absorber by $\Delta|\psi(0)|^2$ the line will be displaced by

$$\Delta E_\gamma = k \frac{\Delta R}{R} (|\psi(0)|^2_{abs} - |\psi(0)|^2_{source}) \tag{2-37}$$

in which k is a constant characteristic of the isotope. If the nuclear radius factor is known, differences in the valency electron density can be derived which again depend on the nature of the bonding of the atoms in the crystal. For example Cu with 0.6 at.% Fe, quenched from 875 °C shows a superposition of two lines I and II (fig. 2.25(a)) which correspond

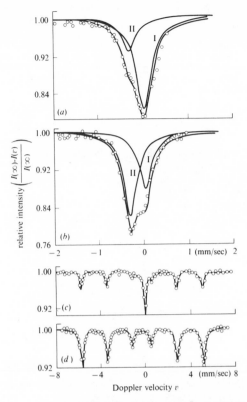

Fig. 2.25. Mössbauer spectra of Cu–Fe alloys (80 K) after various heat treatments (a)–(c). (d) shows the spectrum of pure α-Fe. The source is a ^{57}CoPt foil. After U. Gonser [2.21].

respectively to dissolved atoms and atoms precipitated as γ-Fe. On tempering at 600 °C the proportion of II increases until finally (fig. 2.25(c)) the spectrum of α-iron appears for the precipitated atoms. This is produced by the interaction of the internal magnetic field with the nuclear magnetic moment as described in the following.

(b) Magnetic dipole interaction (hyperfine splitting)

The hyperfine magnetic field H_{eff} effectively acting at the nucleus causes splitting of the energy levels of the ground and excited states corresponding to a nuclear Zeeman effect. Fig. 2.26 shows the six allowed transitions for dipole radiation in ^{57}Fe which explain the spectrum of α-iron in fig. 2.25(d). If the nuclear magnetic dipole moment is known, H_{eff} can be calculated from the splitting. This has been done here for antiferromagnetic

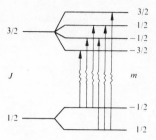

Fig. 2.26. Nuclear Zeeman splitting of the states of ^{57}Fe. The six allowed transitions for dipole radiation are marked.

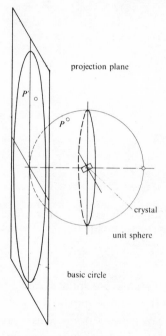

Fig. 2.27. Stereographic projection of the crystallographic direction P to the point P' on the plane of projection. After [2.1].

γ-iron (see section 6.1.2) and also for ferromagnetic α-iron. The interpretation of H_{eff} in terms of electron spin and orbital contributions cannot be discussed here. The way the carbon affects the effective hyperfine field of neighbouring iron atoms is particularly instructive. It depends on whether the carbon occupies sites characteristic of martensite or of various carbides (cf. section 6.1.2). The martensitic transformation (including the fraction of residual austenite, γ-Fe) and the reverse transformation

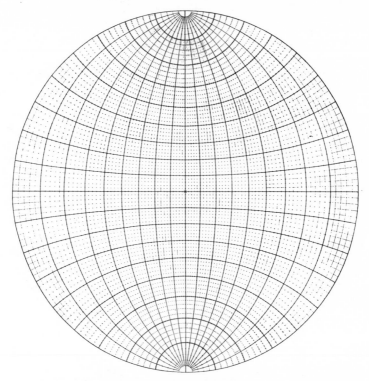

Fig. 2.28. Wulff stereographic net with 2° graduations. After [2.1].

involving decomposition into ferrite and carbide can thus be studied quantitatively (see chapter 13).

In addition to the two types of interaction described above, an electric quadrupole interaction exists which is of interest especially in non-cubic environments. Line broadening occurs during diffusion when the Mössbauer nucleus averages over several kinds of neighbourhood arising during atomic or spin movement. For further details see [2.21].

2.9   The stereographic projection [2.1], [2.2], [2.3]

Angular relationships between crystal faces and directions cannot be visualized satisfactorily from perspective drawings or algebraic equations. An indispensable tool in this context is the stereographic projection. A small crystal is imagined lying at the centre of a large sphere. The axes and faces of the crystal are then imagined to extend until they intersect the surface of the sphere as points or great circles. The sphere surface and the markings on it are then projected from a point on the

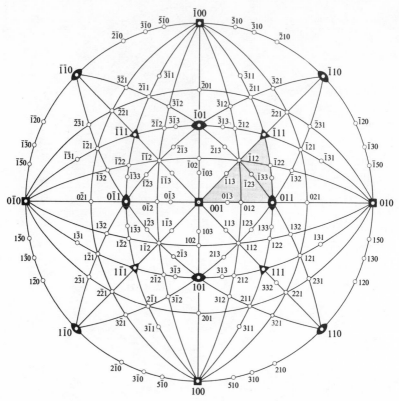

Fig. 2.29. Standard (001) projection for the cubic lattice. After [2.1].

surface on to the tangential plane diametrically opposite (fig. 2.27). It is sufficient to project one hemisphere because each crystallographic direction intersects the surface of the sphere at two points. The external boundary of the projection is the equator. The stereographic projection has the following properties:

(*a*) A circle on the sphere becomes a plane circle on the projection but the centres of the two are not generally coincident.

(*b*) Great circles on the sphere become circular arcs cutting the basic (primitive) circle at two diametrically opposite points.

(*c*) Angles are preserved in projection. It is therefore possible to measure on the projection the angle of intersection of great circles.

(*d*) Areas are not preserved in projection. Point distributions on the sphere appear on the projection with a distorted distribution density.

In practical work involving the stereographic projection, a Wulff stereographic net is used (fig. 2.28). The stereographic projections of crystallographic planes and directions are plotted on a piece of tracing paper laid over the net and attached to it by a pin through the centre of the primitive circle so that paper and net can rotate relative to one another. The following operations can be undertaken with the stereographic net:

Measurement of the angle between two crystallographic directions. The paper is rotated until the projections of the two directions both lie on the same great circle. The angle between them can then be read off.

Rotation of the projection about an axis. With the axis of rotation coincident with the north–south axis of the Wulff net, each point on the projection is moved an equal angle along the small circle on which it lies.

Plotting the standard projection of a crystal. In this case the plane of projection is a low index plane, e.g. the (001) plane of the cubic lattice in fig. 2.29. For reasons of symmetry every possible crystal direction in this lattice can be plotted in the 'standard stereographic triangle' [001]–[011]–[$\bar{1}$11].

Unknown crystallographic directions can be indexed by comparing the angles between them and known crystallographic directions with the angles between indexed directions in the standard projection.

3

Microstructure and phase, grain and phase boundaries

3.1   Definitions

The materials encountered in metallurgy are usually crystalline. Melts and vapour and, very recently, glassy materials (supercooled, frozen melts) are considered only as limiting cases. We start by assuming that we know the crystal structure of every substance. It can be determined by the well known X-ray methods and described crystallographically in terms of atomic positions, symmetry, unit cell, etc. (A structural description can be given for noncrystalline substances also, but the atoms do not adhere to it as strictly as in a crystal.) As stated in the introduction, bulk metal does not normally consist of a single crystal but of many crystal 'grains', in other words it possesses a microstructure. The grains differ from one another in orientation and sometimes also in crystal structure or composition. In the former case the system is described as homogeneous, in the latter heterogeneous. The constituents of the latter, which are in themselves homogeneous, are called phases. The crystal structure, composition and volume fraction of each vary in such a way that the free enthalpy of the system in equilibrium is a minimum. Phases as defined here transform into one another by first order transformations (defined thermodynamically by discontinuities in the first derivative of the enthalpy with respect to temperature) such as melting (see chapter 5).

The clearer understanding of the microstructure of metals and alloys gained by means of the high resolution methods described in the previous chapter, requires that the concepts of microstructure and phase be defined more precisely (cf. [3.1]). Consider first a homogeneous system consisting of crystals of different shapes and sizes but with macroscopically identical composition and structure. The grain boundaries which link the differently oriented grains have a positive interfacial energy, $\tilde{E}$. Their existence does not therefore correspond to a state of minimum energy but has been determined by the history of the metal. For example, when a melt solidifies, crystallization generally proceeds from many differently oriented 'nuclei' which then grow together into a crystalline aggregate. The same is true of crystallization from the vapour phase, electrolytic deposition and 'recrystallization' of a deformed metal when heated (in the solid state) which will be discussed in chapter 15. In thermodynamic equilibrium (chapter 5), a metal would have no grain boundaries. The same is true for one-dimensional lattice defects, i.e. dislocations. Their energy $E_L b$ per atomic spacing is much greater than kT so they are not in thermodynamic equilibrium either but have arisen by kinetic processes such as crystal growth. (This will be discussed in more detail in chapter 4.)

We thus define microstructure as all the existing lattice defects in a metal which are not in thermodynamic equilibrium (according to type, number, distribution, size and shape).

There are lattice defects which are in thermodynamic equilibrium and which do not therefore count towards the microstructure as defined above although they represent deviations from the ideal crystal structure. These are zero-dimensional defects or point defects such as lattice vacancies, the formation of which in an equilibrium concentration (mole fraction)

$$c_V(T) = \exp[S_{VF}/k] \exp(-E_{VF}/kT) \qquad (3\text{-}1)$$

represents a gain in free energy for the crystal although an energy of formation E_{VF} must be expended per vacancy. The numerous ways in which this (small) concentration of vacancies can be distributed over the lattice sites also represent a large increase in entropy which reduces the overall free energy for $T > 0$ (see chapter 5). S_{VF} is the entropy of formation of a vacancy which arises essentially from changes in the lattice vibrations in its vicinity, see section 6.2.2. These thermal vacancies therefore fall within the definition of a (crystalline) phase as a crystal in thermodynamic equilibrium with its environment.

In addition to thermal vacancies, so-called chemical vacancies occur in equilibrium in a number of intermetallic compounds characterized by a particular electron concentration (see section 6.3). NiAl for example is stable with three valency electrons per two atoms. If as a result of some Ni atoms being missing the composition is not perfectly stoichiometric, then as far as the valency electron concentration e/a is concerned, these Ni atoms can be replaced by vacant lattice sites without $e/a = 3/2$ being altered. Chemical ('structural') vacancies like this belong to the phase $Ni_x Al$ for $x \neq 1$ in thermodynamic equilibrium. If on the other hand vacancies are produced by irradiation with high energy particles (see chapter 10), these do not belong to the phase but are part of the microstructure. The latter vacancies can be made to disappear by annealing, the former cannot. The distinction is therefore thermodynamically meaningful.

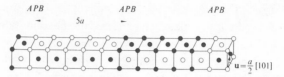

Fig. 3.1. Crystal structure of CuAu II (open and full circles represent Cu and Au atoms respectively).

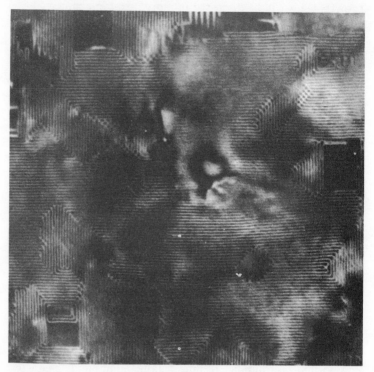

Fig. 3.2. Transmission electron micrograph of CuAu II with a regular arrangement of antiphase boundaries (J. Pashley, Cambridge, England). 25 000 × .

Similar problems arise with antiphase boundaries, APB, in ordered alloys (chapter 7). This lattice defect is caused by a change in the occupation of lattice sites. For example in the ordered alloy CuAu II the cube faces are occupied alternately by Cu and Au atoms (fig. 3.1). After every fifth cube face the occupation changes, i.e. an antiphase boundary arises. This regular succession of antiphase boundaries is actually observed in electron transmission (fig. 3.2). (Corresponding to section 2.2.1.2(a) a shift $\mathbf{u} = (a/2)\langle 101 \rangle$ becomes visible in the superlattice reflections at the antiphase boundary.) Additional antiphase boundaries can be introduced by plastic deformation, i.e. by dislocations with a Burgers vector $\mathbf{b} = \mathbf{u}$. These belong to the microstructure whereas the former belong to the phase CuAu II, i.e. to the structure.

A distinction must also be made in the case of *stacking faults*. The most closely packed planes in the face centred cubic (fcc) structure are stacked according to the pattern ABCABC and those in the hexagonal close-packed (cph) structure according to ABABAB. A regular arrangement of stacking faults gives rise to new stacking patterns or structures (phases) whereas a stacking fault left by a Shockley partial dislocation (see section 11.3.3.1) in the course of plastic deformation of the fcc lattice belongs to the microstructure.

The concept of phase does not embody homogeneity of composition and properties (e.g. that of microhardness). A concentration gradient in the direction of the phase boundary may well be set up in thermodynamic equilibrium (whereby the (partial) free energy is homogeneous, see chapter 5). The phenomenon of clustering in an alloy also falls into this category. This is a preliminary stage in the decomposition of a solid solution (i.e. of a homogeneous alloy) into two phases, in some cases by means of metastable intermediate states. In this stage regions which are still of atomic thickness in one dimension at least become enriched in atoms of one constituent. Alternatively periodic, weak concentration fluctuations are set up throughout the whole alloy as in the case of spinodal decomposition (see chapter 9). Only when a phase boundary has actually formed is it permissible to talk about two phases each of which should be more than a few atoms thick in all three dimensions. This is because even the phase boundary possesses a structure and must hence be assigned a certain thickness as will become clear in the following paragraphs.

3.2   Structure of grain boundaries

An element of a grain boundary between two crystals with the same structure is described by five orientation parameters (Euler angles). A rotation by an angle θ about an axis $\mathbf{u}$ determined by two angles on the unit sphere converts one crystal into the other. The boundary element is determined by a normal $\mathbf{n}$ defined by two angles. The distance separating the two crystals is virtually the atomic distance and the last atoms on either side of the boundary are subject to interaction with neighbours on both sides. A transition structure will therefore be formed. (In the search for the energetic minimum small relative displacements of the two grains must be admitted.) How does this transition structure depend on the five grain

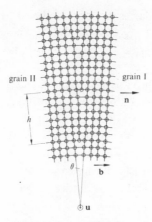

Fig. 3.3. Symmetrical tilt boundary corresponding to a vertical array of edge dislocations at distances h apart.

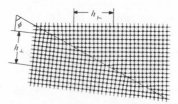

Fig. 3.4. Asymmetrical tilt boundary made up of two sorts of edge dislocations with mutually perpendicular Burgers vectors.

boundary parameters? This question can be answered most easily for a low-angle grain boundary for which $\theta \ll 1$.

3.2.1 *Structure of low-angle grain boundaries* [3.2]

The simplest case, that of a symmetrical tilt boundary $\mathbf{n} \perp \mathbf{u}$, fig. 3.3, is identical to an array of parallel edge dislocations (parallel also to $\mathbf{u}$, section 11.1.1) separated by $h = b/\theta$ where $b =$ the Burgers vector (here the lattice parameter of the primitive cubic lattice). The optimum matching of the two grains in the manner illustrated can actually be used to define the edge dislocation. A less symmetrical tilt boundary for which the boundary surface makes an angle φ with the plane of symmetry is illustrated in fig. 3.4 after W. T. Read and W. Shockley. In this case two sorts of edge dislocations with mutually perpendicular Burgers vectors with separations

$$h_{\vdash} = \frac{b}{\theta \sin \varphi} \quad \text{and} \quad h_{\perp} = \frac{b}{\theta \cos \varphi}$$

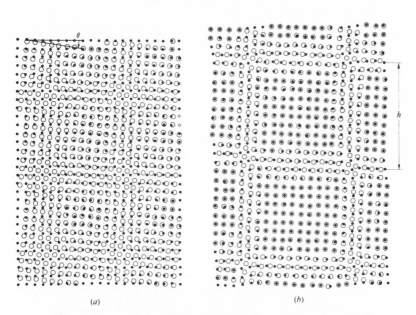

(a)   (b)

Fig. 3.5. Two lattice planes rotated with respect to one another (open and full circles) (a) and the resulting quadratic network of screw dislocations (separation h) (b).

are clearly required to describe the grain boundary. F. C. Frank has described a geometrical method of determining the dislocations contained in any low-angle grain boundary. The resulting arrangement is not always unique and also not always stable, see [11.4].

Another particularly simple limiting case is the symmetrical twist boundary, $n \parallel u$, fig. 3.5. In this lattice it corresponds to a square net of two sets of screw dislocations. The mesh size is again $h = b/\theta$.

This sort of dislocation structure for the low-angle grain boundary is in good agreement with observations on soap bubble models which simulate the lattice structure with a plane array of equally sized soap bubbles. Fig. 3.6 shows a tilt boundary with dislocation structure (and also several high-angle grain boundaries and vacancies) [3.2].

The low-angle grain boundary structure has also been confirmed by direct observation of metals in the TEM (section 2.2.1.2), by metallographic examination (section 2.1) of etch pits, which mark the points of intersection of the dislocations with the specimen surface and by FIM (section 2.4). The field ion micrograph in fig. 3.7 shows that the atoms occupy their normal lattice sites right up to the boundary layer on either side of a 20° ⟨110⟩ tilt boundary in tungsten. This case makes excessive demands on the

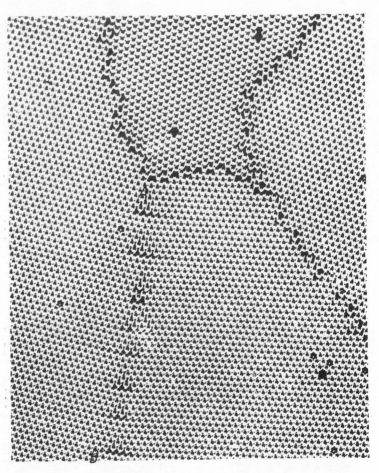

Fig. 3.6. Soap bubble raft with a tilt boundary *AB*, several high-angle grain boundaries and vacancies. After C. S. Smith [3.2].

dislocation model of the low-angle grain boundary, however, because at $\theta \approx 20°$ the dislocation spacing h is of the same order as the distance between the atoms so that the dislocations lose their identity. Some parts of the boundary illustrated here are straight and crystallographically oriented, e.g. right, parallel to $\{100\}$ below, left parallel to $\{110\}$ above. This will be explained in the following.

3.2.2   *Structure of a high-angle grain boundary* [3.3], [3.4], [15.6]
        Less is known about the structure of ordinary grain boundaries than about that of low-angle grain boundaries. The soap bubble model

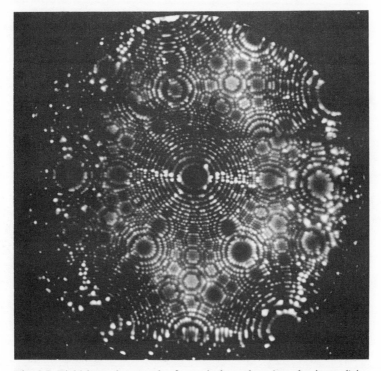

Fig. 3.7. Field ion micrograph of a grain boundary (top, horizontal) in tungsten at 4 K (J. M. Galligan, Brookhaven Nat. Lab., U.S.A.).

indicates that high-angle grain boundaries also consist of a periodic arrangement of 'structural units', of regions of good and bad matching between the two grains (fig. 3.6). The concept of the coincidence lattice due to M. L. Kronberg and F. H. Wilson [3.4] and recent computer calculations by M. Weins, B. Chalmers and H. Gleiter [3.3] represent a considerable advance. The coincidence lattice consists of the sites common to both grains if the atomic configurations of both are imagined to extend beyond the grain boundary. Fig. 3.8 shows two fcc crystals rotated with respect to one another by 38.2° about a ⟨111⟩ axis. One seventh (in general $1/\Sigma$) of the lattice sites belong to the coincidence lattice. It may be assumed that the grain boundary elements possess a low energy when they are 'specially' oriented, i.e. when they contain a high density of coincidence sites.

A limiting case of this is the coherent twin boundary which consists entirely of coincidence sites. It provides the link between two crystals which

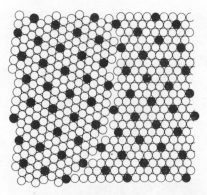

Fig. 3.8. Coincidence lattice (full circles) of two crystals rotated 38.2° about ⟨111⟩ (perpendicular to the plane of the diagram).

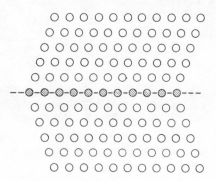

Fig. 3.9. Coherent twin boundary.

are mirror images of one another with respect to this boundary plane (fig. 3.9).

The numerical calculations mentioned above start with these special grain boundaries and attempt to minimize the interaction energy of all the atoms involved by displacements (relaxations and translations) of the two grains. (The dependence of the grain boundary energy on these translations shows that the five parameters given above are insufficient to describe a grain boundary completely.) The interaction is described approximately by central forces between nearest neighbours using a model potential, the parameters of which were fitted to the conditions existing in gold. Fig. 3.10 shows the result of the calculation for a two-dimensional close-packed lattice and a symmetrical 38° tilt boundary. The periodicity of the atomic arrangement in the direction of the grain boundary is clearly visible. The structural unit is made up of seven atoms (six cross-hatched and one in the

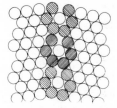

Fig. 3.10. 37.8° symmetrical tilt boundary with structural units (similarly shaded atoms) [3.3].

centre). The cross-hatched zone can be considered as a transition structure between the two grains. The two-dimensional calculations further show that grain boundaries which do not correspond to special orientation relationships (high density of coincidence sites) are made up of structural units belonging to the two nearest special boundaries, e.g. a 28° boundary consists of 50% units of the special 38° boundary and 50% of the special 18° boundary. Since the structural units of the different boundaries do not match exactly, lattice distortions are to be expected in the vicinity of such non-special boundaries. The distortions lead to interactions with impurity atoms of a different size not observed in special grain boundaries and these interactions affect the grain boundary mobility (see section 15.3.1). Asymmetric grain boundaries can also be derived from the structural model for special, symmetrical boundaries by the incorporation of steps. The mean position of the grain boundary is inclined to the plane of symmetry of the two grains and consists of segments of symmetrical boundaries separated by steps. These steps are visible in the electron microscope as fine lines in the grain boundary as observed by H. Gleiter [3.5] in Al alloys.

3.3   Energy of grain boundaries and its measurement

The calculations of the structure of special high-angle grain boundaries mentioned in the previous section also yield values of the interfacial energy $\tilde{E}$ (at $T = 0$) which are of the order of 0.9 J/m² for gold. This should be compared with the surface energy of gold at 0 K which is 2 J/m². The energy of low-angle grain boundaries is composed of the self energy of the constituent dislocations and the energy of interaction between them. When they are widely spaced the interaction energy is naturally proportional to the density $1/h$ and hence to θ. At high values of θ a saturation value of $\tilde{E}(\theta)$ should be reached interrupted only by steep troughs corresponding to the special grain boundaries. These grain boundaries, for example twin boundaries, should have extremely low energies (see fig. 3.14).

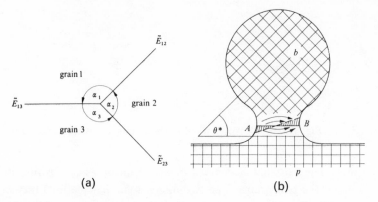

Fig. 3.11. (*a*) Equilibrium of the grain boundary tensions $\tilde{E}_{ik}$ at the common vertex of three grains. (*b*) A sphere (*b*) rotates during sintering onto a plate (*p*) by diffusion from *A* to *B*.

Although as part of the microstructure grain boundaries are generally not in thermodynamic equilibrium, a localized mechanical equilibrium can be established at the vertices of the grains which can be used to measure the grain boundary energy $\tilde{E}$. Consider the situation in fig. 3.11(*a*). Let the boundaries between grains 1, 2 and 3 be in equilibrium at their line of intersection as a result of annealing at high temperatures. To a first approximation, mechanical equilibrium means equilibration of the grain boundary tensions $\tilde{E}_{12}$, $\tilde{E}_{23}$, $\tilde{E}_{13}$ acting in the direction of the grain boundaries at their midpoint. Free grain boundary energies in J/m^2 can be expressed as grain boundary tensions in N/m if $\tilde{E}$ is unaffected by an increase in the grain boundary area when equilibrium is established. Resolving the tensions into their horizontal and vertical components we then have for the equilibrium of forces

$$\frac{\tilde{E}_{12}}{\sin \alpha_3} = \frac{\tilde{E}_{23}}{\sin \alpha_1} = \frac{\tilde{E}_{31}}{\sin \alpha_2}. \tag{3-2}$$

This relationship permits relative measurement of the energies of different grain boundaries, e.g. with respect to the theoretically well known energy of a low-angle grain boundary or the energy of a free surface (in equilibrium with the vapour).

A certain amount of caution must be exercised in the case of equation (3-2) because it does not contain the dependence of $\tilde{E}$ on the orientation of the grain boundary. In other words, not only are the tangential forces $\tilde{E}$ involved, as in fig. 3.11(*a*), but also moments $\partial \tilde{E}/\partial \alpha$. These are of vital importance if one of the boundary energies is small relative to the others as

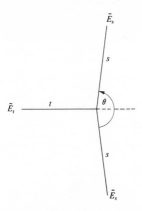

Fig. 3.12. A twin boundary (t) of low energy $\tilde{E}_t$ intersects a surface (s) where it is in equilibrium with the surface tension $\tilde{E}_s$.

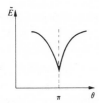

Fig. 3.13. Grain boundary energy corresponding to fig. 3.12.

for example in the case of a coherent twin boundary (t) which intersects a free surface (s) (figs. 3.12 and 3.13). The horizontal equilibrium condition must then be given [3.6] by

$$\tilde{E}_t - 2\tilde{E}_s \cos\frac{\theta}{2} - 4\left|\frac{\partial\tilde{E}_s}{\partial\theta}\right| \sin\frac{\theta}{2} = 0. \tag{3-3}$$

Since $\theta \approx \pi$ the last term cannot be neglected. $\tilde{E}_t$ can be determined by exact measurement of the surface groove which characterizes the line of intersection of a twin boundary with the surface after a high temperature anneal. Similar equilibria have been measured in the interior of crystals using TEM and it has been demonstrated that special boundaries really do possess a smaller energy than ordinary high-angle grain boundaries. Fig. 3.14 illustrates results obtained on copper which do not take into account the dependence of $\tilde{E}$ on the orientation of the grain boundary.

Gleiter *et al.* [3.8] have developed an elegant method for finding relative orientations of two grains which have a low grain boundary energy. They sinter small single crystal spheres (100 μm dmr) onto a single crystal plate

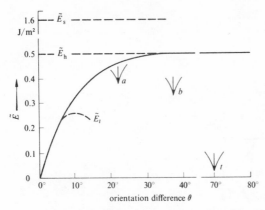

Fig. 3.14. Dependence of the boundary energy $\tilde{E}$ on the angle θ between the grains shown schematically for copper after [3.7]. Limiting values are given for low-angle grain boundaries ($\tilde{E}_l$), high-angle grain boundaries ($\tilde{E}_h$) and surfaces ($\tilde{E}_s$). a and b show the position of special low energy boundaries and t of coherent twin boundaries.

(cf. section 8.4.4). Fig. 3.11(b) shows how the sphere rotates into a special orientation with respect to the plate on annealing, during which diffusion, as indicated by the arrow, moves a wedge from A to B. The X-ray diagram shows a sharper orientation distribution of the spheres after the anneal.

The preceding discussion applies to plane sections perpendicular to the line of intersection of three grains. If all the $\tilde{E}$ are equal, all the α_i equal 120° as is observed approximately in metallographic sections. Four grains meet at a point. At equilibrium (for constant $\tilde{E}$) the angle between the four lines of intersection between any three grains is 109° 28'. This can be illustrated by a three-dimensional soap bubble model produced in a glass bottle by shaking. In the two-dimensional case the required angle of 120° leads to an array of equally sized hexagons. There is, however, no three-dimensional body which when continually repeated would fill the space completely, with simultaneous balancing of the surface tensions [3.2], [3.6]. A regular 14-sided body (known to the physicist as the first Brillouin zone of the fcc lattice) fills the space completely (fig. 3.15) but does not possess quite the correct angles (unless doubly curved boundary surfaces are introduced as in the 'Kelvin tetrakaidecahedron'). A regular dodecahedron with pentagonal faces almost satisfies the mechanical equilibrium conditions but does not fill the space. It can be shown empirically using soap bubbles or metal grains that the mean number of sides per grain surface is somewhat greater than 5 and the mean number of faces $12\frac{1}{2}$.

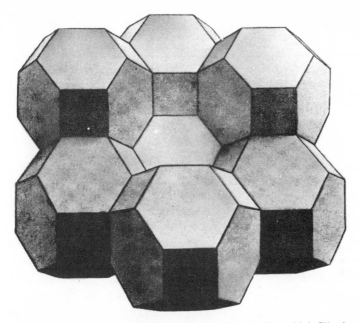

Fig. 3.15. An arrangement of regular 14-sided bodies which fills the space completely with approximate equilibrium of the surface tensions.

By a combination of topological and mechanical arguments, Cyril S. Smith [3.2] was able to show that the mean number of sides per grain surface should be $5\tfrac{1}{7}$ (instead of 6 in the plane!). Deviations from this number indicate a tendency to grain growth (see chapter 15). Topological approaches are extremely useful in attempts at quantitatively determining the three-dimensional morphology of microstructures by the analysis of two-dimensional sections ('quantitative metallography' or 'stereometry') or also at constructing a phase diagram for an n-component system ($n > 3$) (see section 5.7) that can be represented graphically only in sections, [3.7].

3.4   **Interphase interfaces** [3.2], [3.6]

If neighbouring grains differ not only in orientation but also in crystal structure and/or composition, that is they represent different phases (α, β), the boundaries between them are called interphase boundaries. These also possess a specific interfacial energy $\tilde{E}_{\alpha\beta}$ which can in principle be calculated and measured like the grain boundary energy $\tilde{E}_{\alpha\alpha}$.

Let us first assume large structural differences between the phases α and β so that we are concerned with the analogy of the high-angle grain boundary between α/α, the 'incoherent' interface. Actually this term is misleading as

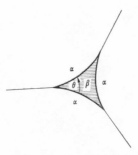

Fig. 3.16. A second phase β at the point of intersection of three α grains.

the two phases will always cohere although perhaps through an atomic layer of a transition structure as recent high-resolution TEM shows in cross-section (F. Ponce, Xerox PARC). Small volume fractions of the second phase (β) will often form in the high-angle grain boundaries of the primary phase (α) because the total boundary energy can thereby be reduced. Fig. 3.16 illustrates a typical arrangement of a β nucleus at the point of intersection of three α grains. The dihedral angle θ measures the equilibrium of the boundary tensions according to (3-3) in this case without the orientation dependence of the (incoherent) boundary surface α/β

$$\tilde{E}_{\alpha\alpha} = 2\tilde{E}_{\alpha\beta} \cos \frac{\theta}{2}. \tag{3-4}$$

The $\tilde{E}_{\alpha\beta}$ are often smaller than the $\tilde{E}_{\alpha\alpha}$. If $2\tilde{E}_{\alpha\beta} \leqslant \tilde{E}_{\alpha\alpha}$, $\theta = 0$ and the second phase spreads along the grain boundaries of the matrix. This can have catastrophic technological consequences if β has a lower melting point than α and the alloy is to be used in a temperature range in which β is liquid. It then disintegrates (e.g. mercury in α-brass).

On the other hand, the wetting of ceramic powder particles by liquid metals can be very useful in sintering e.g. hard metals such as tungsten carbide with cobalt, used for cutting tools (section 8.4.4).

Sometimes the analogue to the low angle grain boundary, a coherent interface, is sufficient to form the junction between structurally similar phases such as often occur in the initial stages of the decomposition of homogeneous alloys. Differences in lattice constant for the same crystal structure can be compensated by dislocations in the interface as shown in fig. 3.17. An interphase boundary in which the coherence is maintained only with the aid of dislocations is known as a semi-coherent interface (example in section 9.1.1).

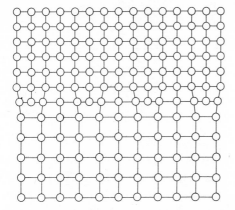

Fig. 3.17. Semi-coherent interface between two cubic lattices.

The cases considered in this chapter show that the quantitative description of grain boundaries is still only in its infancy. On the experimental side, high resolution microscopy such as FIM (section 2.4) and quantitative metallography will provide further data. On the theoretical side, a detailed description of grain boundaries is possible with the help of the computer and a more exact knowledge of interatomic forces (chapter 6).

4

Solidification [4.1], [4.4]

Most crystalline metals and alloys are produced by the process of solidification from the liquid phase. Mixing of components and purification of metals are both best undertaken in the molten state and casting can often be used to produce a desired shape. The microstructure is determined largely by the process of solidification. If solidification is sufficiently rapid the material remains in the state of an undercooled and frozen melt, called a *glass*. Crystallization begins with the *formation* of solid *nuclei* which then *grow* by consuming the melt. These two processes generally govern the formation of new phases. When a homogeneous alloy melt solidifies, inhomogeneous distributions of the components are set up in the resulting solid often leading to the formation of several crystalline phases. The *phase* or *equilibrium diagram* which will be introduced empirically in this chapter shows which phases exist under given conditions. The thermodynamic basis for the phase diagram follows in chapter 5. Pure metals will be considered first of all.

4.1 Homogeneous nucleation [1.3]

As soon as the temperature T falls below the melting point T_m a crystal has a smaller free energy f_S per unit volume than the melt f_L. (In reality we should compare free enthalpies, but it makes no difference at atmospheric pressure.) To a first approximation, the difference $\Delta f_v = f_L - f_S$ can be considered proportional to the undercooling $\Delta T = T_m - T$ ($\Delta f_v = \alpha \Delta T$). At a small degree of undercooling, however, the small crystals which have arisen as a result of thermal fluctuations (assumed here to be spherical nuclei with radius r) are unstable and redissolve because there is a positive free energy change $\Delta F_0 = \tilde{E}_{SL} 4\pi r^2$ associated with the necessary phase boundary. Nuclei capable of growth must satisfy the condition $d(\Delta F_{total})/dr < 0$. The total difference in free energy between a spherical

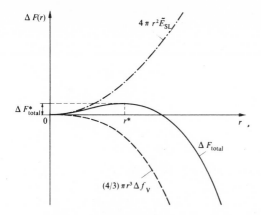

Fig. 4.1. Free energy contributions associated with the formation of nuclei of different radii.

nucleus and the melt is

$$\Delta F_{\text{total}}(r) = -\Delta f_{\text{v}} \frac{4\pi}{3} r^3 + \tilde{E}_{\text{SL}} 4\pi r^2.$$ (4-1)

The function is plotted in fig. 4.1. It goes through a maximum at

$$r^* = \frac{2\tilde{E}_{\text{SL}}}{\Delta f_{\text{v}}} = \frac{2\tilde{E}_{\text{SL}}}{\alpha \Delta T}$$ (4-1a)

of height

$$\Delta F^* = \frac{16\pi \tilde{E}_{\text{SL}}^3}{3\Delta f_{\text{v}}^2} = \frac{16\pi \tilde{E}_{\text{SL}}^3}{3\alpha^2 \Delta T^2}.$$

Larger nuclei gain energy by growth while r^* decreases as the undercooling increases. How large must ΔT be for thermal fluctuations to produce a nucleus capable of growth? In thermal equilibrium, a Boltzmann factor describes the occurrence of fluctuations in the energy ΔF^* according to

$$N^* = N \exp\left(-\frac{\Delta F^*}{kT}\right).$$ (4-2)

N^* is the probable number of critical nuclei per number N of atoms in the melt at the temperature T. The critical undercooling ΔT_{C} for $N^* = 1$ is obtained from (4-1) and (4-2) as

$$\frac{\Delta T_{\text{C}}^2}{T_{\text{m}}^2} = \frac{16\pi \tilde{E}_{\text{SL}}^2}{3T_{\text{m}}^2 \alpha^2 kT \ln N}.$$ (4-3)

Fig. 4.2. Nucleation at a wall.

The factor α can be estimated from the entropy of fusion which is related to the heat of fusion per cm^3 L_v according to

$$\frac{d}{dT}(\Delta f_v) = \Delta S = \frac{L_v}{T_m}, \quad \text{therefore} \quad \alpha = \frac{L_v}{T_m}. \tag{4-4}$$

Unfortunately $\tilde{E}_{SL}$ is not sufficiently well known theoretically. D. Turnbull has measured values of $\Delta T_C/T_m$ up to 0.2 in dispersions of mercury with a droplet size of several μm. From this the interfacial energy per atomic area can be calculated using (4-3) as $\tilde{E}_{SL}a^2 \approx 0.4L_v a^3$ which is approximately equal to half the heat of fusion per atom. The critical nucleus contains about 200 atoms. Naturally the estimates given above are only rough approximations. It has not been taken into account that critical nuclei of density n^* leave the distribution at a rate $j = vn^*$ where v is the frequency of the thermally activated jumps into the nucleus. The free energy of the nucleus ΔF^* can also be estimated more accurately [4.2].

4.2   Heterogeneous nucleation

Unless the melt is finely dispersed as in the experiments on Hg described above, it is generally not possible to undercool it by more than a few degrees. This is because only very few foreign particles are necessary to affect the crystallization of the whole melt. Even the crucible wall can assist nucleation. Figs. 4.2 and 4.3 show that the work needed for nucleation is much less there than elsewhere. The critical volume of a nucleus (and hence ΔF^* and ΔT) decreases to zero with decreasing angle of contact θ for constant radius r^*. The angle of contact θ is again determined by the equilibrium between the boundary tensions according to

$$\tilde{E}_{Lu} = \tilde{E}_{Su} + \tilde{E}_{SL} \cos \theta. \tag{4-5}$$

The efficiency of a substrate (u) in catalysing crystallization is given by the degree of coherence of its crystal lattice with that of the nucleus, its chemical nature and its topography [4.2].

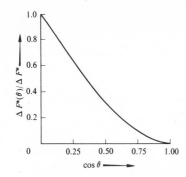

Fig. 4.3. Ratio of the work done in heterogeneous nucleation (section of a sphere) to that done in homogeneous nucleation (complete sphere of the same radius) as a function of the contact angle θ.

4.3   Crystal growth

The atomic stages in the growth of crystals from the melt are related to those of grain growth in the solid state and, like these, are governed primarily by the structure of the interface. Because of the high temperature and the disorder of the melt, the solid–liquid interface is probably more loosely structured than a high angle grain boundary. The experimental value $\tilde{E}_{SL} \approx 0.4 L_V a$ corresponds to the 'semi-molten state' of the atoms in the interface. The roughness of the interface is essential for its movement as crystallization proceeds. New atoms attach themselves at edges in the interface before new edges are formed on perfect close-packed crystal faces. Crystal growth is thus slowest at right angles to these 'flat' crystal planes which are often found in practice to form the surfaces of as grown crystals. In the case of small crystals like the virtually spherical nuclei assumed above there are always enough edge atoms available, whereas on virtually flat interfaces all the edges may run to the outside edge leaving flat crystal faces necessitating further nucleation. This situation is avoided if screw dislocations (section 11.1.1) intersect the interface where they form a self-propagating edge. If the atoms become attached at the same rate all along the edge, the angular velocity at the point of intersection is greater than at the outside and a growth spiral is formed (fig. 4.4). Such spirals are obviously essential, especially for crystal growth at small degrees of undercooling, and have in fact been observed. The rate of crystal growth is then proportional to ΔT.

The shape of the solid–liquid interface is also considerably influenced by the local temperature gradient, which itself is determined by the way in which the heat of crystallization is dissipated. It can be assumed that the

Fig. 4.4. Development of a growth spiral on a crystal surface intersected by a screw dislocation [1.1].

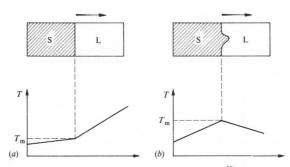

Fig. 4.5. Solidification at an advancing S̈L interface for different temperature profiles.

temperature of the interface itself is always T_m. The velocity of the interface is then determined by the rate of loss of heat. There are two possibilities for this, shown in fig. 4.5. In case (a), the heat flowing from liquid to solid plus the heat of crystallization is dissipated through the solid. The interface is stable because more heat would flow in through a bulge into the liquid, less would be conducted away and the solid bulge would melt again. In case (b) heat is dissipated through the liquid as well as through the solid, and this proceeds more efficiently the further the bulge projects into the liquid. The flat interface is therefore unstable and branched crystalline projections, so-called dendrites, are formed (fig. 4.6). These dendrites grow in crystallographic directions, e.g. in the direction of the cube edges in fcc metals. This is presumably linked with the anisotropic mobility of the solid–liquid interface discussed above. Orientation differences of up to several degrees nevertheless exist between the dendrite arms. Thus on one hand dendritic growth implies a regularity in the microstructure (preferred orientation) or 'texture' and on the other it produces a pronounced substructure of low-angle grain boundaries. In the stationary state, the length of the dendrites $l = vt$ increases proportional to the square of the undercooling ΔT of the melt. This can be explained in qualitative terms. On one hand the heat flow I through the tip of the dendrite (a hemisphere of

Fig. 4.6. A dendrite.

radius r) is proportional to $2\pi r\,\Delta T$. On the other hand the volume of material solidifying per second is $\pi r^2 v \propto I \propto 2\pi r\,\Delta T$. At equilibrium, the radius r must be the critical radius $r^* \propto (\Delta T)^{-1}$ according to (4-1a). Thus v is proportional to $(\Delta T)^2$ and of the order of 1 cm/sec. The residual melt between the dendrites is heated by the heat of crystallization and solidifies slowly.

In order to produce a controlled microstructure it is necessary to regulate the cooling conditions of the melt. On the other hand, heterogeneous nucleation offers possibilities of avoiding coarse grained microstructures which are often technologically undesirable.

4.4   Growth of single crystals and the origin of dislocations

Single crystals are a limiting case of a controlled microstructure. There are a number of techniques by which they can be grown, based on the principles of nucleation and crystal growth described above. Growing crystals does not only involve the application of these principles, it is an art. This is especially true of the production of perfect crystals, required not only for scientific research but also used technologically on a very large scale for semiconductor devices, lasers, magnetic storage units in computers, etc.

4.4.1   *Crystal growth from the melt* [4.3]

In the Bridgman method, a crucible (e.g. graphite) containing molten metal is lowered from the furnace. The crucible has a pointed end in which a single nucleus is formed or which contains a single crystal seed on to which the melt can crystallize. In the Chalmers method the furnace is horizontal and a boat is drawn out of it.

In the Czochralski technique, widely used for semiconductors and ionic crystals, a crystal seed is dipped into the melt and slowly withdrawn, rotating at the same time. If this is done at the same rate as heat is conducted out through the growing crystal, the diameter of the drawn crystal remains

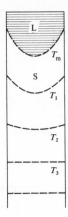

Fig. 4.7. Temperature distribution in a solidifying cylinder. Surfaces of constant temperature $T_m > T_1 > T_2$.

constant. In crucible-free growth from a 'floating' zone a short length of polycrystalline rod is melted by induction or electron beam heating. The melt is held between the adjoining rod ends by surface tension. As the heating element and with it the molten zone traverses the rod the material solidifies as a single crystal.

Other methods utilize transformations in the solid state (see chapter 15) or from the vapour phase. In the growth of single crystals the aim is often to achieve a particular orientation of the crystal lattice with respect to the specimen surface, a single crystal with a particular shape, an alloy crystal with a definite composition (see below) and a high degree of perfection of the crystal lattice.

4.4.2   *Origin of dislocations during crystal growth* [4.2]

In the vicinity of the melting point only very small stresses ($\leqslant 1$ MN/m^2) are required to deform crystals plastically. Such stresses can arise for a variety of reasons during crystal growth producing dislocations which effect plastic deformation. The trivial causes of plastic deformation such as external mechanical stresses due for example to the weight of the single crystal or differences in thermal expansion with respect to the crucible material will be ignored here. Dislocations are frequently produced by:

(*a*) *Thermal stresses.* During crystal growth and cooling of the crystal a radial temperature gradient is often set up in the specimen. In fig. 4.7 the surface regions of the crystal exert a pressure on the inner regions whereas they themselves are under tension. A temperature difference of several K is

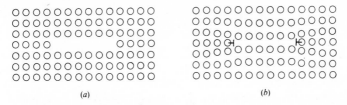

Fig. 4.8. (*a*) Vacancies condensing on a lattice plane. (*b*) Subsequent formation of a prismatic dislocation loop.

enough to produce plastic deformation and therefore also dislocations in a specimen with a radius of 1 cm if dislocation sources are present (see chapter 11).

(*b*) *Constitutional stresses.* Variations in the composition of a solid solution or not absolutely pure metal create stresses if the solute atoms have an atomic radius different from that of the matrix atoms. These stresses can be compensated by the simultaneous incorporation of edge dislocations [4.2].

(*c*) *Supersaturation of vacancies.* As the crystal cools from its melting point the concentration of thermal vacancies in equilibrium decreases exponentially according to (3-1). If the excess vacancies do not find a sink (dislocation, interface) quickly enough they are precipitated as so-called prismatic dislocation rings, fig. 4.8 (section 10.2). Slow cooling produces fewer dislocations.

Dislocations can of course extend into the crystal from a seed or from the crucible wall. The statements made above are also valid for dislocation walls, i.e. low angle grain boundaries. In order to reduce the density of both dislocations and dislocation walls, constrictions or 'necks' are built into the crucible into which the defects run and at which they end. A long anneal a little below T_m possibly with periodic temperature variation (to 'pump' the vacancies) can be valuable. The dislocations are then able to 'climb' and annihilate one another, see section 12.4. In this way the normal grown-in dislocation density in metals, $N_0 \approx 10^7$ cm/cm^3 can be reduced to 10^5/cm^2 and less.

4.5   Distribution of dissolved solute atoms on solidification

Even small additions of solute atoms have a profound influence on the solidification process. Normally solute atoms tend to remain in the melt rather than crystallize out with the solvent atoms. One reason for this is that the melt can accommodate atoms of a different size more easily than can the crystal. The result of this enrichment of the melt in solute atoms is a lowering of the freezing point. The presence of even positive temperature gradients in the melt as a result of heat dissipation can lead to instability of the solid–liquid interface due to so-called *constitutional undercooling* (section 4.5.4). Naturally, enrichment of the melt by solute atoms cannot continue indefinitely. It can, however, no longer be treated as a complication in the solidification of the pure metal but must be taken into account in a complete temperature–concentration diagram showing the solid and liquid phases. Such *phase* or *equilibrium diagrams* are discussed in the next section in connection with slow solidification during the course of which equilibrium is maintained. Solidification processes during which thermodynamic equilibrium is not established are then considered in detail. A more complete analysis and a thermodynamic foundation for the phase diagram will be given in chapter 5.

4.5.1   *Phase diagrams and the solidification of alloys*

The phase diagram of an alloy consisting of components A and B is a 'map' showing the regions in which various phases exist (in thermodynamic equilibrium) over a wide range of temperature T and concentration c of B atoms (in a matrix A). The construction of the diagrams follows specific rules, as will become obvious from the thermodynamics in chapter 5. For the time being these can be summarized in the Gibbs Phase Rule (at a constant pressure of 1 atm.), which will be derived later. According to this, the extent to which independent changes in the system, i.e. in T or c, known as the f degrees of freedom of the system, can take place is governed by the difference between the number n of the components and the number r of phases existing in equilibrium, according to

$$f - 1 = n - r. \qquad \text{fixed } p \qquad (4\text{-}6)$$

In a single phase system ($n = 1$) two phases (e.g. solid and liquid) can exist in equilibrium only at a fixed temperature (T_m), in other words $f = 0$. For $n = 2$, a binary system, a single phase $r = 1$ can exist in a finite (c, T) range ($f = 2$). Two phases ($r = 2$) can exist together in equilibrium for a whole series of $T_i(f = 1)$. The corresponding compositions c_i are then fixed by the diagram and can no longer be chosen independently. On the other hand two phases can exist in equilibrium over a whole range of composition but the choice of temperature fixes the compositions. Thus in the binary diagram there are two-phase regions. The concentrations of the two phases are determined by the boundary curves $c(T)$ of the regions dependent upon the temperature.

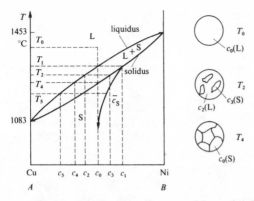

Fig. 4.9. Cu–Ni phase diagram with spatial distribution of the phases and concentrations at three temperatures. Path followed by the mean concentration $\bar{c}_S$ on cooling if equilibrium is not established in the solid phase.

4.5.1.1   Complete solid solution

The simplest case is shown in fig. 4.9 for the system Cu–Ni. Two curves divide the equilibrium diagram into three regions, the liquid phase region L, the solid phase region S and the two phase region (S + L) between the two curves. At a temperature T_2 a solid with composition c_3 and a liquid with composition c_2 co-exist in equilibrium whereby the mean composition c_0 of the system can vary between c_2 and c_3. The ratio of the quantities M_L of the melt (composition c_2) and M_S of the solid (c_3) at a mean composition c_0 is then inversely proportional to the ratio of the lengths $(c_0 - c_2)$ and $(c_3 - c_0)$ on the straight line $T = T_2$ (the so-called tie-line or conode). This is clearly a *lever rule* and can be derived by calculating the number of A and B atoms determined by c_0 existing in the two phases

$$M_L c_2 + M_S c_3 = (M_L + M_S)c_0$$

$$M_L/M_S = (c_3 - c_0)/(c_0 - c_2).$$

(4-7)

We shall now trace the solidification with falling temperature of a melt with composition c_0 under two sets of conditions.

(*a*) *Equilibrium fully established in* L *and* S. (This assumption forms the basis of the phase diagram but is often not justified because the equilibrium composition is reached slowly in the solid phase.) At T_1 the first solid material crystallizes from the c_0 melt with a composition of c_1. Since $c_1 > c_0$, the remaining melt is depleted in B (Ni). As cooling proceeds, its composition shifts along the so-called *liquidus* curve until it reaches c_4. The crystal also contains correspondingly less B as the temperature falls from

Fig. 4.10. Crystals with a smaller Ni-content growing round the first crystals to solidify in Cu–40% Ni (coring). 100×.

$T_1 \rightarrow T_2 \rightarrow T_4$. At T_2, its composition is $c_3 < c_1$ and, according to the assumption made above, is uniform throughout. The solid crystallizing originally as c_1 has also modified its composition to maintain equilibrium. The last remaining liquid of composition c_4 is thus in equilibrium with a solid which again has an overall composition c_0; its composition follows the *solidus* curve. This description of the process is, however, not very realistic because equilibrium is rarely achieved in the solid.

(b) *Equilibrium established in* L *but not in* S. Again the first material to solidify at T_1 in equilibrium with the c_0 melt has a composition c_1. The melt becomes correspondingly richer in A. At T_2 the melt has a composition c_2 while at the solidification front a solid layer of composition c_3 grows on to the original crystal c_1. Since equilibrium is not established in S the mean concentration $\bar{c}_S$ of the solid can only approach c_0 asymptotically as shown in fig. 4.9. The residual liquid has concentrations $c_4, c_5 \ldots$ and the range of solidification is considerably extended. The microstructure arising as a result of this process is characterized by layers of different concentrations growing around the original crystal; the solid is said to be 'cored' (fig. 4.10). This mechanism can be exploited in a one-dimensional arrangement for the purification of metals and will be treated quantitatively in section 4.5.2.

If the melting point is lowered by alloying at both sides of the binary diagram, the case of the Cu–Au system can arise (fig. 4.11). The situation is the same as in the Cu–Ni system, that is there is a continuous series of solid

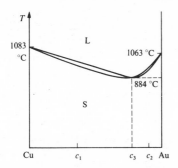

Fig. 4.11. Copper–gold phase diagram.

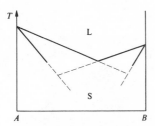

Fig. 4.12. The intersection of two solid solution phase diagrams leads to a eutectic diagram.

solutions in the solid state, except that the liquidus and solidus touch at the common minimum. The alloy c_3 thus solidifies like a pure metal ('congruently') whereas c_1 and c_2 behave as described above.

4.5.1.2   Eutectic system

Fig. 4.12 illustrates another possibility. At the intersection of the two liquidus lines the melt is in equilibrium with two solid phases. According to the Gibbs Rule $f = 0$, no solidification range is possible, and the melt must solidify at exactly T_E, as shown in fig. 4.13 for the Pb–Sn system. The system is called 'eutectic' and is characterized by a uniform fine-lamellar microstructure arising on solidification at point E. The production of this microstructure will be discussed in more detail later. For other initial concentrations of the melt, e.g. c_0, crystals first appear below T_1 corresponding to the left-hand solidus line AB, whilst with decreasing temperature the melt approaches E along the corresponding liquidus line AE. Here the eutectic reaction takes place yielding solid phases with concentrations c_1 and c_2 at a temperature T_E. The microstructure then consists of primary solid solutions with the compositions of the solidus CB,

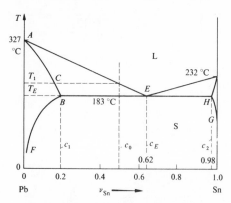

Fig. 4.13. Lead–tin phase diagram.

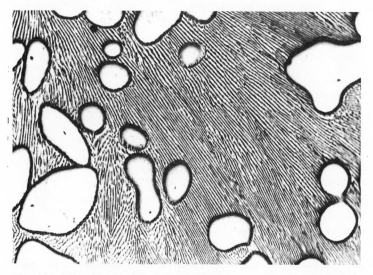

Fig. 4.14. Micrograph of Zn–8% Al showing primary Zn solid solution and eutectic. Cooled in the furnace. 145 ×.

which in the limiting case of complete equilibrium would be the composition c_1 corresponding to the point B, and of secondary solid solutions of composition c_1 and c_2. Thus in equilibrium there are only two phases present (c_1 and c_2) although, as a result of varying degrees of dispersion, a metallographic section appears to show three (fig. 4.14). The Gibbs Rule does not say anything about the actual distribution of the phases. On further cooling beyond T_E the solid phases in equilibrium should transform in the solid state according to the *solubility lines BF* and *HG*. Kinetic problems arise here which will be discussed in section 4.6.

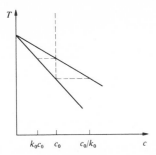

Fig. 4.15. Definition of a distribution coefficient k_0 in the case of linear liquidus and solidus.

4.5.2   *Unidirectional so-called 'normal freezing'*

We shall attempt a quantitative treatment of the redistribution of solute atoms which occurs during the solidification of solid solutions with simple phase diagrams firstly in order to explain the methods for purifying metals which are based on this principle and secondly in order to understand the microstructures which arise during these processes. It is necessary to make the following simplifying assumptions:

(*a*) The solid–liquid interface coincides with the cross-sectional plane of a long narrow crucible and moves at a constant velocity R.

(*b*) In the region of the phase diagram of interest here, the liquidus and solidus are straight lines. The ratio of their slopes is a constant given by

$$c_S/c_L = k_0$$

(fig. 4.15). (The distribution coefficient k_0 can be greater than 1, cf. fig. 4.9 on the Cu rich side.)

(*c*) Diffusion is negligible in the solid phase but is entirely responsible for concentration equilibration in the melt (rather than convection).

(*d*) Equilibrium always exists at the interface, i.e. the concentration of the solidifying crystal at the interface is always k_0 times that of the melt at that point.

Under these conditions the solidification of a melt with a concentration c_0 of solute atoms begins with the crystallization of solid of composition $k_0 c_0$. If $k_0 < 1$ the melt becomes enriched in solute atoms and the material crystallizing out is more concentrated. Finally, in the stationary state, solid and liquid have the concentration c_0 while a concentration peak c_0/k_0 has built up in the liquid ahead of the interface (fig. 4.16). The shape of this peak can easily be described anticipating the kinetics of diffusion explained in chapter 8. The diffusion current (number of impurity atoms diffusing per sec

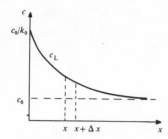

Fig. 4.16. Stationary solute atom distribution in the melt ahead of an advancing SL interface (at $x = 0$).

through unit cross-section) is proportional to the concentration gradient and defines a diffusion coefficient for the impurity atoms in the liquid $j(x) = -D\,\mathrm{d}c/\mathrm{d}x$. Thus a layer between x and $x + \Delta x$ loses $\mathrm{d}j/\mathrm{d}x\,\Delta x$ solute atoms per sec but gains a total of $\mathrm{d}/\mathrm{d}x(Rc)\,\Delta x$ in the same time as a result of displacement of the profile at a velocity R. The differential equation for $c(x)$ in the stationary state is therefore

$$\frac{\mathrm{d}}{\mathrm{d}x}\left(-D\frac{\mathrm{d}c}{\mathrm{d}x}\right) = R\frac{\mathrm{d}c}{\mathrm{d}x}. \tag{4-8}$$

The diffusion coefficient D (cm^2/sec) in the liquid is assumed here to be constant (see chapter 8). Applying the correct boundary conditions, the solution of the equation is

$$c_\mathrm{L} = c_0\left[1 + \frac{1 - k_0}{k_0}\exp\left(-\frac{R}{D}x\right)\right]. \tag{4-9}$$

Typically, the width of the peak D/R is about $1/10$ mm. The material in the peak is taken from the first part of the specimen to solidify and is deposited in the last part to solidify. Fig. 4.17, curve b, shows the profile after solidification (compared with the case of complete equilibrium in the solid and in the liquid as a result of diffusion, curve a).

In reality, however, convection in the melt must be taken into account apart from in a boundary layer of thickness d at the interface. The impurity must diffuse through this layer as before (C. Wagner). In this case it is useful to define a net distribution coefficient k_E as the ratio of c_S at the interface to c_L^∞, the mean concentration in the liquid outside the diffusion layer. In the case considered above, $D/R < d$, in which convection does not extend to the peak and can therefore be neglected, it is clear that $k_E = 1$ at the centre of the specimen. In the other limiting case, $D/R > d$, the 'diffusion peak' is broken down by convection and c_S is less than c_0 at the centre. In the most

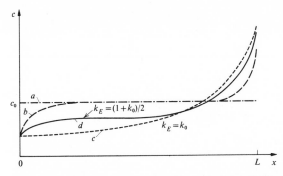

Fig. 4.17. Solute atom profile after uniaxial solidification from 0 to L under different conditions; (*a*) unrestricted diffusion in S and L; (*b*) diffusion in L; (*c*) convection in L; (*d*) partial convection in L.

favourable case of pure convection $k_E = k_0$ (although there is no horizontal portion to the concentration profile in this case). The general relationship between k_E and the parameters determining the redistribution of solute atoms is

$$k_E = \frac{k_0}{k_0 + (1 - k_0)\exp(-Rd/D)}. \tag{4-10}$$

Curves c and d in fig. 4.17 represent profiles in which convection was involved. In the case of perfect convection a large part of the specimen is purified by a factor $k_0 < 1$. (If $k_0 > 1$ it is enriched.)

4.5.3   *Zone refining* [4.5]

More efficient purification can be achieved according to W. Pfann if, instead of melting the whole specimen at once, a narrow molten zone parallel to the plane of cross-section is transported along the specimen at velocity R. The advantage of this is that the solidification and hence the purification process can be repeated an unlimited number of times without the purification effect being lost on remelting as would be the case with normal freezing. This introduces an additional parameter, the length l of the molten zone relative to the length of the specimen, L. Fig. 4.18 shows the progressive purification after n passes. The smaller the distribution coefficient the greater the purification. The closer k_E is to one, the more passes (n) must be made to achieve a given degree of purification. If $k_0 > 1$ there is naturally an enrichment in solute at the front end of the specimen. The technique of zone refining, which can also be carried out without a crucible on a vertical rod, relying on the surface tension to support the

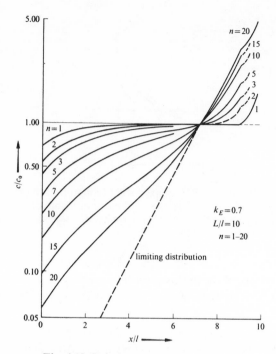

Fig. 4.18. Relative concentration distributions after n passes. After [4.5].

molten zone (cf. section 4.4.1) is of great importance in the production of pure semiconductors.

4.5.4   *Constitutional undercooling* [4.1]

The rapidly decreasing slope of the solute concentration peak ahead of the solid–liquid interface which was calculated in section 4.5.2 raises the liquidus temperature T_L according to the phase diagram and can thus give rise to undercooling leading to instability of the interface similar to that discussed in section 4.3 (fig. 4.19). The condition for this is derived from a comparison of T_L with the actual temperature at a distance x ahead of the interface. According to assumption (*b*) of section 4.5.2, $T_L(x) = T_m - mc_L(x)$ where T_m is the solidification temperature of the base metal and m the slope of the liquidus line. $c_L(x)$ is the concentration curve according to (4-9). The actual temperature at point x in the liquid is

$$T(x) = T_m - m\frac{c_0}{k_0} + \frac{\mathrm{d}T}{\mathrm{d}x}\bigg|_L x. \tag{4-11}$$

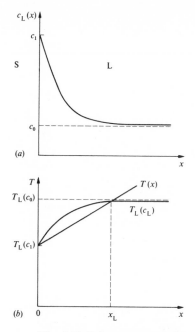

Fig. 4.19. Solute concentration c_L ahead of the LS interface (*a*) and corresponding liquidus temperature T_L compared with the actual temperature profile (*b*).

The first two terms together give the equilibrium temperature at the interface based on assumption (*b*) in section 4.5.2. So-called constitutional undercooling occurs in the region $T_L \geqslant T$, $0 \leqslant x \leqslant x_L$. There is a critical temperature gradient at the interface at which the undercooled region just vanishes for a given growth rate ($x_L = 0$), namely

$$\frac{dT}{dx}\bigg|_L = R\frac{mc_0(1 - k_0)}{Dk_0}. \tag{4-12}$$

A similar situation arises for $k_0 > 1$. As a result of the undercooling, the interface is again unstable to slight perturbations. A bulge in the interface gives rise to transverse diffusion of solute atoms, parallel to the interface. Accumulation of these atoms at P and Q (fig. 4.20) results in localized lowering of the liquidus temperature so that effectively the material is no longer undercooled. It remains liquid and a cellular structure is produced as indicated in Fig. 4.20(*c*). If $k_0 < 1$ the cell walls are rich in solute atoms and, based on what was said in section 4.4.2, in dislocations also. The distance λ between the cell walls is determined by the rate of diffusion of solute atoms parallel to and in the interface in competition with the growth rate R

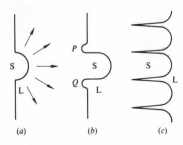

Fig. 4.20. Formation of cell walls in solid solutions.

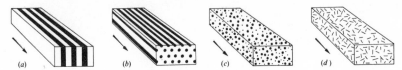

Fig. 4.21. Schematic representation of different eutectic microstructures. After [4.4].

perpendicular to the interface (i.e. $\lambda^2 = Dt$ according to chapter 8 and $t \propto R^{-1}$, i.e. $\lambda \propto \sqrt{(D/R)}$ decreases with increasing growth rate R). Quantitative investigations of the instability are difficult because infinitesimal perturbations in the plane interface are insufficient. A concentration profile must first of all build up ahead of the interface [4.2]. A hexagonal cellular microstructure of 'pencil-like' cells roughly parallel to the growth direction can be demonstrated experimentally. If $k_0 < 1$ the cell walls, which are often crystallographically oriented, become enriched in the alloying element. If $k_0 > 1$ they become depleted in solute. In any event a necessary condition for the occurrence of a cellular microstructure is that for the existence of constitutional undercooling (4-12). This has been confirmed experimentally. If the concentration of solute atoms ahead of the solid–liquid interface is very large, individual projections can extend so far into the melt that branches can be produced. This leads to the formation of dendrites if $dT/dx|_L \cdot k_0/c_0\sqrt{R}$ falls below a limiting value [4.1].

4.6   Eutectic solidification [4.6], [4.7]

Eutectic alloys can solidify in a multitude of different microstructures even when the solidification front is advancing macroscopically in a given direction at a velocity R. Fig. 4.21 shows several examples of continuous (lamellae or rods) and discontinuous (spheroidal or irregular arrangements of the two solid phases α and β) microstructures. This morphological diversity indicates that the growth of the eutectic phase

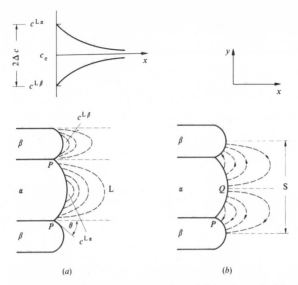

Fig. 4.22. (a) Concentration profile of B in the melt (L) ahead of the phases α, β of the growing eutectic. (b) Diffusion current of B ahead of the interface which is advancing to the right.

mixture is determined by a whole series of factors, the interface energies between α and β or between these and the melt, diffusion in the melt, volume fractions of α and β in the eutectic, preferred orientations for growth and interfaces, the magnitude of the growth rates of the two phases, etc. The repeat distance S between the lamellae or rods as function of the mean growth rate R is of particular interest. According to C. Zener, a particular value S* can be expected representing a compromise between the total interface energy between α and β, which increases with $1/S$, and the mean diffusion path ahead of and parallel to the solidification front, which increases with S.

We want to estimate S* for lamellar eutectics. The geometry is shown in fig. 4.22. First of all the boundary tensions must be in equilibrium at the points P, i.e. $\tilde{E}_{\alpha\beta} = 2\tilde{E}_{L}\cos\theta$ for the symmetrical case of equal boundary tensions $\tilde{E}_{L}$. This results in characteristic curvature of the solid–liquid interface, which, however, in the normal case $(\tilde{E}_{L} \gg \tilde{E}_{\alpha\beta})$ is not very pronounced. In as much as the crystallographic orientations of the α and β phases are closely related, the interface is semi-coherent. (For $\tilde{E}_{\alpha\beta} > 2\tilde{E}_{L}$, lamellar growth is not possible and a discontinuous microstructure is produced.) In the former case the saving in free energy density on solidification is reduced due to the incorporation of αβ interfaces

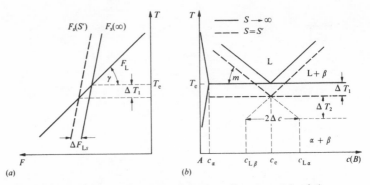

Fig. 4.23. (a) Undercooling ΔT_1 of the eutectic as a result of the expenditure of free energy due to the formation of interfaces. (b) A further undercooling ΔT_2 results from the concentration difference $2\Delta c$ ahead of the interface.

according to

$$\Delta f_{LS} = \Delta f_{LS}(S \to \infty) + \frac{2\tilde{E}_{\alpha\beta}}{S}. \tag{4-13}$$

This energy change is equivalent to an undercooling ΔT_1 of the eutectic solidification as shown in fig. 4.23, where $\Delta T_1 = \gamma(\tilde{E}_{\alpha\beta}/S)$, ($\gamma = $ const). A further undercooling ΔT_2 is necessary to effect equilibration of the concentration ahead of the solid–liquid interface. As shown in fig. 4.22 B atoms are rejected by α and build up ahead of α at Q (and conversely A ahead of β). Only at P is the composition really that of the eutectic c_e. These accumulations of 'wrong' atoms ahead of α and β must in the stationary state be removed by diffusion. (The concentrations $c_{L\alpha}, c_{L\beta}$ in the liquid ahead of α and β respectively lie on the extension of the liquidus lines on the wrong side of the eutectic composition as shown in fig. 4.23.) The diffusion problem (assuming of course a planar interface) has been solved by K. A. Jackson and J. D. Hunt [4.8]. We can justify the essential result by the following estimate. The gradient in B atoms ahead of the interface is $2\Delta c/(S/2)$ assuming a symmetric eutectic system. The diffusion current parallel to the interface is therefore $j_y \approx 4D_L \Delta c/S$, ($D_L$ is the diffusion coefficient in the liquid). In the stationary state j_y must remove exactly as many B atoms as are rejected by α from the eutectic melt, i.e. $2j_y = (c_e - c_\alpha)R$. This gives

$$\Delta c \approx (c_{L\alpha} - c_e) \approx (c_e - c_\alpha)\frac{RS}{8D_L}. \tag{4-14}$$

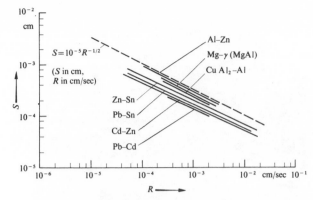

Fig. 4.24. Distance between the lamellae S as a function of the rate of solidification R for different eutectics in comparison with theory, equation (4-16).

Multiplying by m, the slope of the liquidus line, we obtain

$$\Delta T_2 = m \, \Delta c = mRS \frac{c_e - c_\alpha}{8D_L}.$$ \hfill (4-15)

We now assume, with Zener, that the value of S is such that $\Delta T = \Delta T_1 + \Delta T_2$ is a minimum (or for a given ΔT that the growth rate of the eutectic is a maximum). The condition $d(\Delta T)/dS = 0$ leads to

$$S^{*2}R = \frac{\gamma \tilde{E}_{\alpha\beta}}{m} \frac{8D_L}{c_e - c_\alpha}.$$ \hfill (4-16)

Such an expression has been confirmed many times by experiment, not only for lamellae (fig. 4.24) but also for rods which grow preferentially at small volume fraction ($< 1/\pi$) of one or other of the two phases. The separation S adjusts itself to the growth rate at any given moment by termination or transverse shearing of the lamellae. Anisotropic $\tilde{E}_{\alpha\beta}$ leads to the formation of lamellae in preference to rods if not in fact to a discontinuous microstructure if the growth direction and the lamellae direction diverge. A discontinuous microstructure is obtained in the case of grey cast iron (eutectic γ-iron–carbon) which forms spheroidal graphite. The interface energy $\tilde{E}_{CL}$ is extremely low parallel to the basal plane of the graphite and this always lines up perpendicular to the melt (fig. 4.25). If the growth rates R_α and R_β are different, α grows over β and the resulting eutectic microstructure is discontinuous. This can be influenced by a third component which is rejected by α and β in different ways. Discontinuous microstructures as formed for example by the addition of Na to Al–Si alloys

(a)

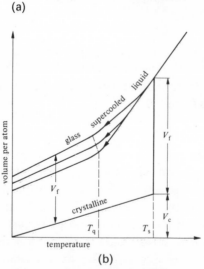

(b)

Fig. 4.25. (*a*) Spheroidal graphite consisting of lamellae, the basal planes of which form the interface to the surrounding γ-iron. International Nickel Co. After [4.1]. 165×. (*b*) Average volume per atom in the crystal, V_c, in the melt and in the glass, the latter larger by the free volume, V_f, as a function of temperature.

(modification of 'silumin') are often technologically desirable on account of their good mechanical properties (see chapter 14). Very finely dispersed microstructures produced by very rapid solidification of alloys with a non-eutectic composition are also technologically important.

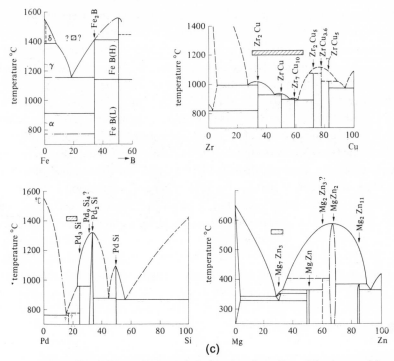

Fig. 4.25. (*c*) Phase diagrams of glass-forming alloys. The composition of existing glasses are hatched.

4.7   Metallic glasses

Recently rapid quenching of liquid alloys has become very important as it leads to non-crystalline, i.e. glassy solids. An undercooled liquid becomes a glass if its viscosity exceeds 10^{13} Poise (10^{12} Pa · sec). Then atoms do not move much any longer. Fig. 4.25(a) shows the decrease of the specific volume on cooling which normally leads to crystallization of the melt at the liquidus temperature T_S. If there is no time to form crystalline nuclei then the metastable state 'glass' appears below a glass temperature T_g. It contracts on further cooling as does the crystal but has about a 1% larger specific volume. That means that 'free volume' has been quenched-in which partly anneals out on tempering below T_g, i.e. during 'relaxation'. Close to $T_g + 20\,^{\circ}\text{C} \approx T_x$ the material crystallizes on annealing.

Rapid quenching (with rates of about 10^5 K/sec) is experimentally done by casting the melt onto a cold wheel which rotates at several thousand rpm. A thin glass band ($\approx 50\,\mu$m thick) forms which can be tens of cm wide nowadays. Also 'metglass'-wire is produced by extruding the melt into a rotating water-filled drum. Furthermore one can make a glassy surface on a massive specimen by scanning it with an intense laser- or electron-beam. Quenching is effected here by rapid heat extraction into the bulk.

The glassy, non-crystalline state is confirmed by X-ray diffraction. While one expects broadened Debye–Scherrer-rings for fine crystalline material as the reflecting lattice planes are of limited dimensions, a glass shows only one or two broad halos around the primary beam. This indicates that the NN of an atom are arranged in a shell at a particular spacing. By Fourier-inversion, equation (2-19), of the curve intensity vs scattering angle one can determine the so-called radial distribution function around a central atom up to three neighbouring shells in the glass. Also the FIM image shows the random arrangement of atoms in the glass very clearly (section 2.4). As only certain alloys, not pure metals, can be transformed into the glassy state one needs partial radial distribution functions for each of the alloy components and obtains these by using different sorts of radiation and substituted alloys. The results show that even in the glass there is chemical short-range-order among the different neighbours, especially after a relaxation anneal.

The most interesting question in the physical metallurgy of glasses is of course which alloys can be obtained in the glassy state. Several classes of such alloys can be distinguished:

(a) $T_{80}M_{20}$ glasses consisting to 80% of a transition metal (like Fe, Co, Ni, Pd) and to 20% of a metalloid (B, C, Si or P). Examples are $Fe_{40}Ni_{40}B_{20}$, $Pd_{80}Si_{20}$.

(b) $T^{(1)}T^{(2)}$ – glasses between a 'late' transition metal (as above) and an 'early' one (like Ti, Zr, Nb), e.g. $Ni_{60}Nb_{40}$, CuZr.

(c) AB-glasses between Mg, Ca ... on the one side and Al, Zn, Cu ... on the other, as in $Mg_{70}Zn_{30}$, but also $Be_{40}Zr_{10}Ti_{50}$.

Many of the (equilibrium) phase diagrams of glass forming alloys show deep eutectics as shown in Fig. 4.25(b) where the compositions of the (non-equilibrium) glasses are also marked. A deep eutectic indicates a tendency towards SRO or compound formation already in the melt as also expressed by a large negative heat of mixing (section 5.6). The 'molecular' units might differ in the molten and crystalline states as exemplified by Fe_3B in the glasses, Fe_2B as the next crystalline phase being observed. The deep eutectic is also favourable for maintaining the molten state at low temperatures.

The question why deep eutectics and easy glass formation often appear at a $T:M$ ratio of 4:1 has been addressed by Polk [4.9]. He assumes a random dense packing of hard spheres for the T atoms and classifies the holes in this structure according to size and frequency. The three largest types of interstices measure between 71 and 82% of the sphere diameter and are found $\frac{1}{4}$ times as frequent as the spheres; so if one fills these holes with the metalloid one accounts for the particularly stable stoichiometry of the TM glasses. Then the M atoms have only T neighbours as is observed in the partial radial distribution function as a special case of chemical SRO. Besides such steric arguments for the glass-forming ability there are electronic ones as with the Hume–Rothery rules for the stability of crystalline alloy phases (section 6.3). Häusler [4.15] observed for example a unique amorphous phase in vapour-condensed alloys of copper or gold with B metals at an electron concentration $e/a = 1.8$. Then the period of the Friedel oscillations coincidences with the NN distance which is energetically favourable. The free volume of metglasses is an essential factor for their properties like the relatively good ductility at extremely high strength (see section 12.7). If the free volume anneals out during a relaxation tempering the glass embrittles. A further factor in embrittlement is the decomposition of the glass in two non-crystalline phases [4.10]. Neutron irradiation at first produces free volume and improves ductility. Later radiation-enhanced diffusion helps decomposition and leads to embrittlement [4.11].

The literature on this new topic grows rapidly and can be cited here only in form of recent reviews [4.12]–[4.14].

5
Thermodynamics of alloys

In previous chapters we have occasionally made reference to thermodynamics and in particular postulated a state of thermodynamic equilibrium. We must now specify quantitative conditions if we are to derive phase diagrams and interpret them in physical terms. We assume an understanding of the basic principles of classical thermodynamics, in particular of state functions such as internal energy, entropy, etc. of a system under different imposed conditions. To interpret these macroscopic state functions on a microscopic scale we use statistical thermodynamics, which averages over the energies and distributions of all the participating atoms. In the case of alloys, which are of the greatest interest in metallurgy, the atomic arrangements are known only for highly simplified model systems, from which then model state functions and phase diagrams can be derived. The matching of these to a real system must be largely empirical and involves thermochemical measurement of the state functions. The discussion of binary and ternary systems which follows is of a more physico-chemical nature. Complementary physical arguments, which have been developed for a few structures, follow in chapter 6.

5.1   Equilibrium conditions [5.1], [5.2], [5.6]

It is well known that in a (thermally and materially) isolated system the *entropy S* has a maximum value at equilibrium. In alloys, equilibrium is more frequently established at constant temperature, that is in contact with a thermostat (furnace). If in a reaction the volume (as well as the number of particles) remains constant, a good approximation in condensed systems, the (Helmholtz) *free energy*, $F = E - TS$, becomes a minimum. E is the internal energy of the system. Strictly speaking it is usually the pressure p which is maintained constant, not the volume. In this case the system does work against the pressure, $W = p\,dV$, on approaching equilibrium if the volume increases by dV. At equilibrium then it is no longer true that F is able only to increase, i.e.

$$\delta F_{T,V} = dE - T \, dS \geqslant 0 \tag{5-1}$$

instead, the external work done by the system cannot be greater than the decrease in F, and the *free enthalpy* (Gibbs free energy)

$$G = F + pV = E + pV - TS$$

at constant pressure becomes a minimum, i.e.

$$\delta G_{T,p} = dE - T \, dS + p \, dV = 0. \tag{5-2}$$

The $p \, dV$ term is small in alloys at atmospheric pressure so that in most cases it is possible to use the free energy. (The enthalpy $H = E + pV$ replaces the energy as state function in systems at constant entropy and at constant pressure but not at constant volume.)

On the other hand we must consider that in many reactions material is added to the system during the reaction (for example that produced by another phase). If dn_i atoms of the component i are added, G changes (at constant T and p) proportional to dn_i by $\mu_i \, dn_i$. μ_i is the *chemical potential* of the component i. If several components are added, the first and second laws state that

$$\left.\begin{array}{rl} & dG = V \, dp - S \, dT + \sum_i \mu_i \, dn_i \\[2mm] \text{or} \quad & dF = -p \, dV - S \, dT + \sum_i \mu_i \, dn_i \\[2mm] \text{or} \quad & dE = T \, dS - p \, dV + \sum_i \mu_i \, dn_i. \end{array}\right\} \tag{5-3}$$

Thus it is possible to define

$$\mu_i = \left.\frac{\partial E}{\partial n_i}\right|_{S,V,n_j} = \left.\frac{\partial F}{\partial n_i}\right|_{T,V,n_j} = \left.\frac{\partial G}{\partial n_i}\right|_{T,p,n_j} \tag{5-4}$$

The μ_i are also called *partial* (molar) free energies (enthalpies). Like pressure and temperature the chemical potentials μ_i are intensive thermodynamic parameters. It will be seen in chapter 8 that a difference in μ_i causes a diffusional flow of the component i (just as a temperature difference causes a flow of heat).

The equilibrium conditions (5-1) and (5-2) are then given by

$$\sum_i \mu_i \, dn_i = 0. \tag{5-5}$$

It is possible to integrate (5-3) for constant T, V so that the ratios of the

$(n_i + dn_i)$ to one another do not change and thus also that the μ_i remain constant in a proportionately enlarged system. Thus

$$F = \sum_i n_i \mu_i = \sum_i n_i \frac{\partial F}{\partial n_i}\bigg|_{T,V,n_j} \tag{5-5a}$$

applies analogous to an interaction-free, purely additive system without mixing of the components i of molar free energies F_i

$$F^0 = \sum_i n_i F_i.$$

Another way of extending an interaction-free system to one in which interaction occurs is to introduce 'mixing terms' as in

$$F = \sum_i n_i F_i + F^M. \tag{5-5b}$$

Example: In a binary system of A and B atoms with the phases α and β in equilibrium at constant T, V, each change in F can be described as the sum of the changes in both phases. Thus in equilibrium

$$\sum_{i=A,B} \mu_i^\alpha \, dn_i^\alpha + \sum_{i=A,B} \mu_i^\beta \, dn_i^\beta = 0 \tag{5-6}$$

with $dn_i^\alpha = -dn_i^\beta$, i.e.

$$(\mu_A^\alpha - \mu_A^\beta)\, dn_A + (\mu_B^\alpha - \mu_B^\beta)\, dn_B = 0. \tag{5-7}$$

In accordance with our assumptions α and β should remain in equilibrium for an interchange not only of A but also of B. The equilibrium conditions are then given by

$$\mu_A^\alpha = \mu_A^\beta \quad \text{and} \quad \mu_B^\alpha = \mu_B^\beta \tag{5-8}$$

or

$$\frac{\partial F}{\partial n_A}\bigg|^\alpha_{T,V,n_B} = \frac{\partial F}{\partial n_A}\bigg|^\beta_{T,V,n_B} \quad \text{and} \quad \frac{\partial F}{\partial n_B}\bigg|^\alpha_{T,V,n_A} = \frac{\partial F}{\partial n_B}\bigg|^\beta_{T,V,n_A}.$$

The latter formulation will be used later in the form of the tangent rule in a discussion of the equilibrium conditions between two phases in the phase diagram. A generalization of the above conditions leads to the *Gibbs phase rule* (for p=const.). In an n-component system there are $(n-1)$ concentration variables to be defined in each of the r phases in addition to the common temperature, altogether $(r(n-1)+1)$ variables. However, $n(r-1)$ equilibrium conditions of the type (5-8) exist between these variables (2 between 2 binary phases). The number of degrees of freedom f of the system is the difference between all the variables determining the system and the number of conditions for equilibrium, therefore

$$f = r(n-1) + 1 - n(r-1) = n - r + 1, \tag{5-9}$$

which has already been used in chapter 4.

5.2 Statistical thermodynamics of ideal and regular binary solutions [5.3], [5.4], [5.5]

Let a system with constant volume be in contact with a (very much larger) heat reservoir which effectively determines its temperature (canonical ensemble). The probability of the system being in the energy state E_i is then given by the Boltzmann distribution

$$W(E_i) = \omega_i \exp[-E_i/kT] \Big/ \sum_i \omega_i \exp[-E_i/kT]. \tag{5-10}$$

ω_i is the number of possible energy states of the energy E_i (in quantum mechanical terms the density of the eigenvalues in the corresponding energy interval or the degeneracy of the state). In the case of a large system $W(E)$ has a sharp maximum, determined by the strong increase in ω with increasing E and corresponding decrease in the exponential function. We shall also use (5-10) for individual atoms in the crystal. In the following let E_i represent the energy of the alloy in contact with the thermostat. $S = k \ln \omega$ is the entropy of the system for a given energy E. The denominator, the partition *function* $Z = \sum_i \omega_i \exp[-E_i/kT]$, contains all the essential information about the system required for thermodynamics. The thermodynamic state functions, in particular the free energy

$$F = -kT \ln Z \tag{5-11}$$

are obtained from Z. E_i, ω_i and Z contain factors which depend only slightly on the arrangement of the A and B atoms in the alloy (for example through the lattice vibrations). These factors are of no interest in the following considerations and yield only an additive term in F. The problem is essentially the calculation of the number $\omega(v_{AB})$ of possible states of an energy $E(v_{AB})$ characterized by the number v_{AB} of AB bonds. This calculation is possible only for simplified models.

5.2.1 Model of the ideal solution

The energy E may be independent of the arrangement of the $Nv_B B$ atoms in the matrix of $Nv_A A$ atoms ($v_A + v_B = 1$). Then ω^M is the number of distinguishable arrangements of the themselves indistinguishable A or B atoms on the lattice sites. The arrangements are obtained by performing all

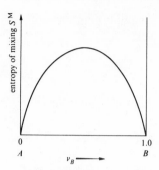

Fig. 5.1. Entropy of mixing of an ideal solution as function of the mole fraction v_B.

possible exchanges for the N atoms excluding those between the A atoms amongst themselves and between the B atoms amongst themselves. Thus

$$\omega^M = \frac{N!}{(Nv_A)!\,(Nv_B)!}.$$ (5-12)

Using Stirling's formula, the *entropy of mixing* is

$$S^M = +k \ln \omega^M = -Nk(v_A \ln v_A + v_B \ln v_B).$$ (5-13)

The form of the curve entropy against concentration v_B is shown in fig. 5.1. It meets the abscissa with an infinite slope at $v_B = 0$ and 1. This is the reason why, from the point of view of the free energy, it is so difficult to obtain materials in a very pure state. The free energy of the ideal solution is $F^{id} = E_0 - TS^M$ and the chemical potential of the component B in the alloy (5-5a) is $\mu_B^{id} = \mu_B^0 + kT \ln v_B$ (μ_B^0 does not depend on v_B).

It should be pointed out that only direct exchanges of A and B atoms (substitutional alloys) have been taken into account and not the occupation of additional sites (interstitial sites) by the dissolved atoms (interstitial alloys). Somewhat different relationships apply for the latter [5.5].

5.2.2   Model of the regular solution

The same alloy is considered using the approximation of pairwise interaction between nearest neighbours. Each atom has n nearest neighbours of the same or a different species. An A atom has nP^{AB} B neighbours and nP^{AA} A neighbours with which it interacts with an energy $\varepsilon_{AB}, \varepsilon_{AA}$ (in the case of attraction $\varepsilon_{ij} < 0$). P^{AB} is thus the probability of an A atom having a B atom as neighbour, which was defined in section 2.3.2. The same applies for a B atom. The total binding energy of a solid solution of given atomic configuration is

$$E = \tfrac{1}{2}Nv_A n\{P^{AA}\varepsilon_{AA} + P^{AB}\varepsilon_{AB}\} +$$
$$+ \tfrac{1}{2}Nv_B n\{P^{BB}\varepsilon_{BB} + P^{BA}\varepsilon_{AB}\}. \qquad (5\text{-}14)$$

Each bond may be counted only once, hence the factor $\tfrac{1}{2}$. The summation rules (as in section 2.3.2) are again applicable

$$P^{AA} + P^{AB} = P^{BB} + P^{BA} = 1; \qquad P^{AB}v_A = P^{BA}v_B.$$

Thus

$$E = \tfrac{1}{2}Nn(v_A\varepsilon_{AA} + v_B\varepsilon_{BB} + 2v_A P^{AB}\varepsilon). \qquad (5\text{-}15)$$

$\varepsilon \equiv \varepsilon_{AB} - \tfrac{1}{2}(\varepsilon_{AA} + \varepsilon_{BB})$ is the exchange energy, which is gained ($\varepsilon > 0$) or lost ($\varepsilon < 0$) when an AB bond is formed from AA and BB bonds. If $\varepsilon = 0$ the solution is ideal.

Equation (5-15) requires a knowledge of the distribution of pairs P^{AB} in the solid solution in order to calculate E even assuming a value for ε. A rigorous solution to this problem still has to be found although in principle, diffuse X-ray scattering, section 2.3.2, offers an experimental method of determining the pair distribution for pairs separated by any arbitrary distance. Using the short-range order coefficients α_m defined in section 2.3.2 we can write

$$E - E_0 \equiv E^M = Nnv_A v_B \sum_{m=0}^{\infty} \varepsilon_m(1 - \alpha_m) \qquad (5\text{-}15a)$$

in which ε_1 (for the nearest neighbour scale) is the ε defined above and the ε_m for $m > 1$ are the corresponding exchange energies for more distant neighbours (cf. section 7.2.2 and [5.12]). Since the α_m (2-19) are effectively the Fourier coefficients of the diffuse X-ray intensity of the solid solution, the energy of mixing can be linked by means of (5-15a) with experimentally determined quantities.

The model of the regular solution assumes that at relatively high temperatures the nearest-neighbour interaction is so weak ($\varepsilon < kT/4$) that it does not influence the statistical distribution of A and B atoms, whereupon $P^{AB} \approx v_B$, $\alpha_1 = 0$, and the ideal entropy of mixing S^M (5-13) can be used. We then have

$$F^M = F - \tfrac{1}{2}Nn(v_A\varepsilon_{AA} + v_B\varepsilon_{BB})$$
$$= Nnv_A v_B\varepsilon + NkT(v_A \ln v_A + v_B \ln v_B). \qquad (5\text{-}15b)$$

The right-hand side describes the total interaction between A and B atoms over and above the simple addition of the pure components. It is designated

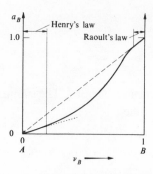

Fig. 5.2. Henry's law and Raoult's law, limiting cases of the curve obtained by plotting the activity of one component of a binary alloy against composition.

the *free energy of mixing* according to $F^M = E^M - S^M T$ (5-5b). The chemical potential, for example of the B component, is obtained directly from (5-5a) using the relationship $v_A v_B = v_A^2 v_B + v_B^2 v_A$ (or by differentiation, although this is more complicated on account of the relationship $v_A + v_B = 1$)

$$\mu_B^{\text{reg}} = \frac{n}{2} \varepsilon_{BB} + n\varepsilon v_A^2 + kT \ln v_B. \tag{5-16}$$

μ_B^{reg} is now compared with μ_B^{id} by defining the *activity* a_B of the component B in the alloy, which replaces the concentration v_B of the ideal solution, with the result that μ_B^{reg} has the same form as μ_B^{id}

$$\mu_B^{\text{reg}} \equiv \mu_B^0 + kT \ln a_B, \tag{5-17}$$

i.e.

$$\ln \frac{a_B}{v_B} = \frac{n\varepsilon}{kT} v_A^2,$$

$$\frac{a_B}{v_B} \equiv \gamma_B$$

where γ_B is the activity coefficient. If γ_B is independent of the concentration (in this case obviously for $v_A \to 1$, $v_B \to 0$) the component B is said to obey Henry's law. If it is unity, which it should be in our model of ideal solutions, $(\varepsilon/kT) \to 0$ or for the almost pure B component, $v_A \to 0$, the component B is said to obey Raoult's law, (fig. 5.2).

5.3   Measurements of the energy of mixing and the activity [5.2], [5.6]

Both these quantities are a measure of the interaction between the alloying constituents. They can be determined in various ways thus permitting us to test the above models which, in the following section, we shall use to make far reaching deductions regarding phase diagrams. Naturally a comparison between a theoretical and an experimentally determined phase diagram is in effect a test of the model used to describe the solution. It is, however, limited by the fact that phase diagrams provide information about the competition between phases rather than about the energetics of a given phase. Using thermochemical measurements, which will be discussed below, information can be obtained regarding the energy not only of realizable alloy states but also of those which are not normally formed because of competition from other phases. Phase diagrams derived from thermochemical measurements are often more accurate than those determined metallographically because thermodynamic equilibrium is attained so slowly at low temperatures†.

The usual thermochemical methods are:

1. The direct measurement of the heat of reaction $H^M \approx E^M$ on the irreversible mixing of components A and B in a calorimeter at constant p, T (ε_{AA}, ε_{BB} can be estimated from the heat of sublimation: $H_A^S = -(nN/2)\varepsilon_{AA}$).

2. The measurement of the partial free energy of mixing $\partial F^M/\partial n_i$ as the work done on reversibly adding 1 mole of the pure substance i to a large quantity of the alloy

(*a*) by isothermal distillation for which $(\partial F^M/\partial n_i) = RT \ln(p_i/p_i^0)$. The partial pressure p_i over the alloy compared with that over the pure component p_i^0 serves as a probe into the energy relationships in the mixture. A comparison with (5-17) shows that the activity $a_i = p_i/p_i^0$ can be measured directly in this way. For ideal solutions $a_i = v_i$; the vapour (p_i) can then be considered as an ideal gas. In consequence of (5-5) and (5-5a) $\sum_i n_i \, d\mu_i = 0$ at equilibrium (Gibbs–Duhem relationship) which if μ_1 is known yields μ_2 for a binary system.

(*b*) by electrolytic transfer in a galvanic cell:

pure metal i | ionic conductor with Z_i-valent | metal i in the alloy.
            | ions of the metal i |

On transfer of one mole of ions (for $Z_i = 1$ this corresponds to a charge of $1\mathscr{F} = 96\,500$ coulombs) an open circuit EMF U is set up between the metal and alloy which does work equal to $\partial F^M/\partial n_i = Z_i \cdot \mathscr{F} \cdot U$.

---

† In contrast to the metallographically determined phase diagrams, those determined from thermochemistry could be called equilibrium diagrams.

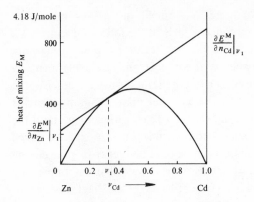

Fig. 5.3. Heat of mixing of molten Zn–Cd alloys at 700 K. The tangent gives the partial heats of mixing for $v_{Cd} = v_1$.

Examples of measured $F(c)$ diagrams

In many cases the measured heat of mixing E^M varies parabolically with the composition, as indicated by (5-15b). The curve for liquid zinc–cadmium alloys is shown in fig. 5.3. The partial molar heats of mixing (chemical potentials) can be determined graphically by the tangent construction as shown. A more accurate analysis of the thermochemical data shows, however, that the regular solution model is at best only qualitatively correct. Asymmetric curves of E^M against concentration are obtained. In particular the measured values of the entropy of mixing S^M, (5-3), deviate strongly from the ideal. The difference is the so-called excess entropy $S^{M,XS}$ which can arise as a result of the nonadditivity of the specific heats of lattice vibrations of the components, contrary to the Neumann–Kopp rule.

More complicated models of solutions are necessary to describe such deviations. Since it was possible to express the energy of mixing in terms of the diffuse X-ray intensity (5-15a) this is proposed also for the excess configurational entropy [5.12]. To a first approximation one obtains by summing over all reciprocal lattice vectors $\kappa = \mathbf{g}$

$$S^{M,XS} \approx \tfrac{1}{2}k \sum_\kappa \ln \alpha_\kappa \tag{5-17a}$$

with

$$\alpha_\kappa \equiv \sum_m \alpha_m \exp[2\pi i \kappa \mathbf{r}_m] = I_{\text{diff}}^{\text{Laue}}(\kappa)/v_A v_B (f_A - f_B)^2$$

according to (2-19). Hitherto, the experimental determination of thermodynamic functions from diffuse X-ray scattering has been limited to

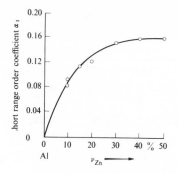

Fig. 5.4. Short-range order coefficient α_1 in AlZn at 400 °C from measurements of the diffuse X-ray scattering by Rudman and Averbach.

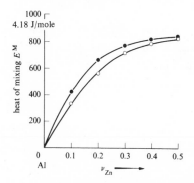

Fig. 5.5. Comparison of the heats of mixing for AlZn from X-ray (●) and calorimetric (○) measurements.

the short-range order coefficient α_1 of nearest neighbours, fig. 5.4, which should be zero assuming $P^{AB} = v_B$, for example in the Al–Zn system. If the P^{AB} data due to P. Rudman and B. Averbach are substituted in (5-15a) good agreement with the experimental values of the heat of mixing is obtained, fig. 5.5. There is a tendency in this system towards precipitation, i.e. a preference for Zn/Zn pairs, see chapter 9. Seen as a whole, the model of the regular solution is of fundamental significance rather than of practical value for describing real systems. We shall consider the above arguments in a little more depth in the next section.

5.4   **More advanced models of solutions** [5.4], [5.5]

First of all we must abandon the assumption that the *pair binding energies* $\varepsilon_{AA}, \varepsilon_{BB}, \varepsilon_{AB}$ in (5-14) are independent of the composition, i.e. of the nature of the neighbourhood. $E^M(v_B)$ is again obtained by a method analogous to that in section 5.2 with $P^{AB} = v_B$. Consequently it is now possible to interpret asymmetric $E^M(v_B)$ curves. Secondly, the lattice parameters of the starting materials (still assuming the same structure) must be matched to those of the alloy, i.e. *lattice distortions* on mixing must be included in the energy calculations.

In order to give a better description of the pair probability P^{AB}, the *quasi-chemical theory* is based on the reaction

$$(A - A) + (B - B) \rightleftharpoons 2(A - B) \tag{5-18}$$

for which the law of mass action (cf. (5-14)) states

$$\frac{(P^{AB} v_A)^2}{(\frac{1}{2} P^{AA} v_A)(\frac{1}{2} P^{BB} v_B)} = K \exp[-2\varepsilon/kT]. \tag{5-19}$$

K must equal 4 in order that the equation reduces to the statistical expression $P^{AB} = v_B$ at $T \gg 2\varepsilon/k$ (applying the summation rule for the P^{ij}). At high temperatures (5-19) can be expanded in ε/kT. To a first approximation the solution is then

$$P^{AB} \approx v_B \left(1 - v_A v_B \frac{2\varepsilon}{kT}\right). \tag{5-20}$$

In this way temperature dependent deviations from a parabolic E^M curve can be explained. (5-20) corresponds to a non-zero short-range order coefficient (according to (2-18)) for $\varepsilon = 0$

$$\alpha_1 = \frac{v_B - P^{AB}}{v_B} = + \frac{v_a v_B 2\varepsilon}{kT} \tag{5-21}$$

as shown in fig. 5.4. Clearly for Al–Zn, ε is positive because $\alpha_1 > 0$, which corresponds to a preference for like nearest neighbours (precipitation).

5.5   Derivation of binary phase diagrams from the model of a solution

5.5.1   *Free energy and phase diagrams*

We first discuss the free energy of mixing F^M *of the regular solution* as a function of v_B and T according to (5-15b). For very high temperatures (or for $\varepsilon = 0$) the shape of $F(v_B)$ is determined by the second term on the right-hand side. Corresponding to the discussion of S^M, fig. 5.1, the result is a parabola open at the top. This shows, however, that the homogeneous mixture of the components is the stable equilibrium state of the alloy. Let us compare the free energy of the homogeneous solution of composition v_B in

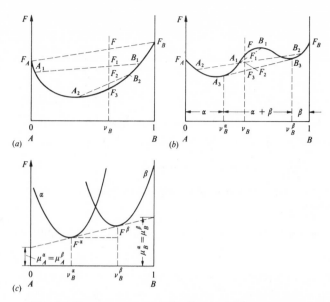

Fig. 5.6. Free energy curves leading to (a) homogeneous, (b) heterogeneous alloys; (c) illustration of the equilibrium between two phases.

fig. 5.6(a) with that of a heterogeneous mixture of two phases of arbitrary composition A_1 and B_1 or A_2 and B_2. The free energy of the heterogeneous mixture is naturally the sum of the free energies of the volume fractions of the constituent phases corresponding to the points F_1, F_2 on the line joining A_1 and B_1, A_2 and B_2. This is expressed by the second term on the left-hand side of (5-15b). In the case of an alloy of composition v_B, the free energies of the mixtures F_1, F_2 are higher than that of the solution F_3. Thus according to (5-1) the homogeneous phase is thermodynamically stable. The situation is different for an $F(v)$ curve as shown in fig. 5.6(b). The v_B alloy can lower its energy from $F \rightarrow F_1 \rightarrow F_2$ by separating into two phases A_1/B_1, A_2/B_2. The state of lowest energy F_3 corresponds to the common tangent to the $F(v)$ curve at the points A_3/B_3 and thus to a separation into two phases with the compositions v_B^α, v_B^β. This *common tangent construction* corresponds to the equilibrium conditions derived earlier (5-8), $\mu_B^\alpha = \mu_B^\beta$, $\mu_A^\alpha = \mu_A^\beta$, for the equilibrium between two phases α, β in a binary system of A and B atoms. This can be seen from the following:

The free energy per atom of the system according to (5-5a) with j α, β

$$F^j = (1 - v_B^j)\mu_A^j + v_B^j \mu_B^j \qquad (5\text{-}21a)$$

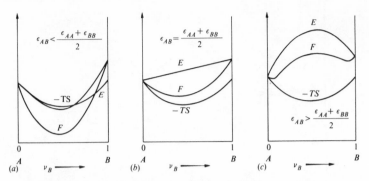

Fig. 5.7. Free energy of solid solutions for different interaction parameters ε.

is differentiated with respect to v_B^j taking into account the Gibbs–Duhem relationship

$$\frac{\partial F^j}{\partial n_B^j} = \mu_B^j - \mu_A^j. \tag{5-21b}$$

The two equations taken together give

$$\mu_A^j = F^j - v_B^j \frac{\partial F^j}{\partial n_B^j};$$

$$\mu_B^j = F^j + (1 - v_B^j) \frac{\partial F^j}{\partial n_B^j}. \tag{5-21c}$$

The geometrical interpretation of this is the common tangent condition (fig. 5.6(c)).

In the case of regular solutions, *for ε > 0 and low temperatures*, we have precisely the situation depicted in fig. 5.6(b) (fig. 5.7(c)). The energy of mixing has the opposite sign to $(- TS^M)$ and thus cannot be compensated by this entropy term at low temperatures, except at low concentrations v_A, v_B. Due to the singular behaviour of

$$\frac{\partial S^M}{\partial v_i} \bigg|_{v_i \to 0}$$

discussed in section 5.2.1 the entropy of mixing term then predominates. $ε > 0$ signifies that $(\varepsilon_{AA} + \varepsilon_{BB})/2$ has a value less than ε_{AB}, i.e. that like atoms form nearest neighbour pairs, which corresponds to the tendency discussed above to decomposition into two phases. We shall discuss the temperature dependence of this case in greater detail in the following. As shown in fig. 5.7(a) the case of $ε < 0$ is qualitatively less interesting. The homogeneous

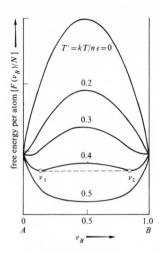

Fig. 5.8. Free energies of a regular solution with $\varepsilon > 0$ for different reduced temperatures.

alloy is stable, nearest neighbours are preferentially of a different species and 'short-range order' is established (chapter 7). $(\partial^2 F/\partial v_B^2) > 0$ over the whole range and though this is a necessary condition it is by no means sufficient for stability once other phases are present as shown by the common tangent region in fig. 5.7(c).

5.5.2   Solubility lines

In fig. 5.8 the free energy per atom $F(v_B)/N$ (5-15b) is plotted for various normalized temperatures $T' = kT/n\varepsilon$ and $\varepsilon > 0$ for the model of the regular solution. It can be seen that for temperatures $T' \leqslant 0.5$ a *miscibility gap* exists as demonstrated by the tangent construction, in this case a horizontal tangent. The miscibility gap is closed at a critical point, the temperature T_c of which is obtained from the maximum energy of mixing for $v_B = \frac{1}{2}$, namely $(2/N)E^M|_{max} = kT_c$. The boundary of the two-phase region is given by the condition $(\partial F/\partial v_B) = 0$; hence the positions $\check{v}_B$ of the $F^M(v_B)$ minima can be calculated from

$$n\varepsilon(1 - 2\check{v}_B) + kT[\ln \check{v}_B - \ln(1 - \check{v}_B)] = 0. \tag{5-22}$$

The boundary $\check{v}_B(T)$ of the two-phase region is plotted in fig. 5.9. We have thus calculated the phase diagram for the alloy at low temperatures. In the two-phase region an α and an α' phase of compositions v^α, $v^{\alpha'}$ given by the intersection of the lines $T = \text{const.}$ (conodes) with the $\check{v}_B(T)$ curve exist together in equilibrium. Whereas the average composition of the alloy

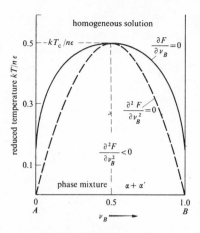

Fig. 5.9. Solubility curve and spinodal according to the model of the regular solution.

varies between $\check{v}_B = v^\alpha$ and $\check{v}_B = v^{\alpha'}$ the compositions of the two phases remain constant for a fixed temperature. It is only the proportion which changes in the manner described by the lever rule. The composition of the two phases corresponds to the saturation concentration of the component A in B and vice versa, the so-called solubility limit. The phase boundary curve can therefore also be viewed as a solubility curve $\check{v}_B(T)$. Its mathematical form, (5-22), can be simplified for $\check{v}_B \ll 1$ to

$$\check{v}_B \approx \exp\left(-\frac{n\varepsilon}{kT}\right). \tag{5-23}$$

A curve of this form is indeed observed, fig. 5.10, although the pre-exponential factor is greater than 1. C. Zener has interpreted this as a factor $\exp(S_V/k)$ where S_V is an additional vibrational entropy or positional uncertainty caused by the lattice distortion due to solute atoms (see chapter 6). The locus $\tilde{v}_B(T)$ of the points of inflection of the $F(v_B)$ curves $\partial^2 F/\partial v_B^2 = 0$ obtained by differentiation of (5-22) is also shown in fig. 5.9. This is the so-called *spinodal*

$$\tilde{v}_B(1 - \tilde{v}_B) = \frac{kT}{2n\varepsilon}. \tag{5-24}$$

In the discussion of the kinetics of precipitation we shall see that the components of the alloy within the spinodal line (i.e. for $\partial^2 F/\partial v_B^2 < 0$) can simply diffuse apart because the sign of the so-called 'chemical' diffusion coefficient changes at the spinodal with the result that diffusion reinforces

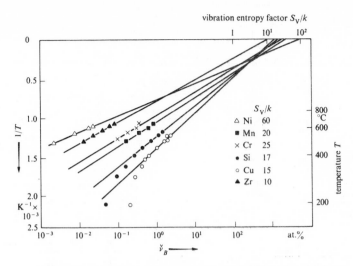

Fig. 5.10. Experimentally determined solubilities in aluminium with Zener's vibrational entropy factor.

concentration differences rather than smoothing them out. Between $\tilde{v}_B$ and $\tilde{v}_B(T)$ on the other hand decomposition must proceed by means of nucleation and growth (see chapter 9). Naturally all the above considerations are valid only qualitatively in the regular solution model which must where necessary be corrected using thermochemical measurements and better theoretical models (see [5.13]).

5.6   Free energies and binary phase diagrams [1.2], [5.7]

The simplest case is a system which is completely if not ideally miscible in both the liquid and solid states. $F^L(v)$, $F^S(v)$ are both parabolae, the relative position of which changes with temperature as shown in fig. 5.11. If the melting points of the pure metals are different the S parabola must be tilted. Intersections and tangent constructions then become possible yielding the well known phase diagram with a liquidus and solidus and otherwise complete miscibility. Such $F(v)$ parabolae have been measured thermochemically for some systems and the corresponding phase diagrams derived. The diagram changes as ε becomes increasingly positive, i.e. $E^M > 0$. First a solubility gap appears, then a congruently solidifying alloy and finally a *eutectic system* as in fig. 5.12. The sequence of $F(v)$ curves for S and L at different temperatures for the eutectic system is shown in fig. 5.13. The eutectic temperature T_E is characterized by coincidence of the tangents to the SL two-phase regions $L + \alpha_1$ and $L + \alpha_2$ (T_4 in fig. 5.13). A

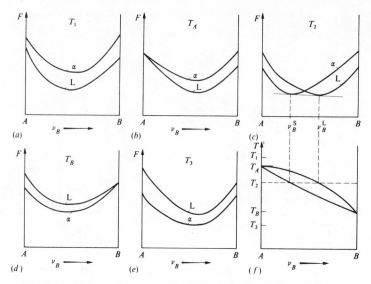

Fig. 5.11. (Melt L, solid phase S = α.) Derivation of a phase diagram from free energy curves for different temperatures.

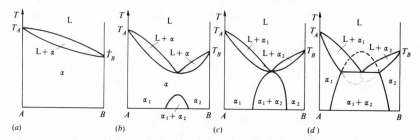

Fig. 5.12. Changes in the phase diagram from (a) to (d) with increasing ε, i.e. increasing energy of mixing $E^M > 0$.

similar construction is obtained if the $F^S(v)$ curve is replaced by two $F^{S\alpha}(v)$, $F^{S\beta}(v)$ parabolae corresponding to two crystals of different structures. In a thermodynamic treatment of an alloy of two elements with different structures it is necessary to take into account the energies and entropies which must be expended to produce a uniform structure, even if it is metastable for one component. L. Kaufman (see chapter 6, [6.4]) has shown how these quantities can be obtained thermodynamically and how with this knowledge and assuming the model of the regular solution it is possible to derive many of the phase diagrams between transition metals surprisingly well.

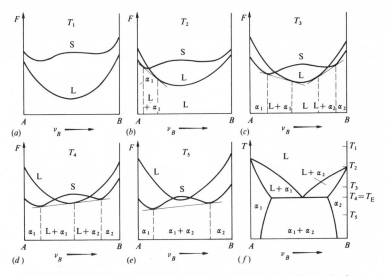

Fig. 5.13. Derivation of a phase diagram with eutectic from the free energy for S and L, dependent on the temperature.

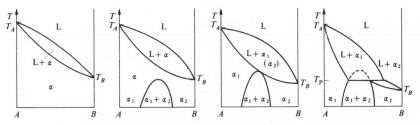

Fig. 5.14. As fig. 5.12 but for the case where the pure metals have widely differing melting points.

If the melting points of the pure components of a binary system are very different, a completely different phase diagram is obtained as the energy of mixing increases, $E^M > 0$, as shown in fig. 5.14. A *peritectic reaction* takes place when the minimum $F^S(v)$ corresponding to α_2 intersects the tangent joining $F^{S\alpha_1}$ and F^L (instead of the F^L curve itself), fig. 5.15. The peritectic diagram in the neighbourhood of T_p is the mirror image of a eutectic diagram, in other words one with an inverted temperature scale. There is a peritectic point at which an α_2 phase of the appropriate composition melts and simultaneously decomposes.

Finally we consider the phase diagrams of alloys which form *intermetallic compounds* (chapter 6) of more or less stoichiometric composition $A_x B_y$.

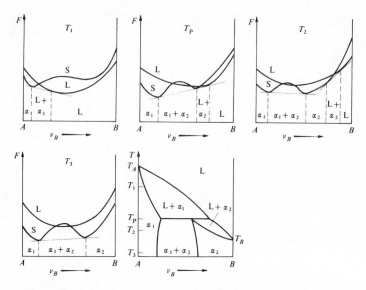

Fig. 5.15. Derivation of a phase diagram with peritectic from $F^L(v_B, T), F^S(v_B, T)$.

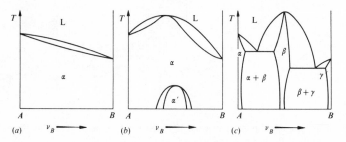

Fig. 5.16. Changes in the phase diagram from (a) to (c) with increasingly negative ε or E^M. β is the intermetallic compound A_xB_y.

Clearly this will be the case when unlike nearest neighbours are energetically favourable, $\varepsilon < 0$. Fig. 5.16 shows how, with increasingly negative energy of mixing, E^M, the phase diagrams change until the intermetallic compound A_xB_y forms from the melt. In spite of its steep $F(v)$ parabola this often occurs over a range of compositions. (Whether or not the composition A_xB_y falls within the homogeneity range of the compound depends on the position of the $F(v)$ parabolae of the neighbouring phases as can be seen from fig. 5.17.) Such intermetallic compounds often form at relatively high temperatures directly from the melt. Liquidus and solidus form a common maximum and the phase diagram divides into two

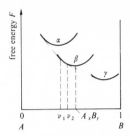

Fig. 5.17. Free energies of three phases α, β, γ. The compound is stable between v_1 and v_2 but not at its stoichiometric composition $A_x B_y$.

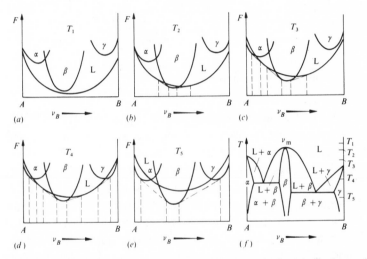

Fig. 5.18. Free energies and the phase diagram derived from them with the intermetallic compound β, which forms directly from the melt L.

diagrams of the usual type, one extending between A and $A_x B_y$ and the other between $A_x B_y$ and B, see fig. 5.18. An intermetallic compound is often formed by a peritectic reaction from L + α, see fig. 5.19. Such a compound melts incongruently, in contrast to the congruently melting compound of fig. 5.18.

If we consider that all the above phase diagrams can degenerate when one or more of the constituent phases expands its field or contracts it to nothing and that up to now we have assumed *complete miscibility in the liquid state* and completely ignored the frequently observed monotectic decomposition $L_1 \rightarrow L_2 + \alpha$, it becomes possible to appreciate the enormous diversity of the binary metal phase diagrams depicted in Hansen

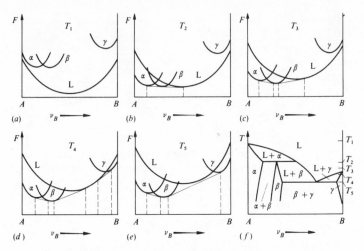

Fig. 5.19. Derivation of a phase diagram with an intermetallic compound β which forms peritectically.

[5.8] and supplementary volumes [5.9], [5.10], [5.10a]. These diagrams are further complicated by reactions in the solid state, which are induced by *allotropy*, i.e. the possibility that the components exist in different structural forms, or by energy relationships between different solid phases similar to those discussed above between S and L. The solid state reactions analogous to peritectic or eutectic decompositions are called peritectoid and eutectoid. These will be discussed in detail later. In order to understand such a variety of diagrams we need to investigate them not only by means of thermodynamics but also to study the atomistic and binding relationships.

5.7   Ternary phase diagrams [5.7]

Alloys with more than two components are quite common in metallurgy. Their composition is indicated on an equilateral triangle, which is subdivided by equidistant lines parallel to the three sides, fig. 5.20. The vertices represent 100% A, B or C. The alloys on the line PR consist of 70% A, on Ta of 10% B and on bS of 20% C. The point of intersection Q of the three lines of equal concentration thus corresponds to an alloy containing the three stated proportions of A, B and C. From elementary geometry the sum of the lengths $bQ + QP + aQ = 100\%$. In order to represent the temperature or free energy as function of the composition the third dimension is necessary. This requires a well developed capacity for visualizing things in three dimensions. Otherwise, as predicted by the Gibbs phase rule, most concepts in binary systems can be applied to ternary

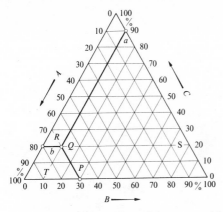

Fig. 5.20. Representation of the composition in a ternary system.

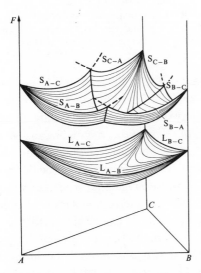

Fig. 5.21. Free energies of a liquid (L) and three solid (S) phases of a ternary system.

systems by raising them one dimension higher. In the following we shall consider one example only of a ternary system, namely one in which the limiting binary diagrams are all eutectic.

Fig. 5.21 shows the free energy surfaces for L and for the minimum of three necessary S phases over the concentration plane. At this temperature the melt is stable at all concentrations. On lowering the temperature, i.e. on lifting the F^L surface, the latter first intersects the F^S surface in the vicinity of A. Using a tangential plane construction, a two-phase region can be

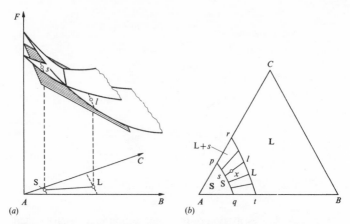

Fig. 5.22. A tangential plane construction to the free energy surfaces
defines equilibrium between *s* and *l* in the ternary system (*a*).
Isothermal section through a ternary phase diagram obtained in this
way with a two-phase region (L+S) and various conodes (*b*). The
quantities of *l* and *s* at point *x* are determined by the lever rule.

obtained for each temperature, see fig. 5.22(*a*). The result is represented in
an isothermal section as shown in fig. 5.22(*b*). In general the tangential
plane touches each of the two energy surfaces at one point only. By rolling
the plane under the two surfaces, all *s–l* points corresponding to coexisting
compositions of the two phases can be obtained. These are shown in fig.
5.22(*b*) joined by tie lines (conodes). (In the binary system the tie lines are
simply the $T = $ const. lines in the two-phase region.) It is necessary to show
the tie lines in the two-phase region of a ternary diagram in order to
describe the metallurgical state completely. It can be shown that tie lines
cannot intersect each other but apart from this their position cannot simply
be guessed. The proportions of the two phases on a tie line can be
determined using the lever rule. On further cooling, i.e. raising the F^L
paraboloid, two-phase equilibria between *A*, *B* and *C*-rich solid solutions
are energetically the most favourable state near the binary systems *AB*, *BC*
and *CA* whereas in the centre of the triangle the F^L paraboloid is still the
lowest. Tangential planes touching the F^L paraboloid and those of the
corner solid solutions likewise yield two-phase regions (fig. 5.23). On the
other hand there are also three tangential planes each of which touches two
of the solid solution paraboloids and the F^L surface (at the points *x, y, z*). In
the triangle *x, y, z* *three phases* are in equilibrium, the compositions of which
are given by the corners of the triangle *x, y, z*. The proportions m_x, m_y, m_z of
these in each alloy *u* within the triangle can again be calculated using the

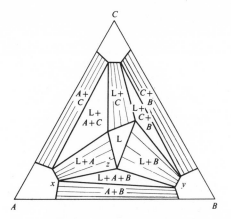

Fig. 5.23. Section through a ternary phase diagram at a temperature above the ternary eutectic temperature but below all the binary T_E.

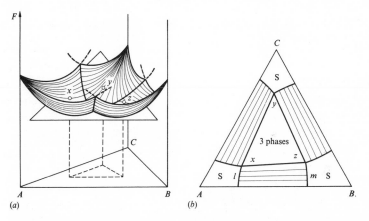

Fig. 5.24. A three-phase equilibrium in the ternary system derived by means of the tangential plane (*a*) and represented in an isothermal section (*b*). $(l-m)$=conode.

lever rule. If the triangle is pivoted at u and loaded with masses m_i at the corners, it should be in mechanical equilibrium. If we reduce the temperature still further, L shrinks to a point at which *equilibrium exists between four phases*: the melt and the three corner solid solutions (according to the phase rule there are no remaining degrees of freedom). This is the ternary eutectic point. At still lower temperatures we have equilibrium between three solid solutions (fig. 5.24). The complete ternary diagram is depicted in fig. 5.25.

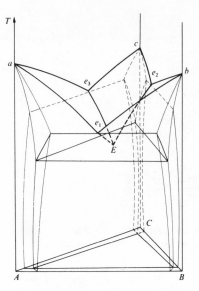

Fig. 5.25. Perspective drawing of a ternary phase diagram with ternary eutectic point E.

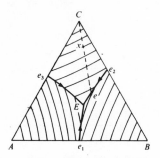

Fig. 5.26. Projection of the liquidus surface of the ternary diagram in fig. 5.25 on to the base.

If we want to follow the course of solidification as we did for binary systems in section 4.5.1.1 (assuming that equilibrium is established) the best method is to project the *liquidus surface* on to the base triangle (fig. 5.26). The contours indicate lines of equal liquidus temperature T_i. The ternary eutectic point E lies in the lowest position. The eutectic valleys $e_i E$ lead to this point. On cooling below T_L C-rich crystals solidify first in an alloy of composition x. The composition of the melt moves along the curve xe until it reaches the eutectic valley $e_2 E$. At this point C and B-rich solid solutions form as in a binary eutectic. Finally the composition of the melt reaches the point E where the remainder solidifies as a ternary eutectic. There are therefore three completely different stages of solidification for the same melt, namely of one, two and three phases successively.

Often vertical sections of ternary diagrams are used, but this has the disadvantage that it is not possible to recognize the composition of the constituent phases. If, however, the binary diagram is subdivided by an intermetallic compound β (fig. 5.18) into one of $A\beta$ and another of βB, then the ternary system is often subdivided along βC. A section along this quasi-binary system is informative and simple. In ternary systems and particularly in quaternary systems, topological considerations and analytical representations are of great importance. R. Vogel and G. Masing have shown that neighbouring regions in ternary systems differ only by one phase, the other phases remaining the same. The rule can be extended to 'poly'nary systems [5.11]. A simplification in phase diagrams where $n > 3$ is that there are probably no quaternary intermetallic compounds so that a knowledge of the thermochemical data for the component binary systems would suffice to determine the diagrams for higher systems. For many years a metallurgist was defined as someone who could interpret phase diagrams. For this, thermodynamics is an indispensable aid.

6
Structure and theory of metallic phases

6.1   Two important binary systems
6.1.1   *Copper–Zinc* (*'Brass'*) [2.1]

Fig. 6.1 shows the phase diagram, which is characterized by a series of five peritectic reactions from the copper to the zinc-rich side. At 73 at.% zinc the δ-phase undergoes eutectoid decomposition into γ and ε on cooling.

The α-phase is close packed cubic, i.e. fcc like copper; the β-phase is body centred cubic (bcc). Below 468–454 °C an ordered arrangement β' of Cu and Zn atoms (present in the ratio of approximately 1:1) forms on the bcc lattice, as shown in fig. 6.2. It is interesting that the β field widens with increasing temperature above the $\beta'\beta$ transition temperature and that consequently the solubility of Zn in Cu is reduced, contrary to (5-22). The gain in entropy on disordering apparently stabilizes β, i.e. lowers the F_β parabola with increasing temperature by more than F_α and F_γ are lowered. The γ-phase is complex cubic with 52 atoms per unit cell. It can be imagined as consisting of $3 \times 3 \times 3$ unit cells of the bcc β'-brass from which 2 atoms have been removed and the remainder slightly displaced. δ also has a very large and complicated cubic unit cell. ε and η, on the other hand, are hexagonal structures, described as close packed hexagonal or cph. The expression cph really implies an 'ideal' axial ratio $(c/a)_{id} = \sqrt{8/3} = 1.633$ whereas that of ε is less than ideal and that of η is greater than ideal, see fig. 6.3 for an idealized brass diagram. In this, a third cph phase ζ with an almost ideal axial ratio has been included instead of the β phase. This phase appears in CuZn alloys only as the result of a martensitic transformation after deformation at low temperatures (see chapter 13), whereas it is stable at the temperatures at which ε and η are stable in analogous systems such as Ag–Cd. The different c/a of these cph structures thus produce different phases.

The primary objective of this chapter is to systematize the many structures which arise even in simple binary phase diagrams and to justify

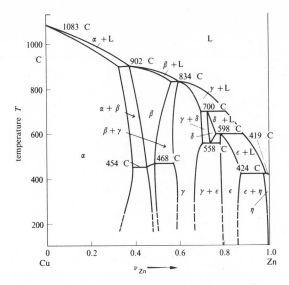

Fig. 6.1. Copper–zinc phase diagram [2.1].

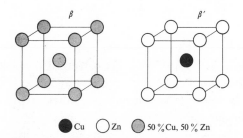

Fig. 6.2. Occupation of lattice sites in β and β'-brass.

their existence in physical terms. We cannot do more here than present the preliminary formulations of a theory of alloy structures or offer structural arguments. First of all, however, we must discuss a second important system in which the pure metal can exist in several stable forms and the solute atom is accommodated on interstitial sites.

6.1.2 *Iron–Carbon ('Steel')* [6.1], [6.2]

Fig. 6.4 shows the phase diagram for iron in equilibrium firstly with graphite and secondly with the reasonably stable compound cementite Fe_3C. (Fe_3C is orthorhombic and has a unit cell containing 12 Fe and 4 C atoms.) The differences in the two phase diagrams are quite small except in the cast iron region, i.e. the region of low melting point Fe–C alloys in the

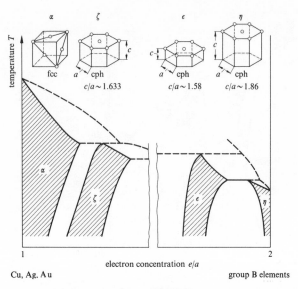

Fig. 6.3. Idealized phase diagram for brasses and related alloys with regions of the fcc and cph phases.

neighbourhood of the eutectic (17.3 at.% C). The eutectic (ledeburite) and the eutectoid reaction (formation of pearlite occurring at 3.61 at.% C as a result of the allotropy of Fe) are the controlling features of this phase diagram. In addition there is a peritectic reaction at high temperatures and low C contents as well as a low temperature metastable phase, martensite, which is not shown. (This will be treated in detail in chapter 13.)

Pure iron is body centred cubic below 911 °C (α-iron, 'ferrite') and above 1392 °C (δ-iron), and face centred cubic (γ-iron, 'austenite') between these temperatures. δ is obviously the high temperature continuation of α. The ferromagnetic transition in α at $T_{CF} = 768$ °C is second-order. It does not itself produce a new phase but from a shrewd analysis of the specific heats, F. Seitz and C. Zener concluded that it necessitates the γ–δ transformation. γ-iron is antiferromagnetic below 80 K, as has been demonstrated with small fcc Fe particles stabilized in Cu. This is astonishing as (unlike the bcc structure) the fcc lattice does not allow an arrangement of antiparallel spins for all nearest neighbours because the nearest neighbours of an fcc atom are nearest neighbours to each other. Both magnetic states stabilize the respective lattice structures at low temperatures; the decrease in energy associated with the ferromagnetic transition is obviously the larger because

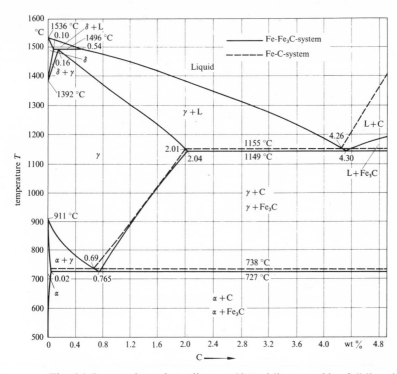

Fig. 6.4. Iron–carbon phase diagram (dotted lines graphite, full lines in equilibrium with Fe_3C) [5.8].

α is stable at $T = 0$. From thermodynamics

$$S = -\partial F/\partial T = \int (c_v/T)\,dT. \qquad (6\text{-}1)$$

At high temperatures the entropy of magnetic disordering, i.e. the extra specific heat c_v which must be supplied at T_{CF}, likewise stabilizes the structure. Fig. 6.5 shows these c_v curves for fcc and bcc iron with extrapolations, which from a knowledge of substitutional iron alloys are quite feasible. The free energies of the two phases F_α, F_γ can be obtained from these and hence $\Delta F^{\alpha\gamma} = F_\alpha - F_\gamma$ as shown in fig. 6.6. γ is stable when $\Delta F^{\alpha\gamma} > 0$ and vice versa. Zener splits $\Delta F^{\alpha\gamma}$ into two contributions: one, 'NM', is obtained by extrapolation from low temperatures and stems partly from the large vibrational specific heat of γ, i.e. as a result of $c_v \propto (T/\Theta)^3$ from its lower Debye temperature Θ. On the other hand the specific heat required to destroy the antiferromagnetic order also stabilizes the γ structure. Although these two effects cause the specific heat of γ *at low temperatures* to be larger

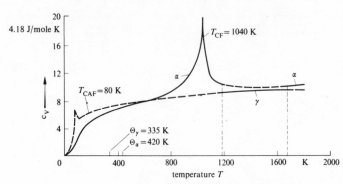

Fig. 6.5. Specific heats per mole for α and γ-iron. After [6.2] and [6.4].

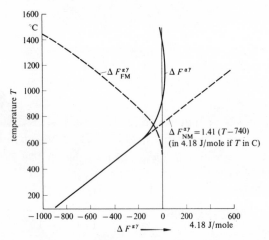

Fig. 6.6. Difference in the free energies for α and γ-iron, resolved according to C. Zener [6.2].

than that of α, due to the double integration they affect F_γ only *at high temperatures*, with the result that iron finally transforms from α to γ at 911 °C. The above entropy considerations give no clue as to why α with a lower specific heat is stable at all at low temperatures. Based on thermochemical considerations, L. Kaufman [6.4] has concluded that were iron not ferromagnetically ordered, it would have a cph structure. Such a structure is observed in iron subjected to a hydrostatic pressure of 15.2 MPa. This demonstrates that pressure is an important additional parameter in the study of the stability of metal phases. It is therefore the ferromagnetism which reduces the internal energy of α at low temperatures

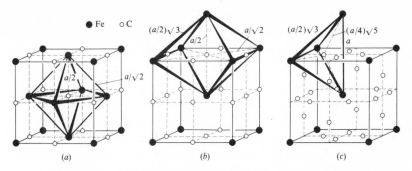

Fig. 6.7. Interstitial lattice sites for carbon in iron. (a) octahedral interstices in γ-(fcc) iron; (b) octahedral (c) tetrahedral interstices in α-(bcc) iron.

to such an extent that (at atmospheric pressure) it becomes the stable low temperature phase.

At the ferromagnetic Curie temperature T_{CF} extra specific heat must again be supplied to disorder the spins. There is a resulting decrease in F_α and the difference $\Delta F^{\alpha\gamma}$ plotted in fig. 6.6 changes in the reverse direction to that at low temperature. Zener explains this by a second ferromagnetic component $\Delta F^{\alpha\gamma}_{FM}$ which he adds to the low temperature branch $\Delta F^{\alpha\gamma}_{NM}$ in fig. 6.6. This stabilizes α (i.e. δ) again at high temperatures as can be demonstrated by the following estimate. On account of the three possible orientations of the atomic magnetic moment in the lattice in the disordered state, the gain in magnetic entropy at the Curie point $\Delta S = k \ln 3 = 2.2\, \text{cal/mol K} = 9.2\, \text{J/mol K}$. This corresponds to the change in slope $\partial \Delta F / \partial T = -\Delta S$ of the curve in fig. 6.6 from 1.41 cal/mol K = 5.91 J/mol K below T_{CF} to -0.7 cal/mol K = -2.93 J/mol K well above T_{CF}. It is therefore the next but one phase change ($\gamma \rightarrow \delta$) rather than the next phase change above T_{CF} which is quantitatively determined by the process occurring at T_{CF}. Also the temperature for the $\alpha-\gamma$ transofrmation is raised by the ferromagnetic transition as shown by fig. 6.6.

The *addition of carbon* obviously extends the γ field and reduces the range of stability of α on the temperature axis. A maximum of 8.9 at.% C dissolves in γ (more usually expressed as 2.0 wt.%) but only 0.095 at.% (0.02 wt.%) C in α. This is due to the size of the *interstices* which are available and occupied in the fcc and bcc structures. Fig. 6.7(a) shows the octahedral interstice in γ, the largest available in the fcc structure. Fig. 6.7(b) shows the octahedral interstice in α. There is a larger tetrahedral interstice (fig. 6.7(c)) in α but it is mostly unoccupied. The radius of the C atom is 0.08 nm, that of the octahedral interstice in γ 0.059 nm, and that of the octahedral interstice in α 0.019 nm (compared with 0.036 nm for the tetrahedral interstice). In all estimates it has been assumed that the iron ions are spherical and just touching (not as shown in fig. 6.7). The advantage of the α octahedral interstice is obviously its anisotropy which means that when it is occupied, only two iron ions need to be displaced as compared with the four iron atoms surrounding the isotropic tetrahedral interstice, fig. 6.7(c).

The anisotropy of the distortion around the C atom in α-iron causes the C atom to changes its site when a uniaxial stress is applied to the lattice (e.g. in the horizontal cube direction in fig. 6.7(b)). In this way, the distortional energy is reduced. A stress applied in the $\langle 111 \rangle$ direction, on the other hand, does not result in a change of site, because all three cube directions are equally stressed and on average equally occupied by C atoms. This carbon site change leads to anelastic effects ('Snoek effect') as described in section 2.7 and is used extensively to measure the carbon content in α-iron (and also the frequency of site change (diffusion) of C in α-Fe, see section 8.2.3). There is naturally no such effect in γ-Fe since the interstices are symmetrical. The local tetragonal distortion of α-Fe by carbon gives rise to high strength, particularly also in the low temperature metastable phase martensite, which as a result of the alignment of all the dipolar distortion fields is in fact body centred tetragonal (see chapter 13).

Finally, it should be mentioned here that ternary iron alloys with interstitial carbon atoms and other substitutional alloying additions are technologically very important, for example as stainless steels. These additions change the γ field in various ways:

(a) Elements which expand the γ field such as Ni, Co, Mn produce austenitic, i.e. non-ferromagnetic, often easily workable, and stainless steels (fig. 6.8).

(b) Elements which restrict the γ field, e.g. Al, Si, Ti, Mo, V, Cr, Nb form carbides and intermetallic phases with α-Fe (fig. 6.9).

γ-Fe and Ni are both fcc and form a continuous solid solution over a wide temperature range, apart from the formation of an ordered phase $FeNi_3$ below 500 °C. α-Fe and Cr are both bcc and they too form a continuous solid solution at high temperatures. At 815 °C, however, the intermetallic σ-phase appears, see fig. 6.10 and section 6.4.3. In addition, thermochemical measurements by O. Kubaschewski indicate a real tendency to decomposition in this system into two bcc phases α, α' which are not detectable metallographically and thus do not appear in the phase diagram, but have been found in a recent FIM study [2.28].

Zener has correlated the change in the γ field by substitutional alloying with the $\Delta F^{\alpha\gamma}(T)$ diagram, fig. 6.6. The addition favours either α or γ, i.e. makes $\Delta F^{\alpha\gamma}$ either negative or positive and to a first approximation linearly

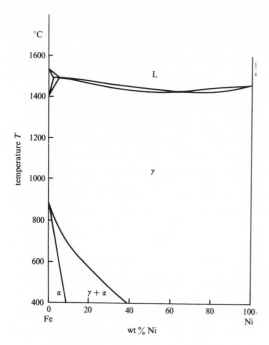

Fig. 6.8. Iron–nickel phase diagram.

dependent on the solute concentration c_{sub}. The abscissa ($\Delta F^{\alpha\gamma}$) in fig. 6.6 can thus be regarded as the c_{sub} axis and thus the $\Delta F^{\alpha\gamma}(T)$ diagram as a (double) phase diagram. The concentration of the γ-restricting elements which favour α should be imagined plotted instead of $\Delta F^{\alpha\gamma} > 0$, whereupon the γ-'belly' appears, e.g. of FeMo in fig. 6.9(a). Instead of $\Delta F^{\alpha\gamma} < 0$ we have the concentration of the γ-widening elements such as Ni, see fig. 6.8. If we take into account the fact that additions of chromium also decrease the Curie temperature of iron, a 'paunch' is obtained, as in the case of FeCr, fig. 6.9(b).

6.2   Structures of pure metals and elastic instabilities

Before attempting to understand why certain structures arise in binary systems we must first provide a physical justification for the structures of the pure metals. This is a problem for the electron theory of metals which is still largely unresolved even for the 'simple' metals of the periodic table. The reason for this is that the binding energy E_0 of a metal is large independent of its structure, i.e. atomic arrangement. For example, the heat of transformation of sodium from the bcc to the cph structure at 36 K is only $E_0/1000$. Very good models and accurate calculations are thus required before it can be predicted, on energy considerations, which

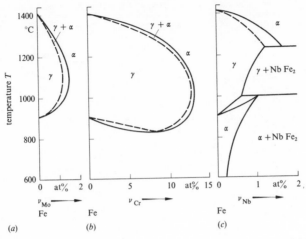

Fig. 6.9. Closed γ regions in the phase diagrams of (a) FeMo; (b) FeCr; (c) FeNb.

structure will be adopted by a given metal. The question is not as academic as many solid state physicists are inclined to assume, 'because the stable structure is known in any case'. The metallurgist wants to know *how stable* the structure is *compared with other* possible structures which might be promoted by alloying. Furthermore, in allotropic transformations, the difference in free energy between the various modifications is directly apparent as the driving force.

At the present time, the *pseudo-potential method* [6.3] is used extensively to calculate the energy of a metal in which the atoms are bound by s and p electrons. In this, the electrons can be considered as virtually free by starting from a screened effective potential due to the metal ions. This is then small compared with the Fermi energy and hence a perturbation calculation is possible. Essentially, the electron energy is dependent only on the specific *volume* of the metal, which is in fact influenced little by changes in the *structure* (see section 6.4). The next smallest contribution to the energy can be expressed by a spherically symmetrical interaction potential between atoms (pair potential), which will be used below. Unfortunately the method fails as soon as d (or f) electrons participate and this is the case in almost all metallurgically interesting elements. Very few electron theoretical estimates have yet been made of the structure dependent energy contributions. They are based on the so-called tight-binding method with atomic d-orbitals [6.15]. Accordingly a systematic trend can be observed in the structures on changing from the typical metallic bond described above to the covalent bond, see Kittel [1.1]. The present discussion must be limited to such qualitative observations. The entropy contribution, in this case of the lattice

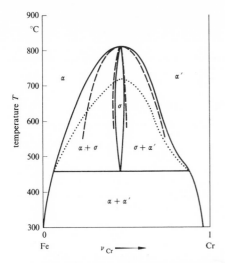

Fig. 6.10. Iron–chromium phase diagram below 900 °C after
O. Kubaschewski. (Full curves obtained thermochemically, dashed
curve obtained metallographically. The dotted curve was calculated
from thermochemical measurements for the decomposition of the bcc
solid solution into $(\alpha + \alpha')$.)

vibrations, must also be considered if we are to investigate the stability of
definite structures at finite temperatures. The spectrum of lattice vibrations
should in principle follow from electron theory since it results from the
reaction of electrons to lattice distortions. We have already encountered the
influence of the entropy of ferromagnetic ordering on the structure of iron
in the previous section. From an empirical standpoint the heats of reaction
for allotropic transformations are particularly interesting because they
demonstrate the structural component of the binding energy.

6.2.1   *Close packing*

Two forms of close packing ($n = 12$ nearest neighbours) exist in
pure metals, fcc and cph. The former is observed in the noble metals in
Group I: Cu, Ag and Au, and also in the neighbouring transition metals of
Group VIII and trivalent aluminium (see table 6.1). Hexagonal close
packing with an almost ideal axial ratio is found only in transition metals
such as Co (Group VIII, below 420 °C) and divalent Mg. The divalent
elements Zn ($c/a = 1.86$) and Cd (1.89) have the cph structure with a non-
deal axial ratio. Here there are essentially only six nearest neighbours (in the
basal plane) and the influence of the $n = (8 - N)$ rule becomes evident.
$N = $ group number in the periodic table, see section 6.2.4. Being close
packed the fcc and cph structures are closely related to one another and
differ only in the stacking sequence of $\{111\}$ planes, see fig. 6.11. In the fcc

Table 6.1. *Periodic system (extract) showing crystal systems and axial ratios*

Group number:						
IA	IIA	IIIA	IVA	VA	VIA	VIIA

Electronic configuration:

s^1	s^2	ds^2	d^2s^2	d^3s^2	d^5s	d^5s^2	d^6s^2
^{3}Li	^{4}Be						
cub 8	hex 12						
hex 12	[1.57]						
[1.64]							
^{11}Na	^{12}Mg						
cub 8	hex 12						
hex 12	[1.62]						
[1.63]							
^{19}K	^{20}Ca	^{21}Sc	^{22}Ti	^{23}V	^{24}Cr	^{25}Mn	^{26}Fe
cub 8	cub 8	hex 12	cub 8	cub 8	cub 8	cub 8	cub 8
	cub 12	cub 12	hex 12			cub 12	cub 12
			[1.59]			complex	cub 8
^{37}Rb	^{38}Sr	^{39}Y	^{40}Zr	^{41}Nb	^{42}Mo	^{43}Tc	^{44}Ru
cub 8	cub 8	cub 8	cub 8	cub 8	cub 8	hex 12	hex 12
	hex 12	hex 12	hex 12			[1.60]	[1.58]
	[1.64]	[1.57]	[1.59]				
	cub 12						
^{55}Cs	^{56}Ba	$^{57-71}$La	^{72}Hf	^{73}Ta	^{74}W	^{75}Re	^{76}Os
		Rare earths					
cub 8	cub 8	cub 12	cub 8	cub 8	cub 8	hex 12	hex 12
		hex 12	hex 12			[1.62]	[1.58]
		[1.61–1.58]	[1.58]				
		^{90}Th		^{92}U		^{94}Pu	
		cub 8		cub 8		complex	
		cub 12		complex			

Key: Crystal systems: cub=cubic; hex=hexagonal; tetr=tetragonal; rh=rhombic; rhd=rhombohedral or complex structure. (With the number of their nearest neighbours.)

Table 6.1—*continued*

Group number: VIII		IB	IIB	IIIB	IVB	VB	VIB
Electronic configuration: d^7s^2	d^8s^2	$d^{10}s^1$	$d^{10}s^2$	s^2p	s^2p^2	s^2p^3	s^2p^4
				^{5}B complex	^{6}C cub 4 Graphite (3)	^{7}N hex 12 [1.65] cub 1 & 7	^{8}O cub 12 rhd 6 complex
				^{13}Al cub 12	^{14}Si cub 4	^{15}P rh 3	^{16}S complex
^{27}Co cub 12 hex 12 [1.62]	^{28}Ni cub 12	^{29}Cu cub 12	^{30}Zn hex 12 (6) [1.86]	^{31}Ga rh 7 (1)	^{32}Ge cub 4	^{33}As rhd 3	^{34}Se complex hex 2 [1.14]
^{45}Rh cub 12	^{46}Pd cub 12	^{47}Ag cub 12	^{48}Cd hex 12 (6) [1.89]	^{49}In tetr 4	^{50}Sn tetr 4 cub 4	^{51}Sb rhd 3	^{52}Te hex 2 [1.33]
^{77}Ir cub 12	^{78}Pt cub 12	^{79}Au cub 12	^{80}Hg rhd 6	^{81}Tl cub 8 hex 12	^{82}Pb cub 12	^{83}Bi rhd 3	^{84}Po complex cub 6

Where the element exists in more than one crystal form, the lattices are given one above the other in order of increasing temperature.
Axial ratio [c/a].

structure the sites occupied in succeeding close packed planes in the $\langle 111 \rangle$ direction are in the order $ABCABC$, and in the cph structure $ABABAB$. A displacement in a $\{111\}$ plane by the vector $(a/6)\langle 112 \rangle$ produces an 'intrinsic' stacking fault in the fcc structure, i.e. a cph layer one atom thick, $ABCAB|ABCABC$. A stacking fault is produced in the cph structure in an analogous manner $ABAB|CAC$. The stacking fault energy γ per unit area of the fault measures (in two dimensions) the difference in (free) energy of the two packing modes. At the temperature of an allotropic transformation fcc–cph (as in Co) $\Delta F^{xy} \approx \gamma/a$ equals zero (a = separation of $\{111\}$) and

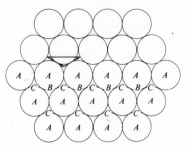

Fig. 6.11. Close atomic packing in plane (A) with two sets of possible sites in the next plane (B or C). The vectors CB or BC introduce stacking faults and combine to an identical translation CC.

Table 6.2. *Stacking fault energy in mJ/m² at 300 K for the stable structure*

Cu	Ag	Au	Al	Ni	Co	Zn
60	20	40	200	130	25	250

indeed stacking faults are very frequently observed. The more the binding depends only on the specific volume of the metal, the smaller should be the value of γ. Directional (covalent) bonding contributions favour certain packing arrangements, and this corresponds to a relatively high γ. Various stacking fault energies measured by electron microscopy (see section 2.2.1.2) are presented in table 6.2.

L. Kaufman [6.5] has deduced the enthalpy $\Delta H^{\alpha\to\zeta}$ and entropy difference $\Delta S^{\alpha\to\zeta}$ for transition metals in the often metastable fcc (α) and cph (ζ) structure from thermodynamical measurements and an analysis of binary phase diagrams. The results are shown in fig. 6.12. It is interesting that ΔH and ΔS depend essentially only on the group number, in rare cases also on the period. The anomaly in Co obviously cannot be attributed to a magnetic energy contribution, since both α and ζ are ferromagnetic. $\Delta H_{\text{Al}}^{\alpha\to\zeta} = 5489$ J/mol and $\Delta H_{\text{Mg}}^{\alpha\to\zeta} = 1948$ J/mol can both be included in fig. 6.12. In the case of Co, ΔH and ΔS have the same sign and hence at a temperature T_u the $\zeta\to\alpha$ transformation must occur according to $\Delta H - T_u \Delta S = 0$. Co becomes fcc above 420 °C. At the beginning of the long periods the cph structure is strongly favoured by ΔH and also but to a lesser extent by ΔS. In the case of Co they oppose one another but otherwise in Groups VIII and IA both tend slightly to promote the fcc structure. If the melting point of Zn were higher than $\Delta H/\Delta S = (-1885$ J/mol)/

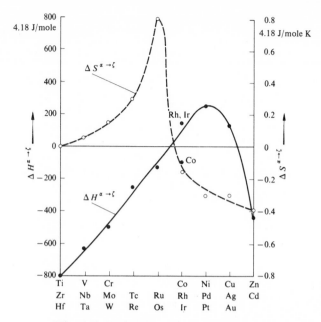

Fig. 6.12. Enthalpies and entropies of transformation necessary to convert one mole of a transition metal from the fcc (α) to a cph (ζ) structure. (The stability of α is guaranteed by (among other things) $\Delta H^{\alpha \to \zeta} > 0$ and $\Delta S^{\alpha \to \zeta} < 0$; that of ζ by the reverse sign.) After [6.4].

(-1.7 J/mol) ≈ 1100 K (instead of 730 K) Zn would adopt the fcc structure on entropy grounds (or the bcc structure according to fig. 6.13).

Fortunately it has been possible recently to derive a curve qualitatively similar to the empirical curve in fig. 6.12 using electron theory, although many assumptions are involved [6.6]. The pseudo-potential method, which does not take into account the d and f electrons, predicts cph (or bcc) as the stable structure for Cu, Ag and Au, i.e. a negative γ (fcc) which is contrary to experience. The binding due to the (inner) d electrons has certain similarities to covalent bonding (see section 6.2.3) although in the case of close packing each atomic wave function participates in several interatomic bonds. There are thus no real energy gaps between bonding and antibonding d electron states in the transition metals but there is a minimum in the middle of the d band which manifests itself in various properties, particularly the specific electronic heat. According to Hund's rule the antibonding half band is first half filled with electrons of one spin direction and then completely filled with electrons of the other. This provides a qualitative explanation of the variation in atomic magnetic moment with the number of holes in the d-band (Slater curve), see [1.1], [6.8].

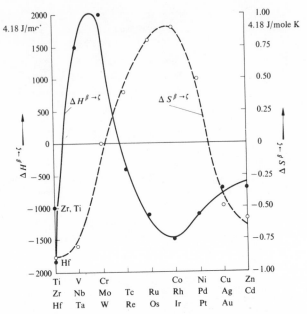

Fig. 6.13. As fig. 6.12 but for the transformation bcc (β) to cph (ζ).

6.2.2   *Body centred cubic structure* (bcc)

This structure is not as densely packed (eight nearest neighbours at $0.866a$ with, however, six next nearest neighbours at a distance of a). It is observed particularly in the transition metals of Groups IV and V over the whole temperature range of the solid phase. There is a series of allotropic transformations from bcc (β) to close packed structures which are the stable low temperature phases (α, ζ). We have already discussed iron. Other examples are Ti, Zr, Hf and also Na ($T_u = 36$ K) and Li ($T_u = 72$ K). The latter transformations are especially satisfying from the point of view of the theory because it is in these simplest metals each with one s electron per atom, that close packing and purely metallic binding, dependent only on volume, are expected. Relative thermochemical stability parameters $\Delta H^{\beta \to \zeta}, \Delta S^{\beta \to \zeta}$ can again be estimated for these structures, fig. 6.13. According to this ΔH and ΔS act in opposition in Hf, Zr and Ti. *The bcc structure in these elements is stabilized at high temperatures only by the entropy contribution.* The same applies for δ-Fe, Na, Li (and for many bcc alloys). We shall now examine the reasons for this.

We consider first a primitive cubic crystal in which the nearest neighbours are connected by springs, fig. 6.14. This crystal can be folded down on to the cube plane without stretching the springs. The lattice is therefore unstable with respect to this shear movement. The same applies

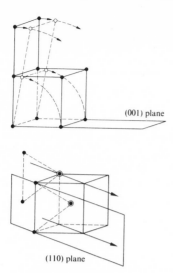

(001) plane

(110) plane

Figs. 6.14 and 6.15. Instability with respect to a shear in the primitive cubic (6.14) and bcc lattice (6.15) assuming nearest neighbour interaction with central forces.

for the bcc lattice on the {110} plane for a shear in the ⟨110⟩ direction, fig. 6.15. As long as the binding is determined by central forces which decline rapidly with distance (a realistic approximation, see [6.3]) so that the binding is restricted essentially to nearest neighbours, the bcc lattice is unstable to shear on the {110} plane. *At low temperatures* it collapses into a close packed structure (martensitic transformation, see chapter 13). *At high temperatures* on the other hand lattice vibrations with little stiffness, i.e. of low frequency, are associated with a high vibrational entropy ('uncertainty of position') which stabilizes the structure thermodynamically. (It can be seen in the theory of lattice vibrations that the vibrational entropy $k \ln(kT/\hbar\omega_j)$ at high temperatures exhibits a singularity when the vibrational frequency ω_j approaches zero [6.7a].) According to J. Friedel [6.7b] the bcc lattice already has a higher vibrational entropy than the fcc because of the smaller number of nearest neighbours which results in smaller vibrational frequencies ω_j.

6.2.3   (8 − N) structures [2.1]

The well known stability of the 8 electron shell (inert gas configuration) leads to characteristic crystal structures for elements in the IV–VI(B) groups in the periodic table. The atoms form covalent bonds as in the Heitler–London H_2 molecule (see [1.1]). They share electrons of opposite spin and thus complete their electron shells. A quadrivalent atom completes the eight-electron shell by covalent bonding with four nearest

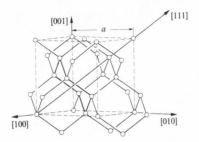

Fig. 6.16. Cubic unit cell of the diamond structure.

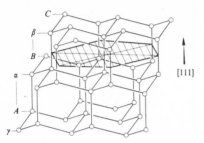

Fig. 6.17. Stacking sequence of the {111} planes in the diamond structure.

neighbours; a pentavalent atom by bonding to three nearest neighbours and a hexavalent atom by bonding to two nearest neighbours. Hume–Rothery's $(8 - N)$ rule therefore states that in covalent bonding between elements of group N, a structure is formed such that each atom has $(8 - N)$ neighbours. Covalent bonding occurs predominantly with the elements of the groups immediately preceding the inert gases. Those in Groups IV–VI(B) are mostly metals and semiconductors. There is often an energy gap between the bonding and antibonding electron states in the crystal (antiparallel and parallel spins for the hydrogen molecule); thus the crystal is a semiconductor if the number of electrons is even (see [1.1]).

In Group IV, C, Si, Ge (and grey tin) possess the diamond structure in which each atom is surrounded by a tetrahedral arrangement of four atoms, fig. 6.16. It can be formed from the fcc structure by double occupation of the {111} planes in the ⟨111⟩ direction, corresponding to a stacking sequence $A\alpha B\beta C\gamma A\alpha B\beta$, fig. 6.17. The structure is relatively open: only 34% of the space is occupied by spheres in contact compared with 74% in close packed metals (Pb, also in Group IV, is an fcc metal and is thus an exception to the $(8 - N)$ rule). The rule also breaks down in Group III for the fcc metal Al, the cph/bcc metal Tl and the metals Ga and In which have complicated structures (Ga seven nearest neighbours, In four nearest neighbours, see table 6.1). Two structures related to diamond, sphalerite and wurzite, are formed from so-called *III–V and II–VI compounds* in which one half of the atoms (from Group III(II)) occupy sites in planes *AB* . . . and the other half (from Groups V(VI)) the sites in the planes $\alpha\beta$ . . . . The average number of electrons per atom, four, is obtained by applying the $(8-N)$ rule overall. (In wurzite the stacking sequence is cph $A\alpha B\beta A\alpha$ . . . .) The electron transfer between the III and the V valent atoms is in reality only partial; the bonds in these compounds have some ionic character, e.g. GaAs, ZnO, see [1.1]. In Group V, As, Sb and Bi have a rhombohedral structure which can be obtained from the primitive cubic structure by distorting it in such a way that each atom has only three nearest neighbours (instead of four). Se and Te in Group VI have trigonal chain structures with two nearest neighbours. It is not intended here to attempt a complete analysis of the structures in the periodic system but only to illustrate the correlation between structure and binding using some metallurgically interesting examples. Such a correlation should assist in categorizing the structures which appear in binary alloys.

6.3   Hume–Rothery phases and electrons in alloys

6.3.1   *Hume–Rothery's rules* [6.9]

Based on their investigations, mainly on copper and silver, W. Hume–Rothery and co-workers have produced a series of rules showing which factors favour the appearance of specific phases in certain concentration ranges of binary *AB* alloys.

(*a*) If the *atomic radius* r_B of *B* differs by more than 15% from that of *A*, r_A, the α-phase (= structure of *A*) is unstable even at low concentrations of *B*. We shall relate this in section 6.4 to a limiting value of the elastic distortion energy relative to kT. In certain cases where the atomic size factor,

$$\delta = (r_B - r_A)/r_A, \tag{6-2}$$

is very large, however, compounds can be energetically favourable (see section 6.4). Also metallic glasses are then easily formed in many systems.

(*b*) If *B* is significantly more *electronegative* than *A*, i.e. if in an *AB* compound *B* attracts electrons more strongly than *A*, then characteristic chemical compounds $A_x B_y$ are formed in which the bonds have some ionic character. It is difficult to quantify the electronegativity of an atom. If two macroscopic pieces of the metals *A* and *B* are brought into contact, the chemical potentials of the electrons (i.e. Fermi energies) find a common level $E_{FB} = E_{FA}$ in that charge is transferred to the more electronegative metal. E_{FA} is thus a measure of the electronegativity EN_A of the component *A* of the compound. According to L. Pauling the exothermic heat of formation of the (ionic) compound is proportional to $\varepsilon = -(EN_A - EN_B)^2$, see section 6.5 for examples of such structures. Recently there have been numerous attempts to describe the ranges of stability of certain intermetallic compounds in form of two-dimensional structure maps [6.15]. Their more or less arbitrary variables are of course functions of the atomic size, valence and electronegativity of the alloy components. Miedema *et al.* [6.14] have recently calculated the heat of mixing of alloys from the EN_i which they obtained from the work functions of the elements. These are proportional to the Fermi energies. One of the coordinates of the structure map thus should be proportional to $-(\Delta EN)^2$ as in Pauling's case. The second, repulsive term to the heat of mixing arises from the electron density on the border between the Wigner–Seitz cells of the A and B atoms. This second Miedema variable relates therefore to ion sizes. The meaning of the Miedema variables in electron theory has recently been studied by Pettifor [6.15].

(c) Specific structures (electron phases) arise preferentially in characteristic ranges of the *valency electron concentration e/a*. This quantity indicates the number of all valency electrons in the alloy (corresponding to the numbers of the group Z_i of the components) per number of atoms, i.e.

$$e/a = Z_A(1 - v_B) + Z_B v_B. \tag{6-3}$$

In the following section we shall discuss this rule and the attempts to explain it in physical terms first made by H. Jones. It is clear, however, after what was said in chapter 5, that the appearance of a phase in the phase diagram depends not only on its own stability but also on that of neighbouring phases.

6.3.2   Electron phases [2.1]

We first consider alloys of Cu, Ag, Au with metals from the Groups IIB, IIIB, etc. ('B metals'). Like the solvent metals the α-phases are fcc. The range of solubility of the α-phases is determined largely by e/a if the parameters δ and ΔEN are favourable, in other words, small (sections 6.3.1(a) and (b)). Fig. 6.18 shows that the solubility of the B metals in Ag, that is the stability of α, extends approximately to $(e/a)_\alpha = 1.4$. The same applies for many binary alloys, see fig. 6.19 (for Au $(e/a)_\alpha$ is about 1.2–1.3). Rapid solidification in many cases produces higher solid solubilities [4.13]. These then agree better with the 'magic' e/a. At higher B concentrations corresponding to $(e/a)_\beta = 1.5 = 21/14$, a bcc β-phase is observed in numerous cases, fig. 6.20, or a cph ζ-phase. These structures and their relative stability have been discussed already in 6.1.1 and 6.2.2 as has also a possible low temperature ordered β' variant (which stabilizes the bcc structure, particularly against shear). The V-form of the β-field at high temperatures and the axial ratio c/a of ζ were also discussed. Finally, at $(e/a)_\gamma = 21/13 \approx 1.62$ the complex γ-phase is stable in many systems and at

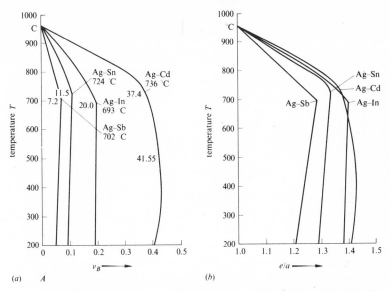

Fig. 6.18. Extent of the α-phase in silver solid solutions (*a*) plotted against mole fraction; (*b*) plotted against electron concentration.

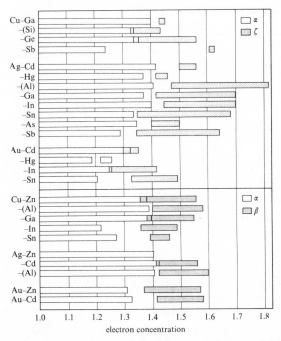

Fig. 6.19. Extent of the phases α (primary solubility), ζ and β for solid solutions of Cu, Ag and Au with *B*-metals. After [2.1].

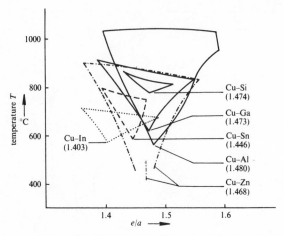

Fig. 6.20. Fields of the disordered β-phase in Cu solid solutions. After [2.1].

$(e/a)_\varepsilon = 21/12 = 7/4 = 1.75$ the hexagonal ε-phase, see section 6.1.1. What is the physical meaning of these magical $(e/a)_{\alpha,\beta,\gamma,\varepsilon}$ values?

It is not obvious that the electron theory of pure metals could be applied to disordered alloys, which lack the basic periodicity of the atomic arrangement. On the other hand the unchanged form of the Bragg reflections shows that the Brillouin zones (BZ) and the energy gaps at their boundaries are still present. The existence of the Fermi surface for the electrons in solid solutions is not in doubt either theoretically or experimentally (perhaps the Fermi surface is less sharp than in the pure metal). Based on this, H. Jones describes a density of states for valency electrons as a function of their energy, similar to that in the pure metal, fig. 6.21(a). The parabolic form is well known for free electrons. On alloying with an element of higher valency the additional electrons should simply fill up further levels in this 'rigid band'. This is the critical assumption made by H. Jones. Due to the energy gaps at the boundaries of the Brillouin zone, the Fermi body does not remain a sphere right up to point C (where a Fermi sphere would touch the Brillouin zone boundary). Already beyond the point A in fig. 6.21 contact occurs at 'necks' on the Fermi body, fig. 6.22. The density of states decreases from point B on; the corners of the first Brillouin zone are filled until at D electrons overlap the second Brillouin zone (see [1.1]). Fig. 6.21(b) shows the resultant electron energy as function of the Fermi energy E_F for a given structure. For another structure with other Brillouin zones contact occurs at a different Fermi energy level, i.e. at another electron concentration. The first contact of a hypothetical Fermi sphere with a Brillouin zone boundary for the fcc structure occurs at

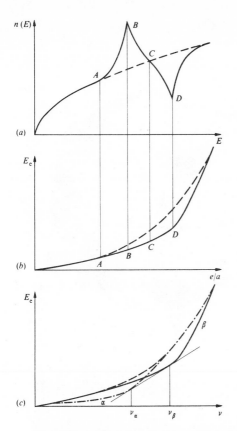

Fig. 6.21. (a) Density of states $n(E)$ and (b) total electron energy
$E_e = \int^{E_F} En(E)\,\mathrm{d}E$ for different occupation $e/a = \int^{E_F} n(E)\,\mathrm{d}E$ of the
electron states for *one* structure and for *two* structures (c). (It can be
shown that the local curvature of the curve in (b):
$\mathrm{d}^2 E_e/\mathrm{d}(e/a)^2 = 1/n(E)$ is equal to the reciprocal of the function $n(E)$
in (a).)

$(e/a)_\alpha = 1.36$ and with a zone boundary for the bcc structure at $(e/a)_\beta = 1.48$.
It is therefore more favourable in the last stages of filling one Brillouin zone,
when the density of states is low, for the electrons to start filling the
Brillouin zone of another structure. In other words, the structure changes
systematically according to e/a. Fig. 6.21(c) shows the resultant appearance
of a two-phase region between the concentrations v_α and v_β by the well
known common tangent construction. In a similar manner the Fermi
sphere comes into contact with a whole series of Brillouin zone boundaries
of the γ-structure at $(e/a)_\gamma = 1.54$ and of cph ζ (depending on axial ratio) at
$(e/a)_\zeta = 1.72$ in apparently good agreement with the experimentally
observed ranges of stability of these phases. Recently, however, this simple

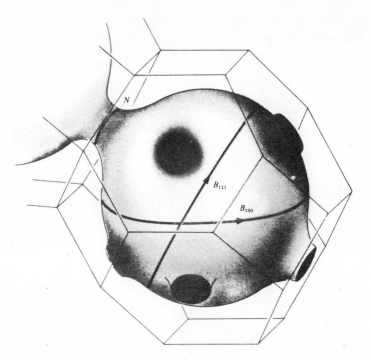

Fig. 6.22. Fermi body of copper in the first Brillouin zone with 'necks'
N and 'bellies' of different circumference (B_{111} and B_{100}) [1.1] after
Pippard.

and successful theory has been the subject of strong criticism. A more
detailed discussion in the following section should permit a better
understanding of electrons in solid solutions, even if agreement with
experiment is sacrificed.

6.3.3   *Electrons in alloys*

It has been known since the measurements made by B. Pippard
that the Fermi body of copper (with $(e/a)=1$) is not spherical but that it
bulges out in so-called necks which touch the $\{111\}$ boundaries of the fcc
Brillouin zone, fig. 6.22. In the Jones model, fig. 6.21(a–c) this occurs for the
pure metal at a point to the right of A.

Jones, however, had considered contact by the α Fermi surface to occur
only at $(e/a)_\alpha = 1.36$ (point C) and identified this point with the
transformation of α into an energetically more favourable structure.
According to fig. 6.21(c) this is not necessary. The fact that the Fermi
surface in the pure metal is already touching the Brillouin zone boundary is
not inconsistent with the continued stability of the α-phase up to a point

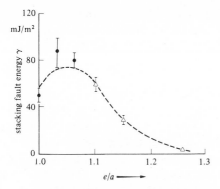

Fig. 6.23. Stacking fault energy of copper solid solutions (with Ga and Ge) of different electron concentration from mechanical measurements (τ_{111}, equation (12-14), full points) and from TEM observations (triangles).

(near to D), which is determined by the common tangent construction, cf. [6.10]. Nevertheless this modified theory is still strongly criticized because it assumes a rigid band. Heine and Weaire [6.3] doubt that there is a noticeable decrease in energy when the Fermi surface crosses a zone boundary, particularly in the case of the α and β-phases, unless this happens simultaneously for many surfaces as is the case in γ-brass (the boundaries of the 'Jones zone', see section 6.3.4). According to V. Heine it cannot be assumed that the energy gaps at the Brillouin zone boundaries are independent of the composition of the alloy. Furthermore, the stacking fault energy γ (fcc/cph), which is a measure of the (free) energy difference α/ζ, decreases gradually to zero with increasing e/a well before the two-phase region is approached, fig. 6.23, and not just at $(e/a)_\alpha$ as predicted by the Jones theory. At the moment, however, there is no better theory (using a 'soft band'). The pseudo-potential theory cannot explain the fcc structure of even the *pure noble metals* with or without involving the participation of d electrons, see [6.10a]. What can it then predict for solid solutions based on these metals?

In fact it gives a good picture of the *spatial distribution of the electrons* in a solid solution [6.10]. A Z-valent atom in a univalent matrix cannot donate Z electrons to the homogeneous electron gas of the metal. Rather the excess charge on the remaining ion core 'binds' $(Z-1)$ electrons as an electrostatic screen. These bound states appear below the occupied energy band. Since the total number of states for N electrons is constant, some states at the top of the band must be missing, i.e. the Fermi surface moves closer to the top of the band on alloying as would be expected from the oversimplified model in which the Brillouin zone is filled up by the valency electrons of the solute atoms (J. Friedel, see [6.8]).

An infinitely wide Fourier spectrum of electron waves is, however, required, according to the Thomas–Fermi method, to screen a point charge, and this is not available. The shortest wave length

$$\lambda_F = 2\pi/k_F = 2\pi\hbar/\sqrt{(2mE_F)}$$

corresponds to the Fermi surface. Thus the screening is incomplete and long-range potential and charge inhomogeneities, the so-called *Friedel oscillations* persist, see [6.10]. The Friedel oscillations lead to an interaction energy between two atoms, which is determined by E_F, i.e. by the electron concentration. (The difficulty of screening at the Fermi surface can also be described by a decrease in the dielectric 'constant' $\varepsilon(k)$ at the wavenumber $k = 2k_F$, see [6.10*a*].) A correlation is thus established between the energy of a specific atomic arrangement (structure) and e/a, which describes the same in real space as the effects discussed above of contact between the Fermi surface and the Brillouin zone boundary in reciprocal space. As we shall now try to show, the description in real space is, for many purposes, more obvious. The screened potential at large distances from a Z_B-valent B atom is given by [6.10]

$$V(r) = \alpha Z_B \frac{\cos 2k_F r}{k_F r^3 (2k_F + 1/a_B)^2} \tag{6-4}$$

($\alpha = $ const, $a_B = $ Bohr radius; the argument of the cos can in principle contain an additional phase angle which we neglect, however, for near-neighbour interaction). The additional electron density $(-\Delta\rho(r))$ and the interaction energy with an A atom $\varphi_{AB} \propto Z_A V(r)$ show the same functional dependence (fig. 6.24).

Let us first assume $Z_A = Z_B$, a homogeneous material, in which case we are concerned with the screening of the pseudo-potential. If within certain limits the A atoms can choose their positions around B, as in the melt, they lie preferentially in the $V(r)$ minima and their density distribution resembles this function. In a crystal on the other hand the sites are fixed (apart from small displacements). In Fig. 6.24 the sites of the nearest neighbours are indicated for a particular Z_A which in turn fixes k_F and thus the period of the cosine. The unfavourable position of the nearest neighbours in the pair potential of the monovalent matrix is compensated by an increase in volume-dependent interaction energy, which is not taken into account in $V(r)$.

We now consider a solute atom in a monovalent matrix at the origin in fig. 6.24 ($Z_A = 1$, $Z_B > 1$). A second solute atom is more strongly repelled from the nearest neighbour position than an A atom. This corresponds to the case $\varepsilon < 0$ in section 5.5.1, i.e. short-range order as observed in *CuZn*, *CuAl*, etc. In addition a lattice expansion $\delta > 0$ is expected for such an alloy. In a trivalent matrix ($Z_A = 3$ like Al) there is an attractive potential for a

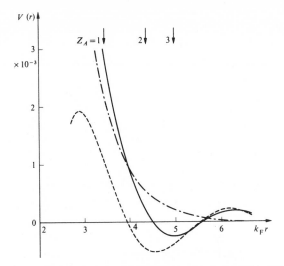

Fig. 6.24. Screened potential (in atomic units 27.21 eV) around a B atom in a matrix of valency Z_A (Fermi momentum k_F). (Full curve: self consistent after Hartree, dashed curve: asymptotic approximation to this, dot–dash curve: classic, after Thomas–Fermi.) The nearest neighbour places are marked for different Z_A.

second solute atom at the nearest neighbour site to a copper atom ($Z_B = 1$). The 'overscreening' at this point means that there are *fewer* electrons here than elsewhere and this is exploited by a second Cu atom which is *negatively* charged with respect to the Al matrix. There is thus a tendency to decomposition which is in fact observed in *Al*Cu and *Al*Ag. This explains why the limit of solubility on the side of the high valency element in binary systems is always found empirically to be lower than on that of the low valency element. The order of magnitude of the heats of mixing and the atomic size factors are also satisfactorily explained by the theory, even if it should be applicable only to very dilute solutions, see [6.10]. The group number in the periodic table, i.e. e/a, is therefore a significant parameter in describing alloying behaviour.

6.3.4 *Changes in structure to maintain a favourable e/a*

Brillouin zones are purely geometrical bodies derived from the lattice structure by drawing the reciprocal lattice and constructing the perpendicular bisecting planes to all reciprocal lattice vectors. Each Brillouin zone can accommodate e/a = two electrons per atom. A Jones zone is bounded by those Brillouin zone surfaces for which the structure factor does not disappear, that is by those at which there really is an energy gap. The difference is significant, for example in the cph structure (and also

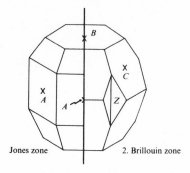

Fig. 6.25. Second Brillouin zone (right) and Jones zone (left) of the cph structure.

in γ-brass). Fig. 6.25 shows the second Brillouin zone for the cph structure (right) and the Jones zone (left). The first Brillouin zone is bounded by $\{0001\}$ and $\{10\bar{1}0\}$ planes (A) and contains a maximum of $e/a = $ one electron per atom (the unit cell of the structure contains two atoms!). There are, however, no energy gaps on the $\{0001\}$ planes. The second Brillouin zone is bounded essentially by two $\{0002\}$ planes (B) and 12 $\{10\bar{1}1\}$ planes (C) (apart from the notches Z formed by A planes); it can accommodate $e/a = $ two electrons. The Jones zone can be derived from the observation of strong X-ray reflections from the A planes. The maximum electron capacity of the Jones zone depends on the axial ratio c/a of the structure according to [6.9]

$$\frac{e}{a}\bigg|_{\text{full}} = 2 - \frac{3}{4}\left(\frac{a}{c}\right)^2\left(1 - \frac{1}{4}\left(\frac{a}{c}\right)^2\right). \tag{6-5}$$

The axial ratio c/a of an alloy for constant atomic volume can vary with concentration in such a way that a favourable e/a is achieved (a high density of states near B in fig. 6.21) as regards the stability of ζ. It is not only a full Jones zone which provides favourable conditions for contact (on all its boundaries) but possibly also the contact of individual sets of planes at electron concentrations quite different from $(e/a)_{\text{full}} \approx 1.75$. Experimentally, there is a universal though not yet well understood correlation between c/a and e/a for many Hume–Rothery phases, fig. 6.26 [6.9]. In ternary alloys, e.g. *Cu*GaGe, with an extensive ζ cph field the c and a lattice parameters change considerably with composition whereas for a section for which $e/a = $ const., c/a is also observed to be constant. The mean volume per atom in the alloy changes steadily through the alloy field despite all the individual changes in the lattice parameters.

In the case of noble metal or aluminium alloys with transition metals it is difficult to quantify e/a. It is often more practical not to include the d

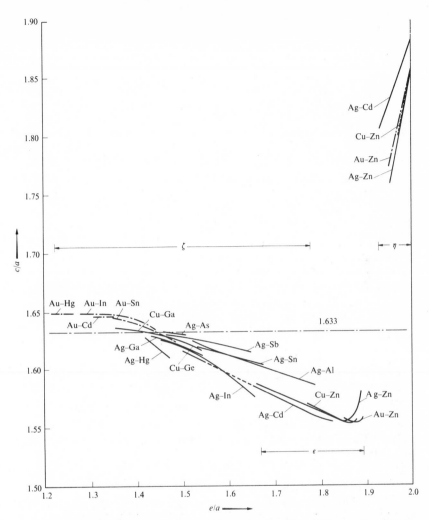

Fig. 6.26. Change in axial ratio c/a of cph solid solutions with electron concentration e/a. After [6.9].

electrons and to assign zero valency to Fe, Co, Ni, Pd, Pt, etc. in Hume–Rothery phases. Interesting examples are the β-phases NiAl and CoAl which are obviously electron phases with $(e/a)_\beta = \frac{3}{2}$, as can be seen from the ternary phase diagram Cu–Ni–Al. Ni with a valency of zero, counts here as a vacant lattice site and in fact a sudden decrease in crystal density and a lattice contraction is observed in non-stoichiometric Al-rich alloys, fig. 6.27. Similar behaviour is found for γ-CuAl and γ-CuGa in which no transition metals are present but in which the electron concentration is maintained

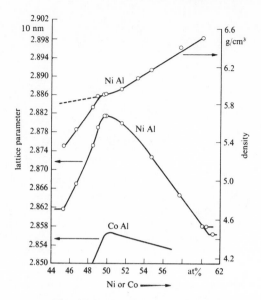

Fig. 6.27. Lattice parameter and density of β solid solutions with chemical vacancies and with more than 50% Al so that e/a is maintained at 3/2. After [6.9].

constant $(e/a) = 1.7$ by chemical ('structural') vacancies if there is an excess of Al(Ga).

The electron concentration or mean group number of an alloy can be used in many other systems as a parameter to predict the stability of certain phases although a convincing electron theoretical explanation is still not available.

6.4   Alloy phases determined by atomic size

It is necessary to define *atomic size* in order to be able to investigate its influence on crystal structure. The Goldschmidt radius r_1, which is simply half the bond length between the atoms, is generally used. This must then be corrected for the appropriate coordination number. The correction implies that the volume per atom remains constant in an allotropic transformation, an assumption which is usually well satisfied. In his 15% size difference rule for good solubility Hume–Rothery used the distance between nearest neighbours d_i obtained from lattice parameter measurements of the pure metals. An empirical parameter, the 'apparent atomic diameter', is often used. This is obtained by extrapolating the lattice parameters of the solid solution to 100% solute. More recently Massalski and King [6.11] have advocated the 'Seitz radius' which is calculated from the mean atomic volume Ω, $r_\Omega = (3\Omega/4\pi)^{1/3}$. Fig. 6.28 shows this atomic size

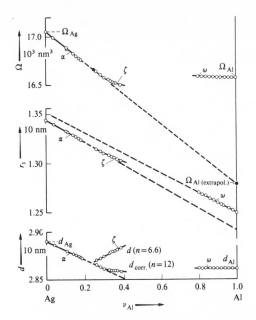

Fig. 6.28. Different atomic size parameters in AgAl: volume Ω per atom, Goldschmidt radius r_1 and distance between nearest neighbours d (dependent on the coordination number n). After [6.11].

parameter for the system AgAl. Table 6.3 gives the differential volume size factor $(1/\Omega)\,d\Omega/dc_B \approx 3\delta$, important in many metallurgical calculations, for the most important base metals and alloying elements (B). A more accurate analysis of the atomic distortion in solid solutions is possible using diffuse X-ray scattering, as described in section 2.3.1. Early investigations of atomic size in alloys were based on Vegard's Law, which corresponds to a linear interpolation of the lattice parameters of the pure components, and attempted to explain deviations from this law.

6.4.1   *Atomic size and solubility*

Fig. 6.29 shows the atomic size (d_i) of various elements with respect to their solubility in iron. The 15% limit for $\delta = \Delta d_i/d$ is shown to be the controlling factor for extensive solubility in this and other matrix metals. According to linear elasticity theory, the (free) enthalpy of distortion for a solute atom with an atomic size ('misfit') factor δ is

$$E_s = G_A \Omega \delta^2 \tag{6-6}$$

(G_A = shear modulus of the matrix). If this energy is equated to the maximum energy of mixing in a symmetrical regular model system (section 5.5.2) and if other energy contributions arising on mixing are neglected, a

Table 6.3. *Volume–size factors of Cu, Ag, Au, Al and Fe solid solutions*

Alloying element	Base metal				
	Cu	Ag	Au	Al	α-Fe
Li			−19.2	−2.1	
Cu		−27.7	−27.8	−37.8	+17.5
Ag	+43.5		−0.6	+0.1	
Au	+47.6	−1.8			+44.2
Be	−26.4				−26.2
Mg	+50.8	+7.1		+40.8	
Zn	+17.1	−13.7	−13.8	−5.7	+21.1
Cd	+67.4	+14.8	+13.1		
Hg	+5.4	+14.0	+18.9		
Al	+20.0	−9.2	−10.2		+12.8
Ga	+24.1	−5.1	−4.3	+4.9	
In	+79.0	+23.5	+20.6		
Tl	+129	+39.4	+23.8		
Si	+5.1			−15.8	−7.9
Ge	+27.8	+1.7	+5.5	+13.1	+16.5
Sn	+83.4	+32.4	+28.8	+24.1	+67.7
Pb		+54.5		−53.6	
P	+16.5				−13.2
As	+38.8	+10.3	+17.7		
Sb	+92.0	+44.9	+34.6		+36.4
Ti	+25.7		−7.7	−15.1	+14.4
V			−8.9	−41.4	+10.5
Cr	+19.7		−16.4	−57.2	+4.4
Mo			−14.8		+27.5
W					+33.0
Mn	+34.2	+0.1	−5.3	−46.8	+4.9
Fe	+4.6		−19.9		
Co	−3.8		−25.2		+1.5
Ni	−8.4		−21.9		+4.6
Pd	+28.0	−17.2	−14.2		+62.2
Pt	+31.2	−20.1	−12.6		

After H. W. King, *J. Mat. Sci.*, **1** (1966), 79.

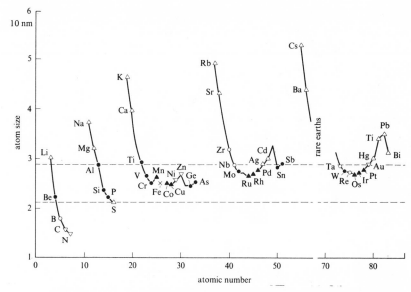

Fig. 6.29. Atomic size of alloying elements in iron, and their solubility: △ insoluble, ▲▽ expand γ field, ●○ restrict γ field. The dashed lines mark a deviation of $\pm 15\%$ from the atomic size of iron. After [6.9].

miscibility gap is predicted with a critical temperature T_c given by $kT_c = 2E_s$. The solubility of the system is considered limited if T_c lies above the solidus temperature, which in this analysis is taken to be the melting point T_{mA} of A. From this, the Hume–Rothery condition for limited solubility is obtained

$$\delta \geqslant \sqrt{\frac{kT_{mA}}{2G_A\Omega_A}}. \tag{6-7}$$

For most metals $G\Omega = 30kT_m$ [6.7] so that a limiting value of $\delta \approx 10\%$ can easily be understood, see [5.2].

6.4.2   *Laves phases*

The Laves phases are the most common intermetallic compounds. They have either the cubic structure of $MgCu_2$ or the hexagonal structures of $MgZn_2$ or $MgNi_2$. The essential condition appears to be that the radii of the participating atoms in AB_2 are in the ratio $r_A/r_B = 1.225$. In this way a very high packing density is achieved, corresponding to an average coordination number of 13.3.

The $MgCu_2$ structure possesses eight formula units per unit cell. The B atoms are situated at the vertices of tetrahedra which themselves are linked at the apices (fig. 6.30(a)). The (larger) A atoms are situated at the midpoints of the tetrahedra and form a diamond lattice. Each A atom has 4 A and 12 B

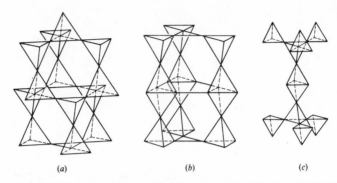

Fig. 6.30. Tetrahedra formed by B atoms in (a) MgCu$_2$, (b) MgZn$_2$, (c) MgNi$_2$.

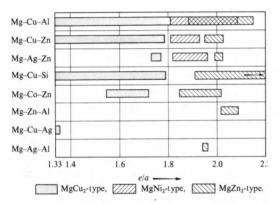

Fig. 6.31. Ranges of existence of different Laves phases dependent on the electron concentration. After [6.9].

nearest neighbours, each B has 6 A and 6 B nearest neighbours. MgZn$_2$ consists of a cell containing four formula units. In this case the tetrahedra are connected alternately at the apices and at the bases (fig. 6.30(b)). The A atoms form the cph wurzite structure. MgNi$_2$ with eight formula units per cell consists of a mixture of the other two (fig. 6.30(c)). The stacking sequence of hexagonally packed double planes of A atoms is $ABCABC$ in MgCu$_2$, $ABAB$ in MgZn$_2$ and $ABACABAC$ in MgNi$_2$. It is interesting to note that in ternary Laves phases the range of existence of the three structures is determined by e/a, fig. 6.31. The strongly diamagnetic behaviour of the ternary MgCu$_2$–MgZn$_2$ compounds in the region of $e/a \approx 1.75$ confirms the correlation of the stability limit with the contact between the Fermi surface and the Brillouin zone boundaries [6.12]. Laves [6.13] points out that in all AB_2 type compounds the AA and BB separations are smaller than the AB separation so that in a hard sphere

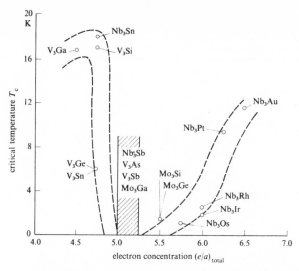

Fig. 6.32. β-tungsten structure A_3B. Superconductivity transition temperature as function of the total electron concentration (no superconductivity in the cross-hatched area). After [6.12].

model like atoms touch each other! The ferromagnetic Co_5–rare earth compounds, interesting on account of their pronounced crystal anisotropy, possess the $CaCu_5$ structure and are derived from $MgZn_2$ by replacing AA by AB contacts [6.12].

6.4.3   *Intermetallic phases with transition metals* [6.12]

A_3B compounds with the 'β-tungsten' (or A 15) structure belong to the group of *stoichiometric phases* containing a significant proportion of transition metals. A is a transition metal from one of the Groups IVA–VIA and B is a Group VIII or a B metal (see table 6.1). The B atoms are arranged on a bcc lattice and have twelve A nearest neighbours. If the e/a ratio is suitable, compounds of this structure exhibit superconductivity with a high transition temperature, see fig. 6.32. The large electronic specific heats of the compounds near $(e/a)_{total} = 4.5$ are indicative of a high density of d electron states.

Among the phases with a *broad homogeneity range*, the sigma phase has attracted special attention (see section 6.1.2). The unit cell is strongly tetragonal ($c/a = 0.52$) and consists of 30 atoms most of which are arranged in two alternating quasi-hexagonal layers and occupy five crystallographically equivalent positions (fig. 6.33). At this stage the packing principle becomes difficult to visualize unless other classification patterns are used such as the *Kasper polyhedron* [6.13]. According to F. C. Frank and J. S. Kasper all the atoms are considered chemically equivalent

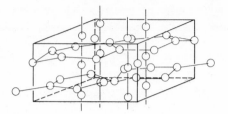

Fig. 6.33. Crystal structure of the σ-phase.

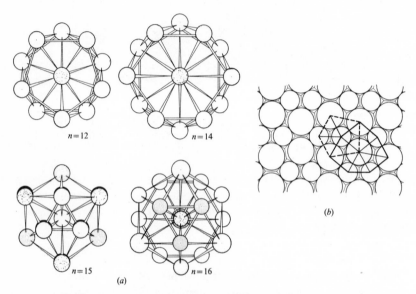

Fig. 6.34. (*a*) The four Kasper polyhedra with different coordination number n [6.13]; (*b*) two-dimensional analogue of the Kasper polyhedron of *A* atoms (small circles, $n=5$) and *B* atoms (large circles, $n=7$) making alternating contact [6.13].

and configurations are sought in these structures for which $(n=)12, 14, 15$ or 16 atoms on *X* sites constituting the vertices of a convex polyhedron enclose one atom on a *Y* site. Five or six triangles with occupied vertices meet in each polyhedron vertex. Each atom site can be considered as a *Y* site. Fig. 6.34(*a*) shows the only Kasper polyhedra to fulfil the above conditions. Fig. 6.34(*b*) shows a two-dimensional analogue with $n=5$ and 7 demonstrating the efficient utilization of space which can be achieved by irregular packing of such tetrahedra. This is a different packing principle from that of close packing (12 nearest neighbours) which contains tetragonal as well as octahedral interstices (see section 6.1.2). Kasper uses only tetrahedra, but they are different, they are arranged irregularly and their bonding distances and angles are not quite identical, all of which helps to accommodate complicated bonding relationships. The Kasper phases are recently used as reference structure for metallic glasses as well as for the newly discovered 'quasi'-crystalline Al_6Mn phase [6.16]. F. Laves has stated the geometrical principles determining the structure in crystal chemistry as: (a) high density of packing, (b) high symmetry and (c) a large number of 'contacts' or bonds between the atoms forming the structure. The principles are competitive and taken together with the other structural arguments discussed earlier explain the variety of crystal structures observed.

6.4.4   *Interstitial compounds* [2.1]

The compounds of greatest interest here are those of small non-metallic elements $X = H(r_X = 0.05$ nm), $B(0.08$ nm), $C(0.077$ nm) and $N(0.074$ nm) with transition metals $M = Zr$, Ti, V, Cr, W, etc. because they are used extensively in cutting tools. G. Hägg has formulated a rule for these compounds based on the radius ratio r_X/r_M. Simple compound structures are obtained only if r_X/r_M is less than 0.59. In these compounds the M atoms generally form an fcc or cph lattice in which the metalloid atoms X occupy the interstitial sites (see section 6.1.2). The compounds lie close to the stoichiometric compositions MX, M_2X, MX_2, M_4X. A NaCl or zinc blend structure is observed for MX as in ZrN, TiC, ZrH. In W_2C the M atoms are arranged hexagonally. The bonding is very strong, probably covalent. Cementite Fe_3C has $r_C/r_{Fe} = 0.61$ and therefore a complex structure. C atoms can be replaced by B but not by N although $r_N < r_C < r_B$. The valency of X appears to exercise some influence.

6.5   **Normal valency compounds**

Normal chemical valency compounds also exist between metals; an electropositive and an electronegative metal form an ionic bond of strength

$$\varepsilon = \varepsilon_{AB} - \frac{\varepsilon_{AA} + \varepsilon_{BB}}{2} = -(EN_A - EN_B)^2$$

measured on a somewhat arbitrary electronegativity scale EN_i, see section 6.3.1. The structures of the 'salts', e.g. NaCl, CaF_2 are often adopted by these compounds, e.g. MgSe, Mg_2Sn or $Mg^{2+}Mg^{2+}Sn^{4-}$. The NaCl lattice can be visualized as two interpenetrating fcc lattices, each occupying the octahedral interstices of the other (see section 6.1.2). If the tetrahedral interstices are occupied, the fluorite structure is obtained. The octahedral interstices of the cph structure can also be filled to produce the NiAs structure, which is adopted by many compounds between transition metals and metalloids. Many of these are conductors, i.e. the binding has a partially metallic character. According to a rule derived by E. Zintl only elements from Groups IVB–VIIB can serve as the electronegative partner in normal valency compounds (perhaps also Ga and In). Further (defect) structures can be produced by partial occupation of the interstices in close-packed structures.

7
Ordered arrangements of atoms

7.1 Superlattices

We have seen in section 6.1 that the ordered β'-brass structure is produced from the disordered β structure when the corners of the bcc unit cell are occupied by one atomic species and the cell centre positions by the other. The crystallographic nomenclature for β is $A2$ and for β' is $B2$ (for the lattice type nomenclature $A, B, \ldots, D, L, \ldots$, see [7.11]).

An ordered distribution of the alloying components over the lattice sites, i.e. the formation of sublattices, results in a larger unit cell or *superlattice*. Often additional X-ray (superlattice) reflections are observed if the waves reflected from parallel A atom and B atom planes do not cancel each other completely. (The structure factor $[\,f_A + f_B \exp(-i\pi(h+k+l))]$ see (2-7) with $\mathbf{u}_n = 0$, is no longer zero for odd values of $(h+k+l)$, if $f_A \neq f_B$, see [2.2].)

Other superlattices based on $A2$ are observed in $Fe_3Al(D0_3)$, fig. 7.1, and the Heusler alloy, $Cu_2MnAl(L2_1)$, which is employed as an iron free ferromagnet. (Al on the sublattice X, Mn on the sublattice Y in fig. 7.1.) The superlattice Cu_3Au I ($L1_2$), fig. 7.2, and the tetragonal superlattice CuAu I ($L1_0$), fig. 7.3, are based on the disordered fcc lattice $A1$. The superlattices Mg_3Cd ($D0_{19}$) with an axial ratio for the ordered cell of 0.8038 is formed on a cph $A3$ lattice.

Various other superlattices, including the long-period superlattice CuAu II, fig. 3.1, are shown in the CuAu phase diagram, fig. 7.4. As was mentioned in section 3.1, this consists of $2M = 10$ tetragonal CuAu I cells stacked in the **b** direction. An *antiphase boundary*, section 7.3, is inserted after every five cells which causes a $\frac{1}{2}(\mathbf{a} + \mathbf{c})$ displacement of copper atoms on to Au sites and vice versa. (The word phase in the context of the antiphase boundary has a different meaning from that in thermodynamics; the order is 'in phase' in the sense of being 'in step' and in antiphase when it is 'out of step'.) The mean separation of two such antiphase boundaries is $b(M + \delta)$, where δ represents the slight expansion of the lattice in the direction of periodicity. $\{110\}$ superlattice reflections are observed for CuAu I. This corresponds to

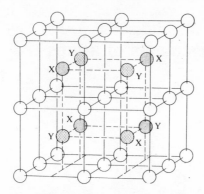

Fig. 7.1. Superlattices of Fe_3Al (DO_3) ($\bigcirc = X = Fe$, $Y= Al$) and of Cu_2MnAl $(L2_1)$ ($\bigcirc = Cu$, $X = Al$, $Y= Mn$).

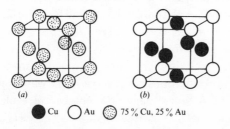

Fig. 7.2. Superlattice of Cu_3Au I $(L1_2)$ (b), which can be derived from the fcc structure $(A1)$ (a).

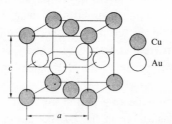

Fig. 7.3. Superlattice CuAu I $(L1_0)$.

CuAu II with $M \rightarrow \infty$. For $M \rightarrow 1$ these reflections must largely merge with those of the disordered CuAu structure because then Cu follows Cu on neighbouring $\{110\}$ planes causing cancelling. Therefore, in the CuAu II structure with a finite period M the $\{110\}$ superlattice reflections are split in the direction of periodicity as shown by the reciprocal lattice in fig. 7.5. The *splitting* is $2\Delta = 1/Ma$ [7.1]. This has been confirmed convincingly on vapour deposited Cu–Au films using electron diffraction. Furthermore, the

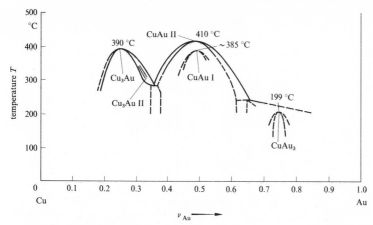

Fig. 7.4. Part of the Cu–Au phase diagram showing superlattices and
critical temperatures.

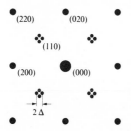

Fig. 7.5. X-ray intensity distribution in the (001) plane of the
reciprocal lattice of CuAu II, splitting 2Δ of the $\{110\}$ superlattice
reflections by two types of antiphase domain periodic in the [010] and
[100] directions. For CuAu I, $\{110\}$ reflections are obtained with
$2\Delta = 0$, for disordered CuAu, no $\{110\}$ reflections.

period M can be measured directly in the transmission electron microscope,
fig. 3.2.

If third elements are added to the Cu–Au film the periodicity changes, in
fact it is controlled by the electron concentration of the alloy, fig. 7.6. This
points to the stabilizing effect of contact between the Brillouin zone and
Fermi surface, already mentioned in section 6.3 but in this case applied to
the superlattice of a particular periodicity M. Fig. 7.7 shows the Brillouin
zone of the disordered fcc structure and of the ordered CuAu I superlattice
with an enlarged unit cell producing a smaller first Brillouin zone bounded
by the (001) and $\{110\}$ planes responsible for the superlattice reflections.
The splitting of the $\{110\}$ reflections on transition from CuAu I to CuAu II
causes a split in the Brillouin zone, as shown in fig. 7.8 in a (001) section.

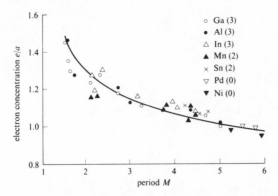

Fig. 7.6. Distances between antiphase boundaries $Ma = 1/2\Delta$ in CuAu II with additions of third elements of the given valencies, which change the electron concentration e/a. After [7.1].

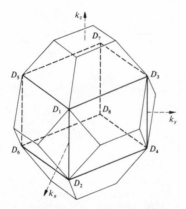

Fig. 7.7. First Brillouin zone of CuAu (disordered, thin outer lines) and of the CuAu I superlattice (thick lines between the points D_i).

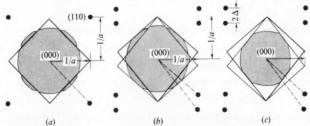

Fig. 7.8. Section parallel to (001) through the first Brillouin zones of CuAu I (a) fig. 7.7, and of CuAu II (b) and (c), and through the Fermi bodies. In the case of CuAu II there are two energetically favourable Fermi bodies which touch Brillouin zones, (b) and (c), instead of overlapping them (a) as in the case of CuAu I. After [7.1].

Instead of the Fermi sphere of CuAu I containing $e/a = 0.86$ electrons touching the $\{110\}$ planes, fig. 7.8(a), two Fermi spheres can be inscribed touching the outer or inner $\{110\}$ surfaces respectively, fig. 7.8(b) and (c). H. Sato and R. Toth [7.1] have calculated the electron content of these two Fermi bodies as

$$\left. \frac{e}{a} \right|_{\{110\}} = \frac{\pi}{12t^3} \, (2 \pm 2(a\Delta) + (a\Delta)^2)^{3/2}, \tag{7-1}$$

where t is a free parameter ($= 0.95$) giving the ratio of the radius of the actual Fermi body in the $\langle 110 \rangle$ direction to that of a sphere. The curve $e/a(\Delta)$ describes the experimental results of fig. 7.6 extremely well (with the $+$ sign corresponding to contact of the external $\{110\}$ surfaces). The stability of a period M corresponding to $(e/a)_{\{110\}}$ arises from the increased density of states for electrons in the neighbourhood of a Brillouin zone boundary with an energy gap although this effect should not be overrated in view of the remarks made in section 6.3. Similar effects have been described for other superlattices [7.1]. Many unanswered questions still remain, however [7.2]:

(a) Based purely on energy considerations the above electronic stabilization favours CuAu II at low temperatures. According to fig. 7.4, however, the CuAu I structure is observed. Tachiki and Teramoto consider the discrepancy to be the result of the additional distortional energy associated with the CuAu II antiphase boundaries. On the other hand if the subdivision by antiphase boundaries is somewhat more irregular, CuAu II possesses an additional configurational entropy due to the number of configurations possible for the antiphase boundaries. This entropy stabilizes CuAu II at higher temperatures.

(b) An antiphase boundary has a specific energy $\tilde{E}$. The energy density of all antiphase boundaries is proportional to $\tilde{E}/M$ and this must be included in the stability considerations. $\tilde{E}$ depends on the composition (even at constant e/a) and on the temperature. It is not clear whether these effects really manifest themselves in the observed M. Nevertheless the Sato and Toth model is an interesting application of the electron theory of alloy structures, as described in section 6.3.

7.2 Incomplete order, degrees of order

In the previous section it was assumed that order was complete and that the superlattice was perfect. This is impossible from entropy considerations, as was shown in chapter 5 in a discussion of the quasi-chemical theory of solutions for the *coefficient of short-range order* α_1 for nearest neighbours (5-21). This coefficient was defined in section 2.3.2 in connection with its measurement by diffuse X-ray scattering as the excess of like nearest neighbours over the random number; for short-range order therefore $0 \geqslant \alpha_1 \geqslant (-1)$ for an AB alloy. Short-range order coefficients α_m are also defined and measured for distant (mth) neighbours, but they are not independent of each other or of α_1, because the next nearest neighbours, NNN, of an atom can be nearest neighbours, NN, of one another (see [7.2]). In non-stoichiometric alloys, complete short-range order ($\alpha_1 = -1$) cannot in any case be established throughout, as shown by (2-18).

7.2.1   Long-range order

Short-range order was obtained from the neighbourhood relationships P_m^{AB} of an atom. Long-range order on the other hand was derived in section 7.1 from the occupation of certain sublattices 1, 2 by one atomic species. Consequently it is necessary to define the degree of long-range order s as the fraction of A atoms on the correct sublattice (1): $P_{A1} = (\frac{1}{2})(1 + s)$ and on the incorrect (2): $P_{A2} = (\frac{1}{2})(1 - s)$. For a random distribution $P_{A1} = \frac{1}{2}$ and $s = 0$. For perfect long-range order $s = 1$ and $P_{A2} = 0$ or $s = -1$ and $P_{A1} = 0$. As is shown by Kittel [1.1], this parameter s can be used to formulate the energy of a regular solution (Bragg–Williams theory of long-range order), as was done in section 5.2.2 using the short-range order parameter P^{AB} (see (5-15), short-range order theory). Once again we use the nearest-neighbour pair interaction parameters between the sublattices (i.e. a short-range order concept). The distribution of the atomic species within each sublattice is considered as random and the entropy is taken to be that for ideal mixing. A comparison of the results for the two theories (long-range order, short-range order) shows that the energy of a 50% alloy is reduced on ordering by

$$\Delta E = E - \frac{Nn}{4} \left(\frac{\varepsilon_{AA} + \varepsilon_{BB}}{2} + \varepsilon_{AB} \right),$$

$$= \frac{Nn}{4} \varepsilon (2P^{AB} - 1) = \frac{Nn}{4} \varepsilon s^2. \tag{7-2}$$

Using the definition of the degree of short-range order α_1 for an alloy AB in (5-21) this implies

$$-\alpha_1 = 2P^{AB} - 1 = s^2. \tag{7-3}$$

This simple relationship between two basically different concepts of order stems in the last analysis from the use of similar approximations in both theories (short-range order, long-range order), namely the calculation of the energy of mixing for nearest neighbours and the use of the entropy of mixing for a random distribution. The presence of only one antiphase boundary in the microstructure is sufficient to cause difficulties in the definition of s. In this case there is a phase step in the occupation of the sublattices by A and B whereby s can become zero although $\alpha_1 \approx -1$. In determining s from the intensity of superlattice lines it is important only that there is no antiphase boundary in the coherently scattering volume. α_1 is determined (2-19) from the diffusely scattered background. In fact they are often found not to be related in the manner described by (7-3).

The temperature dependence of s is obtained by minimizing

$$(E(s) - TS^M(s))$$

with respect to s (S^M is the entropy of mixing), cf. [1.1]. The same result is obtained from a quasi-chemical theory for the reaction of an AB alloy

$$\binom{A}{1} + \binom{B}{2} \rightleftharpoons \binom{A}{2} + \binom{B}{1}, \tag{7-4}$$

for which the Boltzmann distribution law yields (with $\varepsilon < 0$)

$$\frac{P_{A2}}{P_{A1}} = \frac{1-s}{1+s} = \exp(+n\varepsilon s/kT), \tag{7-5}$$

$n\varepsilon s$ is the energy difference between the two positions of an A atom, expressed by that of all its bonds. Thus the disordering energy itself decreases with the degree of order which is described as a 'demoralizing effect'. Equation (7-5) can also be expressed as

$$\tanh x = \frac{2kTx}{n|\varepsilon|} \equiv s \tag{7-5a}$$

which suggests a graphical solution, fig. 7.9. (7-5a) yields the $s(T)$ curve shown in fig. 7.10. In the limiting case a straight line of unit slope is a tangent to the tanh curve at the origin. It defines a critical temperature T_c by $kT_c = n|\varepsilon|/2$. As predicted by improved theories (see [7.3]), the measured $s(T)$ for CuZn is somewhat steeper. A particularly suitable quantity for comparison with experiment is the specific heat necessary to produce disorder $c_v = (dE/ds)(ds/dT)$. Fig. 7.12 shows experimental results for CuZn. It can be seen that disordering processes take place even above T_c, i.e. that short-range order still exists there, as shown also by diffuse X-ray scattering, section 2.3.2. The c_v curve for Cu_3Au corresponds to a first-order phase change. The degree of long-range order changes discontinuously at

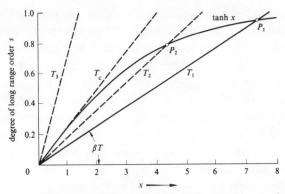

Fig. 7.9. Graphical solution of the equation tanh $x = \beta T x$ for different $T_1 < T_2 < T_3$. At temperature T_c the point of intersection P coincides with the origin.

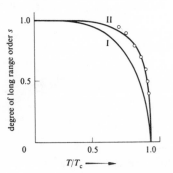

Fig. 7.10. Temperature dependence of the degree of long-range order s according to measurement by B. E. Warren and co-workers on β'-CuZn compared with theory: curve I equation (7-5a); curve II section 7.2.2 with $\varepsilon_2 = \varepsilon_1/3$.

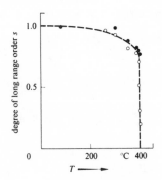

Fig. 7.11. $s(T)$ for Cu_3Au (● specimens quenched to room temperature; ○ superlattice lines measured at the annealing temperature).

T_c, fig. 7.11. This is due to the fact that the function corresponding to the left-hand side of (7-5a) for the A_3B lattice does not have a constantly decreasing slope like tanh x but is intersected, possibly several times, by a straight line through the origin (see fig. 7.13). The stable graphical solution (0 or B) is found by comparing the cross-hatched areas. If the area above the intersecting straight line ($2kT_2x/n|\varepsilon|$) is greater than the area below it, B is stable and vice versa. At the critical temperature the areas are equal (dotted). In the case of CuZn, the transformation is of higher order and the specific heat remains finite. The approach to the critical point and the

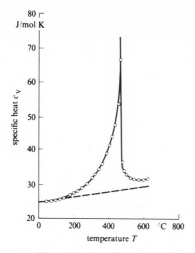

Fig. 7.12. Specific heat of disordering of CuZn.

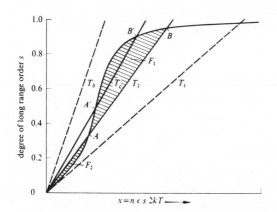

Fig. 7.13. Graphical determination of the degree of long-range order for an A_3B lattice. Of the three solutions for temperature T_2, A is unstable and B stable according to a comparison of the shaded areas F_1 and F_2. At T_c, B' and 0 are stable solutions, i.e. s increases with T with an infinite slope. At T_3, $s = 0$. After [7.3].

associated fluctuation effects (critical opalescence) are the subject of current research. Sometimes a deformation of the lattice accompanies ordering which changes the lattice symmetry, as in the case of CoPt alloys (bcc becomes tetragonal face centred). Then the transformation is first order with a corresponding two phase region between it and the disordered phase.

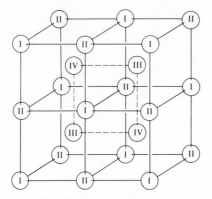

Fig. 7.14. Subdivision of the bcc lattice into four sublattices.

7.2.2   Ordering and pair potential

It was shown in section 2.3.2 that the short-range order parameter α_i for various neighbourhoods (ith shells) can be obtained from diffuse X-ray scattering. For example, for Cu_3Au at 450 °C (above $T_c = 390$ °C) values were measured of $\alpha_1 = -0.19$; $\alpha_2 = +0.21$; $\alpha_3 = 0.00$; $\alpha_4 = 0.08$; $\alpha_5 = -0.05$, etc. (see [7.4]). The fact that the sign alternates with i is to be expected from the superlattice, see fig. 7.2. In order to give a physical explanation of these more distant preferred neighbourhoods the interaction parameters ε_{ij} are required not only for nearest neighbours, as used in the exchange energy ε ($= \varepsilon_1$) in the theory of regular solutions, but also between next nearest neighbours, which lead to an exchange energy ε_2, cf. (5-15a). ε_i can then be determined from the measured α_i. These have been used by Inden and Pitsch [7.5] for superlattices in the system Fe(Co)Si. Fig. 7.14 shows the subdivision of the bcc lattice into four sublattices employed by these authors. The sublattices I and II are in the nearest neighbour (8) position to III and IV, but the relationship between I and II and again between III and IV is that of next nearest neighbours (6). Fig. 7.15 illustrates qualitatively the different possible ways of filling these sublattices by Fe and Si atoms for various Si concentrations and various ε_i. If $|\varepsilon_i| \ll kT$ or $\varepsilon_i = 0$ a random distribution of atoms obtains, see fig. 7.15(a); if $\varepsilon_1 < 0$, $\varepsilon_2 \geqslant 0$ the resulting arrangement has the highest possible number of pairs of unlike nearest neighbours (b); if $\varepsilon_1 < 0$ and $\varepsilon_2 \lesssim 0$, but $\varepsilon_2 > 2\varepsilon_1/3$ a tendency to order is apparent even for the next nearest neighbours (c); if the tendency of the next nearest neighbours to order predominates, i.e. $|\varepsilon_2| > |2\varepsilon_1/3|$ the atomic arrangement depicted in (d) is expected; finally (e) shows a tendency to decomposition among the nearest neighbours ($\varepsilon_1 > 0$) and a tendency to order among the next nearest neighbours ($\varepsilon_2 < 0$). The distribution of atoms

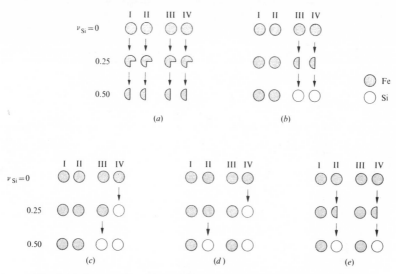

Fig. 7.15. Occupation of the sublattices in fig. 7.14 with two atomic species Fe, Si for different proportions of Si, ν_{Si}, and different interaction parameters ε_1 and ε_2 (cases (a) to (e)). After [7.5].

can be calculated quantitatively from the thermodynamics of a regular solution (up to next nearest neighbours). For FeSi (and FeAl) in particular, agreement is obtained with the measured enthalpy of mixing given by $(8\varepsilon_1 + 6\varepsilon_2)$, see [7.5] and [5.3], and with the measured energy of disordering or the critical point, section 5.5.2, which are proportional to $(8\varepsilon_1 - 6\varepsilon_2)$ if $\varepsilon_1, \varepsilon_2$ are chosen to be negative and $\varepsilon_1/\varepsilon_2 = 2$. In the case of FeCo agreement is obtained if $\varepsilon_1 < 0$, $\varepsilon_2 > 0$ and $|\varepsilon_1| \approx |\varepsilon_2|$. This analysis thus permits an excellent description of phase diagrams, scattered neutron intensities and Mössbauer spectra and also of ordering reactions in ternary systems. The ordering reactions $A2 \to B2 \to D0_3$ in FeSi above the Curie temperature as a function of Si content appear not to be of the first but of higher order, with the result that no two phase regions are observed (fig. 7.16).

In the CuAu system S. Moss and P. Clapp [7.4] have carried out a similar thermodynamic analysis based on a formulation resembling the quasi-chemical theory, section 5.4, (5-21) extended to next nearest neighbours. In this the interaction of an atom with its next nearest neighbours can be expressed in terms of that with the nearest neighbours between them (see [7.2]). The interaction energy of the ith neighbour pair $V_i^{AB}(r_i)$ is assumed to be dependent on the distance r_i instead of being of constant strength $\varepsilon_{AB,i}$ as in section 5.2.2. This has the advantage that we obtain the changes in lattice parameter with the degree of order, which in many cases cause a first

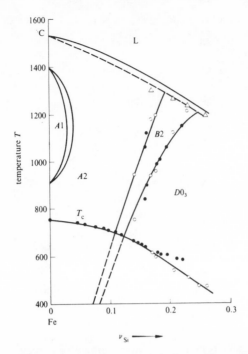

Fig. 7.16. Section of the iron–silicon phase diagram with two
superlattices and a ferromagnetic region (below T_c). Calculated curves
[7.5] and experimental values from neutron scattering ($\bigcirc\triangle$) and
specific heat ($\bullet$).

order transformation [7.10]. From the α_i Moss and Clapp are able to
determine the magnitude and sign of the interaction potentials V_i between
nearest neighbours, next nearest neighbours and next next nearest
neighbours! The curve in fig. 7.17 shows an interaction which decreases
with distance while it oscillates in sign, which is not unlike the Friedel pair
potential following from the pseudo-potential, section 6.2. Different
superlattices arise depending on the sign of V_i, and for $V_1 < 0$ (and not too
large $V_2 < |V_1|$) there is a tendency to decomposition as already found in
connection with fig. 6.24. A link with the results of electron theory (chapter
6) is thus established, according to which the pair interaction (plus a binding
energy dependent only on the volume of the lattice cell) is sufficient to
describe the energy relationships in an alloy. Many-body interactions are
involved in very few of the observed cases [7.10].

Finally, the energy contributions of lattice distortions due to atomic size
differences have to be taken into account. These always favour order rather
than decomposition as was seen in the case of intermetallic (Laves) phases.

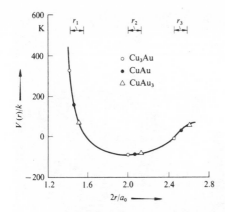

Fig. 7.17. Interaction potential as a function of the distance between atoms from X-ray measurements of the short-range order parameter for neighbours at different distances (r_i) in CuAu alloys. (a_0 is the lattice parameter for Cu_3Au.) After [7.4].

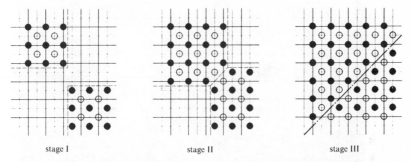

stage I   stage II   stage III

Fig. 7.18. Formation of antiphase boundaries (dashed) when ordered regions, in which A (●) *and* B (○) atoms occupy different sublattices, grow together.

The lattice distortions can be determined from the (diffuse) Huang scattering (see section 2.3.2).

7.3   Ordered domains and their boundaries

7.3.1   *Domain structure and types of antiphase boundary*

In order to describe long-range order, we introduced the concept of the sublattice occupied by one or other of the atomic species. The sublattices are generally equivalent, so that at any given nucleation site either A or B atoms can occupy the sublattice I (fig. 7.18). As the ordered regions grow and meet, antiphase boundaries inevitably appear in the *microstructure* as described earlier in sections 3.1 and 7.1 (in that case as a

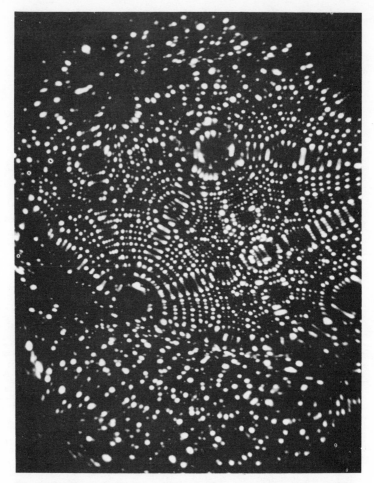

Fig. 7.19. Boundary between ordered (above) and disordered CoPt (below) in the field ion microscope (T. T. Tsong and E. W. Müller, Penn. State Univ.).

constituent of the CuAu II *phase*). In this model a small degree of order is described by small reasonably perfectly ordered domains in a disordered matrix. Above, on the other hand, the short-range order parameter α_1 was interpreted by a homogeneous distribution of AB pairs, in other words by a rather different model. Experimental observations on CoPt with the field ion microscope (section 2.4, fig. 7.19) and on Fe_3Al using transmission electron microscopy support the heterogeneous model, as also do electron diffraction and computer simulation of a Cu_3Au alloy above T_c.

An antiphase boundary is produced by an antiphase vector **u** which

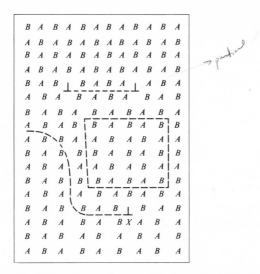

Fig. 7.20. *AB* superlattice with thermal antiphase boundaries, one of which ends (at *X*) in a normal lattice dislocation (partial dislocation of the superlattice). A dissociated superlattice dislocation is also shown of which the partial dislocations bound an antiphase boundary.

brings about the displacement of one atomic species from one sublattice to another. This displacement is a manifestation of the phase jump which can arise at the boundary when ordered domains grow together (thermally generated antiphase boundary). The displacement can also be effected mechanically by a dislocation with a Burgers vector smaller than the translation vector of the superlattice (partial dislocation of the superlattice with $\mathbf{b}=\mathbf{u}$, see section 11.3.3). Each antiphase boundary ending in the crystal terminates at such a dislocation, see fig. 7.20. An antiphase boundary is sufficiently described by $\mathbf{u}$ and the normal $\mathbf{n}$ to the boundary if this can be assumed to be locally plane. If $\mathbf{u}$ lies in the boundary, as is the case for an antiphase boundary produced by a translation, a relatively small antiphase boundary energy $\tilde{E}_{APB}$ is expected because no atoms have been added or removed (type 1 antiphase boundary). In the other case, $(\mathbf{u} \cdot \mathbf{n}) \neq 0$, $\tilde{E}_{APB}$ is often larger (type 2 antiphase boundary), see fig. 7.22. Fig. 7.21 shows the sublattices I–IV for the $L1_2$ structure (Cu$_3$Au) with the three associated vectors $\mathbf{u}_i = (a/2)[110]$, $(a/2)[101]$, $(a/2)[011]$. For $\mathbf{n} = [001]$ a type 1 antiphase boundary is produced by $\mathbf{u}_1$, the others are of type 2. Thus four different ordered domains are possible. This is also the least number necessary to maintain a metastable structure of antiphase boundaries (analogous to a soap lather) as was explained in section 3.3 for grain

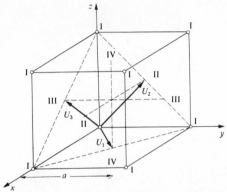

Fig. 7.21. Sublattices I (Au) and II–IV (Cu) of Cu_3Au and the three vectors U_i which generate antiphase boundaries.

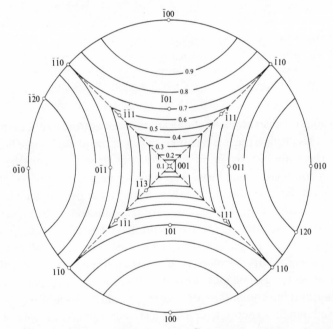

Fig. 7.22. Lines of constant energy $\tilde{E}_{APB}$ for an antiphase boundary created by $\mathbf{u} = (a/2)[110]$ in Cu_3Au as function of the plane normal $[hkl]$. (Nearest neighbour interaction.)

boundaries. Unlike the $L1_2$ superlattice, $B2$ (fig. 7.20) has only two possible sublattices and domains, one possible vector $\mathbf{u}$ and no possibility of stabilizing an antiphase boundary structure. In this case the antiphase boundaries cannot end except in dislocations [7.9].

7.3.2   Energy of antiphase boundaries

The (free) energy $\tilde{E}_{APB}$ can be estimated as a function of the orientation of the antiphase boundary, i.e. of $\mathbf{n} = (n_1, n_2, n_3)$ using the approximation of pair interaction (section 5.2.2) between nearest neighbours and next nearest neighbours. In the case of $L1_2$ and using the known pair exchange energy ε_1 we obtain

$$\tilde{E}_{APB}\left(\mathbf{u} = \frac{2}{2}[110]\right) = 2\varepsilon_1 n_1 s^2 / a^2 \sqrt{(n_1^2 + n_2^2 + n_3^2)} \qquad (7\text{-}6)$$

assuming $n_1 \geqslant n_2$ [7.6]. For $\mathbf{n} = [001]$, i.e. the type 1 antiphase boundary, $\tilde{E}_{APB} = 0$, in other words the antiphase boundary in Cu_3Au lies preferentially in the cube planes. (If next nearest neighbours are taken into account $\tilde{E}_{APB}$ remains finite for this orientation). The orientation dependence of $\tilde{E}_{APB}$ for $L1_2$ is shown in fig. 7.22 by lines of constant energy in a stereographic projection of the possible $\mathbf{n}$. Besides the minimum at $\mathbf{n} = [001]$, there are maxima at $[100]$, $[010]$ and saddle points at $[110]$, $[1\bar{1}0]$. $\tilde{E}_{APB}$ is, however, reasonably isotropic for the $D0_3$ superlattice in Fe_3Al due to the strong influence of next nearest neighbours, and is of the order of $0.1\ J/m^2$.

These estimates have been confirmed by transmission electron microscopy. The remarks on contrast formation at stacking faults in section 2.2.1.2 apply here to antiphase boundaries when $\mathbf{g}$ is a reciprocal lattice vector of the superlattice. Fig. 7.23 shows a domain structure in partially ordered Cu_3Au after a 75 h anneal at 380 °C. The domain size is about 75 nm. (It should be noted that about $\frac{1}{3}$ of the antiphase boundaries are out of contrast for the superlattice reflection used to form the image.) The cube orientation of the antiphase boundaries confirms the strong anisotropy of $\tilde{E}_{APB}$ predicted by (7-6). Occasionally a comparison of thermally generated and mechanically generated antiphase boundaries shows that the contrast at the latter is sharper. In the former case the phase jump in the occupation of the sublattices is spread by diffusion to a profile of finite width (several lattice constants wide for $T < T_c$). This is especially true for non-stoichiometric alloys in which the excess atoms of one species can form type 2 antiphase boundaries without any expenditure of energy. Using the field ion microscope, these concepts of the structure of antiphase boundaries can be examined closely. A domain structure in bulk alloy material, which becomes distorted tetragonally on ordering (as mentioned for CuAu I), is associated with a defect microstructure which compensates the distortion (see chapter 13).

Fig. 7.23. Antiphase boundary microstructure in partially ordered Cu$_3$Au (R. Fischer, U.S. Steel Corp., Monroeville).

7.4   Kinetics of ordering [7.7]

The time required for order to be established in an alloy is determined by atomic movements, diffusion. We shall discuss these phenomena in the following chapters. Here it is possible to give only a general qualitative picture of the kinetics of ordering, which relates this process to more general topics discussed individually later on. In ordering, unlike in precipitation and decomposition processes, no large concentration differences arise which necessitate transport of material over large distances. Here on the contrary it is much more a question of atomic exchange between neighbouring sublattice sites. The preferential occupation by one atomic species can start in a *nucleation process* (*a*). The resulting ordered domains then extend outwards into the disordered material until they meet other *growing domains* (*b*) (fig. 7.18). Finally the metastable domain structure coarsens until an ordered crystal free of antiphase boundaries (*c*) is obtained (Ostwald ripening). Concurrently with these processes, the *degree of order* within each domain is changing with time (*d*). This case is treated first, with reference to the work of G. Dienes. (Recently de Fontaine [7.13] has proposed a process called *spinodal*

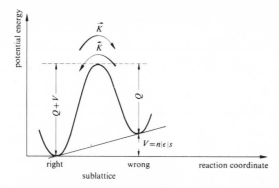

Fig. 7.24. Potential energy of an atom in the superlattice and atomic jumps with rate constants $\vec{K}, \overleftarrow{K}$.

ordering. Far below the ordering temperature he envisages periodic modulations of the degree of order which grow steadily with ageing time – similar to the spinodal decomposition process dealt with in chapter 9.)

(*d*) *Homogeneous ordering*. We start from the quasi-chemical ordering reaction for an *AB* alloy, (7-4), the rate of which, using the notation of section 7.2, is

$$\frac{\mathrm{d}P_{A1}}{\mathrm{d}t} = \overleftarrow{K} P_{A2} P_{B1} - \vec{K} P_{A1} P_{B2}$$

(7-7)

or $2\dfrac{\mathrm{d}s}{\mathrm{d}t} = \overleftarrow{K}(1-s)^2 - \vec{K}(1+s)^2.$

The rate constants $\overleftarrow{K}, \vec{K}$ are described as in fig. 7.24 by Arrhenius factors (with an atomic vibration frequency v_0).

$$\overleftarrow{K} = v_0 \exp\left(-\frac{Q}{kT}\right); \qquad \vec{K} = v_0 \exp\left(-\frac{Q+V}{kT}\right).$$

(7-8)

The results obtained using expressions (7-7) and (7-8) are shown in fig. 7.25. At low degrees of order s the rate of change $\mathrm{d}s/\mathrm{d}t = 0$. At medium values of s, $\mathrm{d}s/\mathrm{d}t$ goes through a maximum which is particularly pronounced just below $T_c = 205$ K. Isothermal $s(t)$ curves are S-shaped as would be expected from nucleation and growth theories, although nothing was assumed here corresponding to the postulates (*a*) and (*b*).

In the case of the A_3B alloys, which undergo a first-order transition (fig. 7.13), the establishment of order is much slower. Fig. 7.25 shows that well below T_c the rate $(\mathrm{d}s/\mathrm{d}t)_{t=0} < 0$. A larger fluctuation must occur leading to a finite degree of order before ordering can proceed further. As is confirmed in

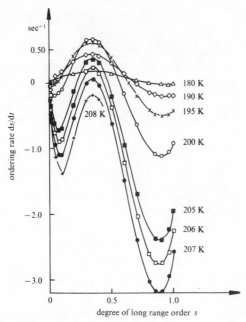

Fig. 7.25. Calculated rates of ordering for an A_3B alloy with $T_c = 205$ K. After G. Dienes.

practice, order is achieved in A_3B superlattices much more slowly than in AB. It is virtually impossible to quench-in the disorder in CuZn but quite easy in Cu_3Au.

(b) *Growth of ordered domains* ('*Johnson–Mehl kinetics*'). As in the discussion of the kinetics of recrystallization which follows later in chapter 15, let us assume that dependent on the temperature, there are initially N domain nuclei per unit volume the size of which increases in all directions at a constant rate v until the domains come into contact. Between t and $t + dt$, therefore, each domain grows by $dV = 4\pi v^3 t^2 \, dt$ and the ordered volume fraction X increases by growth into that not yet ordered $(1 - X)$ by

$$dX = 4\pi v^3 N t^2 \, dt (1 - X),$$ (7-9)

i.e. $-\ln(1 - X) = \tfrac{4}{3}\pi v^3 N t^3$,

or $X = 1 - \exp(-\tfrac{4}{3}\pi v^3 N(T) t^3).$ (7-10)

For a linear growth rate we put

$$v = \frac{D(T)}{kT} \cdot \frac{\Delta E(s(T))}{a},$$ (7-11)

where $D(T)$ is the diffusion coefficient, a measure of the number of atomic movements per second (see chapter 8), and ΔE is the gain in energy per atomic volume associated with ordering, see (7-2). (In the discussion of recrystallization this postulate 'boundary migration rate equal to mobility times driving force' will be justified in more detail.) The temperature dependence of v passes through a maximum below T_c because at low temperatures D decreases exponentially and on approaching T_c the factor ΔE disappears. Qualitatively the time dependence of the observed degree of ordering $s_{\text{eff}} = \int sX(s)\,dV/V$ predicted by this theory does not differ essentially from that described in (d). For short times X increases proportional to t^3 whereas after long times saturation is approached exponentially.

(c) *Ostwald ripening of the domains.* The growth of domains into material which is still disordered was compared above with the process of (primary) recrystallization, i.e. the absorption of a deformed microstructure by new strain-free crystals, see chapter 15. These processes end when $X = 1$, the disordered volume has all been consumed, and the material is therefore made up completely of ordered domains. This does not correspond to a state of stable equilibrium, since the antiphase boundaries contain additional energy. The next kinetic process is domain growth (to be compared with grain growth or secondary recrystallization in chapter 15). Individual domains increase at the expense of their neighbours so that the total antiphase boundary area is reduced. Attention has already been drawn to the delicate nature of the equilibrium of the antiphase boundary tensions in A_3B alloys, which contain four possible domains. (In AB alloys with two sublattices and domains there is no mechanically stable antiphase boundary configuration.) Fig. 7.26 shows a statistical arrangement of four domain types (A, B, C, D) in A_3B together with the first stage of their growth as a result of the existing mechanical instability of the $\tilde{E}_{\text{APB}}$.

The mathematical description of the process is again based on (7-11) in which now the driving force for domain growth is the specific energy of all antiphase boundaries (referred to the atomic volume a^3), thus for a mean domain diameter L

$$\Delta E \approx \frac{\tilde{E}L^2}{L^3}\,a^3 \quad \text{and} \quad v \equiv \frac{dL}{dt} = \frac{D(T)}{kT}\frac{\tilde{E}a^2}{L} = \frac{K}{L}. \tag{7-12}$$

Integration yields the time law for '*Ostwald ripening*'

$$L^2 - L_0^2 = 2Kt, \tag{7-13}$$

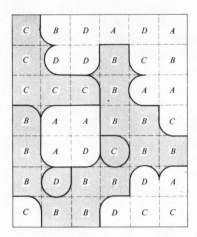

Fig. 7.26. Domain growth (shaded) resulting from the non-equilibrium of antiphase boundary energies in a random initial distribution of four types (A, B, C, D) of square domains in an A_3B superlattice.

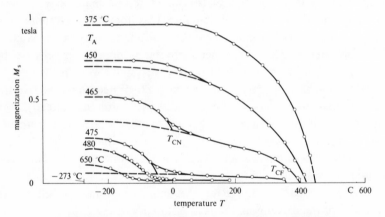

Fig. 7.27. Magnetization–temperature curves for Ni_2Mn after annealing at T_A and quenching. After Marcinkowski and Brown, *J. Appl. Phys.*, **32** (1961), 375.

which is confirmed for Cu_3Au for L between 10 and 80 nm if the actual initial domain size distribution L_0 is taken into account [7.8].

Although the existence of process (c) is beyond question, and although thanks to short-range order sufficient nuclei are always present (a), it is still not possible to decide between the mechanisms of homogeneous ordering (d) and of domain growth (b). On the contrary, it appears that these two processes, which work towards short-range order (d) and long-range order (b) respectively, proceed side by side and partly in competition. This is, for example, apparent in the system Ni_3Mn which is ferromagnetic in the ordered state. A Curie temperature T_{CL} of about 480 °C is observed for the long-range ordered state with the usual temperature dependence of the spontaneous magnetization M_s (fig. 7.27). Specimens quenched from above 450 °C yield a $M_s(T)$ curve with a Curie temperature T_{CS} less than 0 °C which must correspond to the short-range ordered state. Both degrees of order increase with time in qualitative agreement with relationships derived above, although at a mean annealing temperature of 425 °C the volume fraction with short-range order first increases and then decreases again to be replaced by that with long-range order [7.12].

8
Diffusion [8.1], [8.6]

The question of how long it takes for certain atoms in a solid alloy to 'diffuse' from one place to another has already arisen several times in the preceding chapters. Since the time of Faraday it has been known that such atomic migration was possible, in contradiction of the rule *corpora non agunt nisi fluida*. A knowledge of diffusion in the solid state is necessary to describe the kinetics of solid state reactions. In addition, the atomistic mechanism of diffusion is itself of interest because it gives an insight into the properties of lattice defects, mainly vacancies, without which it would be impossible for atoms in a close packed metal to change places. We shall consider these lattice defects in chapter 10. In the following paragraphs the course of diffusion in an alloy with time will be described, using essentially a continuum model.

8.1 Isothermal diffusion with a constant diffusion coefficient

We imagine a dilute single-phase alloy (B in A, or an isotope A^* in A or AB). Let the distribution of B atoms in the x-direction be inhomogeneous. Then, according to Fick's first law, a diffusion coefficient D_B (cm^2/sec) for component B is defined as the number j_B of B atoms migrating through unit cross-section per sec in the x-direction divided by the local concentration gradient: $j_B = -D_B \, \partial c_B/\partial x$. $c_B = N_v v_B$ is the number of B atoms per unit volume (N_v the total number of all atoms per unit volume). The change with time in the number of B atoms situated between x and $x+dx$ is then

$$\Delta x \frac{\partial c_B}{\partial t} = (j_B(x) - j_B(x + \Delta x)) = \Delta x D_B \frac{\partial^2 c_B}{\partial x^2} \tag{8-1}$$

if D_B *is independent of position, i.e. of the concentration* c_B. As long as N_v can be assumed constant, Fick's second law has the form

$$\frac{\partial c_B}{\partial t} = D_B \frac{\partial^2 c_B}{\partial x^2}. \tag{8-2}$$

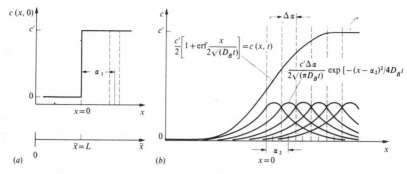

Fig. 8.1. (a) Initial distribution of solute and (b) distribution after diffusion time t, represented as the sum of thin film solutions for films of thickness $\Delta\alpha$. After [8.1].

The physicist will recognize this differential equation as that for the conduction of heat, also in three dimensions, in which case $(D_B \Delta c_B)$ is on the right hand side. In the present case it is often convenient to use spherical polar coordinates for which the equation takes the form

$$\frac{\partial c_B}{\partial t} = D_B \left(\frac{\partial^2 c_B}{\partial r^2} + \frac{2}{r} \frac{\partial c_B}{\partial r} \right). \tag{8-3}$$

Numerous solutions of (8-2) and (8-3) are given in [8.2]. The most important start from the thin film solution or the Fourier solution. In the first case a thin layer of material B is introduced between the end faces of two long rods of material A which are then welded together. If x is measured along the rod axis starting from the welded joint the concentration at time t is given by

$$c_B = \frac{m_B}{2\sqrt{(\pi D_B t)}} \exp\left(-\frac{x^2}{4 D_B t} \right). \tag{8-4}$$

$m_B = \int_{-\infty}^{+\infty} c_B(x, t)\, dx$ equals the initial amount of B in the layer; $c_B(x, 0)$ has the form of a δ-function. The half peak width of the Gaussian distribution, i.e. the width of the layer traversed by the B atoms on average in time t, is $2\sqrt{(D_B t)}$. Other solutions corresponding to different initial conditions can be derived from the solution (8-4) by superposition, i.e. integration. The integral

$$\frac{2}{\sqrt{\pi}} \int_0^{x/2\sqrt{(Dt)}} \exp(-\xi^2)\, d\xi \equiv \mathrm{erf}\left(\frac{x}{2\sqrt{(Dt)}} \right)$$

is tabulated as the error function. Fig. 8.1 shows an example.

In contrast to the thin film solution (8-4) the second type of solution for the differential equation (8-2) can be separated into x and t variables, $c_B(x, t) = A(x)B(t)$. This separation leads to a Fourier series for $A(x)$ the amplitudes of which decrease exponentially with time.

$$c_B(x, t) = \sum_{n=1}^{\infty} A_n \sin(\lambda_n x) \exp(-\lambda_n^2 D_B t). \tag{8-5}$$

This type of solution is particularly useful if the boundary conditions have to be satisfied at a finite distance: for example it describes the degassing of a plate of thickness h, for which the boundary conditions are $c_B = c_0$ for $0 < x < h$, $t = 0$, $c_B = 0$ for $x = 0, h$; $t > 0$. Then $\lambda_n = n\pi/h$ and $A_n = 4c_0/n\pi$ for odd values of n and $A_n = 0$ for even values of n. The higher Fourier terms, which describe the rectangular shape of the initial profile, decrease rapidly with time. The mean gas content at time t (for $\bar{c}_B < 0.8c_0$)

$$\bar{c}_B(t) = \frac{1}{h} \int_0^h c_B(x, t) \, dx \approx c_0 \frac{8}{\pi^2} \exp(-\pi^2 D_B t/h^2) \tag{8-6}$$

is of practical interest. Clearly the half peak time τ for degassing is given by the relationship $h = \pi\sqrt{(D_B\tau)}$.

8.2   Atomic mechanisms of diffusion

8.2.1   *Vacancy mechanism*

From the theory of one-dimensional Brownian motion it is known that after m random steps of length a along the x-axis, a B particle is on average a distance $\sqrt{(\overline{x_m^2})} = a\sqrt{m}$ from its starting point $(m \gg 1)$. $\overline{x_m^2}$ is the mean square displacement. It can be compared with the mean square width of the distribution (8-4):

$$\overline{x^2} = \int c_B(x)x^2 \, dx \bigg/ \int c_B(x) \, dx = 2D_B t.$$

If Γ_B is the jump probability per sec, i.e. $m = t\Gamma_B$, the comparison yields

$$D_B = \tfrac{1}{2}a^2\Gamma_B. \tag{8-7}$$

In the case of a three-dimensional random walk only $\frac{1}{3}$ of the jumps in $\overline{r^2}$ contribute to $\overline{x^2}$ so that the numerical factor in (8-7) is $\frac{1}{6}$.

An atom in a solid can change its place only if the neighbouring lattice site is vacant. The *vacancy mechanism* is the predominant diffusion mechanism for lattice atoms in close packed metals. Vacancies are present in thermal equilibrium at finite temperatures. According to (3-1) the

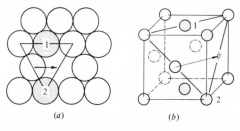

(a)   (b)

Fig. 8.2. When an atom jumps into a neighbouring vacancy, the neighbouring atoms (shaded) have to make way for it.

probability of a lattice site being vacant is given by $c_V(T)$. We therefore expect $\Gamma_B = n \cdot c_V \cdot v_V$ is the jump frequency of the atom under consideration into the vacancy (or the jump frequency of the vacancy itself). The thermal vibrations of the atom about its equilibrium position have a frequency v_0 (approximately equal to the Debye frequency $\approx 10^{13}/\text{sec}$). They do not normally suffice to transfer the atom to a neighbouring vacant site because other neighbouring atoms hinder this movement, see fig. 8.2. The transfer can be achieved only by the expenditure of (free) distortional enthalpy, F_{VM}. This is provided by thermal fluctuations as described by a Boltzmann factor

$$v_V = v_0 \exp\left(-\frac{F_{VM}}{kT}\right). \tag{8-8}$$

$F_{VM} = E_{VM} - TS_{VM}$, where E_{VM} is the migration energy of the vacancy, and the migration entropy S_{VM} comes from the change in lattice vibration during the transfer†. We consider the factor $\exp(S_{VM}/k)$, which generally varies very little, to be incorporated into v_0 to give v_0' and a similar factor $\exp(S_{VF}/k)$ to give c_∞ in (3-1). S_{VF} is the entropy of formation of the vacancy. Then

$$D_B = \frac{n}{6} a^2 v_0' c_\infty \exp\left(-\frac{E_{VF} + E_{VM}}{kT}\right) \equiv D_0 \exp\left(-\frac{E_{VD}}{kT}\right). \tag{8-9}$$

This exponential dependence of the diffusion coefficient on temperature has been observed experimentally in numerous cases. As an example, fig. 8.3 shows the *isotope diffusion coefficient* D_{A*} of the radioactive isotope $A*$ in A

† A more thorough analysis of the process of site exchange due to St Rice [8.15] starts with a consideration of the thermal vibrations of the atoms in the vicinity of the saddle point. The frequency v_V with which these reinforce one another to give displacements which make the saddle point position accessible to a jump can be estimated quantitatively. The contestable assumption (8-8) is thereby rendered unnecessary.

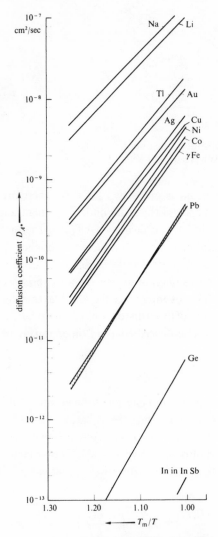

Fig. 8.3. Isotope diffusion coefficient in pure cubic metals plotted against the reciprocal temperature.

measured for various cubic metals in a reduced plot. Values of the pre-exponential factor D_0 and the activation energy E_{VD} for diffusion obtained from this plot will be discussed in more detail. In chapter 10 it will be shown how the splitting of D into its components c_V and v_V can be checked in detail and hence also that of the activation energy $E_{VD} = E_{VF} + E_{VM}$.

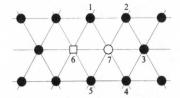

Fig. 8.4. Close packed plane with vacancy (□) and isotope (○).

8.2.2 *Correlation effects*

We must first return to the mean square displacement in the case of an isotope A^* diffusing in A by means of vacancies. The situation after A^* has *changed place once* (from site 6 to 7) is shown in the close packed plane in fig. 8.4. Next to A^* is a vacancy which has jumped from 7 to 6. Its next jump will with equal probability be an exchange with any one of its nearest neighbours (the vacancy has the same interaction with A^* as with A). The situation is quite different for A^*. Its next jump will with the greatest probability be back to 6 rather than to 1–5 simply because the vacancy which made the earlier jump possible is still there. The preferred reverse jump which is directly correlated with the preceding jump falsifies the calculation of the mean square displacement which is made statistically without correlation. It also demolishes the earlier assumption that the 'self diffusion coefficient' in A defined by $D_A = (a^2/6)c_V v_V n$ is identical with the actually measurable isotope diffusion coefficient D_{A^*}.

A *correlation factor* f is defined quantitatively as

$$f = \lim_{m \to \infty} \frac{\overline{r_m^2(A^*)}}{\overline{r_m^2(V)}} = \frac{D_{A^*}}{D_A}. \qquad (8\text{-}10)$$

In our example (fig. 8.4) the probability of a second change of position between 6 and 7 is exactly $\frac{1}{6}$ (or in general $1/n$) of the vacancy jump rate (for n nearest neighbours). If it occurs, A^* has made two jumps which do not contribute to $\overline{r_m^2}(A^*)$. As an approximation, therefore, we can expect from (8-10) that $f \approx 1 - 2/n$. A more accurate calculation for cubic metals gives $f = (1 + \overline{\cos \theta})/(1 - \overline{\cos \theta})$, where θ is the angle between the preceding and succeeding jump vector for A^* averaged over all possible jumps according to their frequency. (For a reverse jump $6 \to 7 \to 6$, $\theta = \pi$, $\cos \theta = -1$.)

Typical values for f are 0.78 for fcc and cph, 0.73 for bcc and 0.5 for the diamond structure†.

If the marked atom used to measure D_B is in fact a solute atom (B in A) and not simply an isotope of A, correlation becomes a decisive process since the vacancy can interact differently with B than with A (see section 8.3.4). It can, for example, be attached firmly to B, because B is an extra large atom or one with a higher valency. In this case reverse jumps are the rule and $f \ll 1$. The jump rate of the B atom into the vacancy, Γ_B, is then much greater than that of an A atom, Γ_A, and instead of (8-7) we expect

$$D_B = D_A f \frac{\Gamma_B}{\Gamma_A} = \frac{a^2}{2} f \Gamma_B \ll \frac{a^2}{2} \Gamma_B. \tag{8-11}$$

The equality on the left of this equation follows from the definition $f = \overline{r_m^2}(B)/\overline{r_m^2}(A)$ in the alloy, see (8-10), as $\overline{r_m^2}(i) = D_i m/\Gamma_i$. f can be given quantitatively for a situation similar to that shown in fig. 8.4, but in which changes in position between 6 and 7 are now more frequent than between 6 and 1 or 5. To a first approximation $\overline{\cos \theta} = -\Gamma_B/(\Gamma_B + (n-1)\Gamma_A)$ and

$$f \approx \frac{(n-1)\Gamma_A}{2\Gamma_B + (n-1)\Gamma_A}. \tag{8-12}$$

Taking other jumps into account (beyond the nearest neighbours) it is found that the statistical weight of Γ_A is smaller than $(n-1)$, namely only 7.15 instead of 11 for the fcc lattice. The situation is even more difficult for concentrated alloys, see [8.3].

8.2.3   *Other diffusion mechanisms*

The movement of an atom (usually a smaller solute atom) from one interstitial site to another (a typical case, carbon in α-iron, will be discussed later) is described as an *interstitial mechanism* of diffusion. The movement again requires thermal activation and there is a migration energy threshold

---

† The diffusion coefficients of two isotopes α and β with atomic weights m_α and m_β of an element A in A differ [8.16] by

$$\left(\frac{D_\alpha}{D_\beta} - 1 \right) = \left(\sqrt{\left(\frac{m_\beta}{m_\alpha} \right)} - 1 \right) f \Delta K. \tag{8-10a}$$

The first term on the right-hand side takes into account the dependence of the jump frequency v_0 of the diffusing atoms on their mass, see (8-8). The last term, ΔK, gives the *fraction of the kinetic energy remaining with the jumping atom* on passing through the saddle point. This factor $0 \leqslant \Delta K \leqslant 1$ allows for the participation of several atoms (different masses) in the lattice vibrations leading to passage of one particular atom through the saddle point.

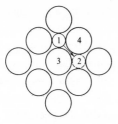

Fig. 8.5. Interstitial atom diffusing in the fcc cube plane.

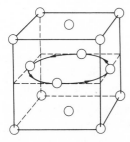

Fig. 8.6. Atoms changing place by a ring mechanism in the fcc cube plane.

E_M corresponding to (8-8) and fig. 8.5. The diffusion is obviously uncorrelated, all nearest neighbour places are equivalent, as long as the number of interstitial atoms is small. Atoms of the base metal can move into interstitial positions (self interstitials) although this requires a considerable amount of energy (see chapter 10). Applying (3-1) to self interstitials, we find that their equilibrium concentration is small and their contribution to diffusion might just be possible because E_M is small too. The existence of a *ring exchange mechanism* of diffusion, as indicated in fig. 8.6, is still unsubstantiated. According to C. Zener, it would be relatively advantageous from the energy standpoint for such a mechanism if the movement of more than two atoms occurred simultaneously. In close packed structures the activation energy for this process is too high for it to make a real contribution to diffusion but possibly not in more open structures such as bcc. In some cases of bcc metals the plot of ln D against $1/T$ is actually observed to be curved instead of a straight line as predicted by (8-9) and this could indicate a second diffusion mechanism.

One of the best documented cases of diffusion which does not take place by means of vacancies is that of *carbon in α-iron*. Not only is a suitable isotope ^{14}C available for self diffusion measurements using the technique described in section 8.1 but also measurements of the anelasticity by the method described in section 2.7 can be used. According to section 6.1.2, a carbon atom stretches one of the three cube edges of the α-iron by $\delta_x \approx 40\%$. If a tensile stress is applied in the direction of a cube edge, the frequency of occupation of this edge by C-atoms will be slightly higher than that of the other cube edges leading to an anelastic relaxation. According to section 2.7 the degree of relaxation is measured by the modulus defect $\Delta\hat{E}/\hat{E}$, which results from the competition between elastic distortion enthalpy (see (6-6)) and the thermal energy of all the C atoms (analogous to the derivation of the Curie law)

$$\frac{\Delta\hat{E}}{\hat{E}} = \alpha\, \frac{\hat{E}\delta_x^2\Omega_{Fe}}{kT}\, v_C, \qquad (8\text{-}13)$$

(in which Ω_{Fe} is the molar volume of iron, v_C the mole fraction of carbon and α is a numerical factor (see [2.20])). The height of the damping peak (fig. 2.22) for the cyclic elastic straining of an α-iron wire is therefore a direct measure of its carbon content, see (2-35). The damping peak (Snoek effect) occurs at a frequency $\omega = 1/\tau$ where τ is the relaxation time or in other words the time needed for a carbon atom to jump from *one* cube edge of the iron lattice to another, see fig. 6.7(*b*). Apart from a factor $\frac{3}{2}$, see [2.20], this relaxation time also determines the mean square displacement according to (8-7) so that the diffusion constant of C in α-Fe can be obtained from anelasticity measurements and (8-8)

$$D_C = \frac{a^2}{6}\Gamma_C = \frac{a^2}{9\tau} = \frac{a_0^2}{36} v_0' \exp\left(-\frac{E_M}{kT}\right) = D_0 \exp\left(-\frac{E_M}{kT}\right). \qquad (8\text{-}14)$$

($a_0 = 2a$ is the length of the cube edge in the iron lattice, a is the jump distance.) A plot of the diffusion coefficients of N in α-iron (fig. 8.7(*a*)) obtained from both anelasticity and chemical measurements obeys (8-14) over fifteen orders of magnitude of D! The appropriate diffusion paths $\sqrt{x^2}$ vary from the lattice constant in the case of the anelasticity measurements covered by the nitrogen in about 1 sec at room temperature to several mm penetration depth measured at $700\,°C$ over anneals lasting several hours. The slope of the plot in fig. 8.7 gives $E_M = 74\,kJ/mol$ and the intercept $D_0 = 0.004\,cm^2/sec$. We shall come back to these quantities in the following paragraph.

8.2.4   *Activation energy and D_0 theory, diffusion in glasses*

The plots of $\ln D$ in fig. 8.3 are parallel straight lines if the temperature is plotted as the reciprocal homologous temperature T_m/T.

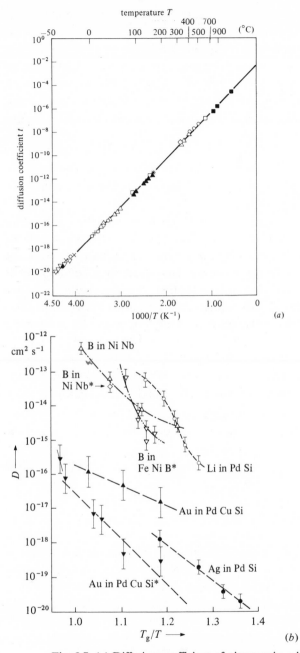

Fig. 8.7. (*a*) Diffusion coefficient of nitrogen in α-iron dependent on temperature according to various measurements. (*b*) Diffusion coefficients in metallic glasses vs temperature. Alloys marked by an asterisk have been annealed before the diffusion experiment.

This implies that $E_{VD} = \text{const } T_m = 16.5L_m$ where L_m is the atomic heat of fusion $\approx 8.4T_m$ if L_m is measured in J and T_m in K. According to the interpretation of N. Nachtrieb, if an atom is to change its place in the lattice, it and its nearest neighbours, on average 16.5 atoms, must be in the 'molten' state. As far as close packed metals are concerned this interpretation is certainly not correct because diffusion takes place by a vacancy mechanism. The equation given above is, however, a useful rule of thumb for estimating E_{VD}. Typical values are

$$E_{VD} \approx 2 \text{ eV} \quad \text{for} \quad T_m \approx 1350 \text{ K (copper).}$$

The activation energy for interstitial diffusion E_M of C in α-Fe is only half this value.

At this point some comments are in order on the diffusion in melts, especially the ones which are frozen, i.e. metallic glasses (section 4.7). These are characterized by a distributed free volume (rather than by vacancies) which is dependant on temperature and quenching conditions. By irradiation with high-energy particles (section 10.4) one obtains more free volume and faster diffusion. The migration of an atom in the glass makes use of this free volume although the details of this process are not yet known. It follows an Arrhenius law with an activation energy which is of the same order as in crystalline alloys but proportional to the glass temperature T_g. Metalloid and transition metal atoms diffuse at different rates in the glass. Diffusion in them can be measured only at low temperature ($T < T_g$), small $\sqrt{\langle x^2 \rangle}$, which accounts for the large scatter of the available data, fig. 8.7(*b*), together with the strong dependence of the free volume on the quenching rate [8.20].

The starting point in an estimate of D_0 for diffusion through interstitial sites according to C. Zener is

$$D_0 = \frac{a^2 v_0}{6} \exp(S_M/k) \tag{8-15}$$

where $S_M = -dG_M/dT|_p$, G_M is the elastic work (at constant T, p) to expand the lattice (by δ_M) when an atom jumps, fig. 8.2. According to (6-6) $G_M \approx \hat{E}\delta_M^2\Omega$. The temperature coefficient of the elastic modulus is

$$\beta = -\frac{d \ln \hat{E}}{dT/T_m} \approx 0.4.$$

Hence

$$S_M = \beta \frac{G_M}{T_m} \approx \left(\frac{G_M}{L_m}\right)\beta k. \tag{8-16}$$

For interstitial atoms in bcc metals, $G_M \approx 6L_m$ and $S_M \approx 2.4k$. Then according to (8-15) $D_0 \approx 10^{-16} \times 10^{13} \times 10 = 10^{-2}$ cm^2/sec which is the correct order of magnitude. For diffusion by means of vacancies, another factor $c_\infty = \exp(S_{VF}/k)$ appears in D_0 which takes into account the entropy of formation of a vacancy (8-9), which according to what has been said about the solubility of solute atoms, (5-23), should be positive and of the same order of magnitude as S_M. In the case of the vacancy mechanism, $D_0 \approx 10^{-1}$ cm^2/sec is in fact observed. From this, the diffusion coefficient at the melting point for many metals is $D(T_m) \approx 10^{-8}$ cm^2/sec.

8.3   Diffusion with a concentration-dependent D

8.3.1   *Boltzmann–Matano analysis*

If D depends on the composition of the alloy and hence on the position in the diffusion couple, which we indicate by the symbol $\tilde{D}$, it can no longer be determined by the methods given in section 8.1. The diffusion equation (8-2) in the one-dimensional case

$$\frac{\partial c}{\partial t} = \frac{\partial}{\partial x}\left(\tilde{D}\frac{\partial c}{\partial x}\right) \tag{8-17}$$

becomes a normal differential equation with the new variable $\eta = x/\sqrt{t}$,

$$-\frac{x}{2t^{3/2}}\frac{\mathrm{d}c}{\mathrm{d}\eta} = \frac{1}{t}\frac{\mathrm{d}}{\mathrm{d}\eta}\left(\tilde{D}\frac{\mathrm{d}c}{\mathrm{d}\eta}\right) \quad \text{or} \quad -\frac{\eta}{2}\frac{\mathrm{d}c}{\mathrm{d}\eta} = \frac{\mathrm{d}}{\mathrm{d}\eta}\left(\tilde{D}\frac{\mathrm{d}c}{\mathrm{d}\eta}\right). \tag{8-18}$$

Equation (8-18) can be integrated for the boundary conditions in fig. 8.1 ($c = c_0$ for $x > 0$, $t = 0$; $c = 0$ for $x < 0$, $t = 0$; i.e. $c = c_0$ for $\eta \Rightarrow \infty$; $c = 0$ for $\eta \Rightarrow -\infty$) giving

$$\tilde{D}(c) = -\frac{1}{2}\frac{\mathrm{d}\eta}{\mathrm{d}c}\bigg|_c \int_0^c \eta\,\mathrm{d}c' = -\frac{1}{2t}\frac{\mathrm{d}x}{\mathrm{d}c}\bigg|_c \int_0^c x\,\mathrm{d}c' \tag{8-19}$$

since we are considering a diffusion profile for fixed t, i.e. we can substitute $\eta = x/\sqrt{t}$. $\tilde{D}$ can be obtained graphically from fig. 8.8. First the 'Matano plane' ($x = 0$) is fixed by the conservation condition $\int_0^c x\,\mathrm{d}c = 0$ (cross-hatched areas above and below the curve are equal). The double cross-hatched area is the integral in (8-19) for $c = 0.2c_0$ and the tangent at this point is the corresponding multiplying factor. The product of these gives the diffusion coefficient at this concentration. Naturally the problem of mutual diffusion of the two (mutually soluble) components is described by *one* coefficient, the so-called *interdiffusion coefficient*, which is designated $\tilde{D}$. There is, however, no doubt that this coefficient represents the combined

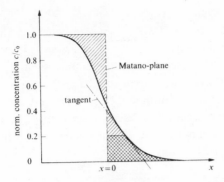

Fig. 8.8. Matano analysis for the determination of the interdiffusion coefficient at $c = 0.2c_0$.

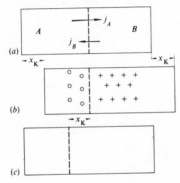

Fig. 8.9. *Kirkendall shift* x_K of weld plane (dashed) because A diffuses more quickly in B than B does in A. Pores are created on the side of the more rapidly diffusing component, while on the other side the material 'swells'.

effect of two individual diffusion coefficients $\tilde{D}_A, \tilde{D}_B$. We would like to determine these. An unexpected difference effect in the $\tilde{D}_i$'s arises in the quantitative analysis.

8.3.2   Kirkendall effect and Darken equations

We again start with the diffusion couple shown in fig. 8.1. The two starting materials are welded together so that small oxide inclusions or small wires of an inert high melting point metal mark the position of the weld plane permanently. If the two components diffuse at different rates, fig. 8.9, then after the anneal, extra material will be found on the side of the slower component measured from the weld plane. The material 'swells' (in all three dimensions). On the other side, more than the equilibrium concentration of vacancies are created as a result of loss of the more rapidly diffusing component. These extra vacancies coagulate to pores and the material shrinks, see fig. 8.12. If the whole specimen is held externally, the weld plane markings move in the opposite direction, fig. 8.9(c). This is the *Kirkendall effect.*

Clearly the system of reference is of great importance in the description of the diffusion experiment (end of specimen or weld plane). According to L. Darken the $\tilde{x}$-coordinates are measured from the left-hand end of the diffusion couple in fig. 8.1 and the x-coordinates from the weld plane. The flux of B through the weld plane in the x-coordinate system is then

$$j_B(x=0,t) = -\tilde{D}_B \frac{\partial c_B}{\partial x}\bigg|_{x=0} = -\tilde{D}_B \frac{\partial c_B}{\partial \tilde{x}}\bigg|_{\tilde{x}=L(t)} . \tag{8-20a}$$

In the $\tilde{x}$-system an additional transport term is involved

$$j_B(\tilde{x}=L,t) = -\tilde{D}_B \frac{\partial c_B}{\partial \tilde{x}}\bigg|_{\tilde{x}=L} + vc_B|_{\tilde{x}=L}. \tag{8-20b}$$

v is the velocity of the weld plane (at the point $\tilde{x}=L(t)$) with respect to the left-hand end of the specimen. Similar equations hold for component A. The continuity equation for the total density of atoms is

$$\frac{\partial c}{\partial t} = \frac{\partial c_B}{\partial t} + \frac{\partial c_A}{\partial t} = -\left(\frac{\partial j_B}{\partial \tilde{x}} + \frac{\partial j_A}{\partial \tilde{x}}\right)$$

$$= \frac{\partial}{\partial \tilde{x}}\left\{\tilde{D}_A \frac{\partial c_A}{\partial \tilde{x}} + \tilde{D}_B \frac{\partial c_B}{\partial \tilde{x}} - cv\right\}. \tag{8-21}$$

If the total density remains constant $\partial c/\partial t = 0$ and the expression in curly brackets is a constant. It must be zero because at the left-hand end of the specimen, $\tilde{x}=0$, the concentration gradients and v vanish. Thus the speed of the weld plane markings (at $\tilde{x}=L$) is given by

$$v = \frac{1}{c}\left(\tilde{D}_A \frac{\partial c_A}{\partial \tilde{x}} + \tilde{D}_B \frac{\partial c_B}{\partial \tilde{x}}\right) = (\tilde{D}_A - \tilde{D}_B)\frac{\partial v_A}{\partial \tilde{x}} \tag{8-22}$$

taking into account that $c = $const, $\partial c_A/\partial \tilde{x} = -\partial c_B/\partial \tilde{x}$ and $v_i = c_i/c$.

Substituting (8-22) in (8-20b) for B, an equation is obtained of the form of Fick's second law

$$\frac{\partial c_B}{\partial t} = \frac{\partial}{\partial \tilde{x}}\left(\tilde{D}_B \frac{\partial c_B}{\partial \tilde{x}} - \tilde{D}_A \frac{c_B}{c}\frac{\partial c_A}{\partial \tilde{x}} - \tilde{D}_B \frac{c_B}{c}\frac{\partial c_B}{\partial \tilde{x}}\right)$$

$$= \frac{\partial}{\partial \tilde{x}}\left(\frac{\tilde{D}_B c_A + \tilde{D}_A c_B}{c}\frac{\partial c_B}{\partial \tilde{x}}\right). \tag{8-23}$$

This equation defines the interdiffusion coefficient obtained from the Matano evaluation as

$$\tilde{D} = v_A \tilde{D}_B + v_B \tilde{D}_A. \tag{8-24}$$

(8-22) and (8-24) are the well known Darken equations. Using them it is possible to calculate $\tilde{D}_A$ and $\tilde{D}_B$ separately after having measured $\tilde{D}$ and v. In the case of a Cu–22% Zn alloy at 785 °C we obtain

$$\frac{\tilde{D}_{Zn}}{\tilde{D}_{Cu}} = 2.3 \quad \text{and} \quad \frac{\tilde{D}_{Zn}(v_{Zn} = 0.22)}{\tilde{D}_{Zn}(v_{Zn} \to 0)} = 17.$$

These changes still have to be investigated more closely, but first it is necessary to consider the assumptions made in the Darken equations.

8.3.3   *Thermodynamic factor*

The absence of concentration gradients $\partial c_i / \partial x$ is in no way a generally valid condition for equilibrium in an alloy, i.e. for the currents j_i to disappear within the meaning of Fick's first law. Constancy of the chemical potentials μ_i is a necessary condition for equilibrium according to section 5.1. Gradients in μ_i are thus expected to produce currents j_i. The following equations are formulated in irreversible thermodynamics specifically for our isothermal binary system AB containing vacancies (V):

$$\left.\begin{aligned}
j_A &= -M_{AA}\frac{\partial \mu_A}{\partial x} - M_{AB}\frac{\partial \mu_B}{\partial x} - M_{AV}\frac{\partial \mu_V}{\partial x}, \\[2mm]
j_B &= -M_{BA}\frac{\partial \mu_A}{\partial x} - M_{BB}\frac{\partial \mu_B}{\partial x} - M_{BV}\frac{\partial \mu_V}{\partial x}, \\[2mm]
j_V &= -M_{VA}\frac{\partial \mu_A}{\partial x} - M_{VB}\frac{\partial \mu_B}{\partial x} - M_{VV}\frac{\partial \mu_V}{\partial x}.
\end{aligned}\right\} \tag{8-25}$$

A whole series of relationships exist between the coefficients M_{ij}. First of all $(j_A + j_B + j_V) = 0$ must hold *outside the zones* in which vacancies are created or destroyed. This condition must hold over wide ranges of the individual gradients (which can of course be independent in the absence of equilibrium). This means, for example, that $(M_{AA} + M_{BA} + M_{VA})$ must also equal zero. Furthermore the Onsager reciprocity relations $M_{ij} = M_{ji}$ must be fulfilled. Equations (8-25) thus become

$$\left.\begin{aligned}
j_A &= -M_{AA}\frac{\partial}{\partial x}(\mu_A - \mu_V) - M_{AB}(\mu_B - \mu_V), \\[2mm]
j_B &= -M_{AB}\frac{\partial}{\partial x}(\mu_A - \mu_V) - M_{BB}(\mu_B - \mu_V).
\end{aligned}\right\} \tag{8-25a}$$

In order to arrive at the Darken equations, the non-diagonal terms must be negligibly small and the vacancies must be in overall equilibrium, i.e. $\partial\mu_V/\partial x = 0$. A comparison of (8-20a) with (5-17) then gives

$$j_B = -\tilde{D}_B \frac{\partial c_B}{\partial x} = -M_{BB}\frac{\partial\mu_B}{\partial x} = -M_{BB}kT\frac{\partial\ln a_B}{\partial x}. \tag{8-26}$$

We now consider an alloy with an infinite dilution of B in A. This brings us back to the diffusion coefficient D_B which we defined in sections 8.1 and 8.2 and which we call the *component diffusion coefficient*. In this case $a_B = v_B$ and (8-26) becomes

$$D_B\frac{\partial c_B}{\partial x} = M_{BB}kT\frac{\partial\ln v_B}{\partial x} = M_{BB}kT\frac{\partial\ln c_B}{\partial x} \tag{8-27}$$

i.e. $D_B = kTM_{BB}/c_B$.

Assuming that M_{BB} represents the same thing in the general case, (8-26), and in the case of the ideal solution, (8-27), and defining the activity coefficient $\gamma_B = a_B/v_B$ we obtain

$$\tilde{D}_B = D_B\frac{d\ln a_B}{d\ln v_B} = D_B\left(1 + \frac{d\ln\gamma_B}{d\ln v_B}\right). \tag{8-28}$$

The expression in brackets is the *thermodynamic factor*. It links the individual diffusion coefficient $\tilde{D}_B$ in an alloy of finite concentration with the component diffusion coefficient (as a measure of the mobility M_{BB}/c_B) by means of a concentration dependent activity coefficient γ_B. According to section 5.2.2, the activity coefficient describes the interaction of the alloy partners in a real solution. As shown in fig. 5.2, γ_B is constant at very small concentrations of B (Henry's law) and equal to unity as $v_B \to 1$ (Raoult's law). In these cases $\tilde{D}_B \approx D_B$ (even if D_B has different values in the two almost pure metals). The thermodynamic factor now appears in both Darken equations (8-22) and (8-24); in fact only for one component. Since the Gibbs–Duhem equation is valid

$$v_A\frac{d\ln\gamma_A}{dv_A} = v_B\frac{d\ln\gamma_B}{dv_B}. \tag{8-29}$$

We thus have the possibility of testing the Darken equations by measuring $\tilde{D}(v_B), v(v_B), D_A, D_B$ and $\gamma_B(v_B)$ for the desired alloy composition. In the case of interdiffusion of AuNi alloys the result largely confirms the theory, see fig. 8.10. The situation is not so unambiguous for the Kirkendall shift. This is directly caused by the creation of vacancies on one side of the weld plane (at dislocations) and their destruction on the other (in pores).

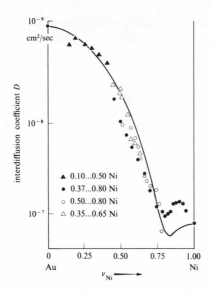

Fig. 8.10. Interdiffusion coefficient for the given AuNi couples at 900 °C. Full curve according to Darken calculated with the aid of the isotope diffusion coefficients measured in the homogeneous alloys.

These processes are related to finite supersaturations of vacancies, i.e. deviations from vacancy equilibrium (contrary to the assumption made above in the derivation of the Darken equations). Fig. 8.11 shows a concentration profile and the component fluxes resulting from differentiation and multiplication by D_A, D_B. The difference between the vacancy flow $j_B - (-j_A) = -j_V$ is not divergence-free, thus the vacancies are not in equilibrium nor is μ_V constant (equal zero) everywhere. (In agreement with fig. 8.12, the porous zone is clearly outside the weld plane!)

It is generally agreed that the Kirkendall effect cannot be explained in terms of a ring exchange mechanism (requiring $D_A = D_B$). H. Schmalzried has, however, questioned such a generalization. If the vacancy concentration depends on the composition, any concentration change, no matter what the cause, means a local change in density and hence some kind of Kirkendall shift.

8.3.4   *Effect of the thermodynamic factor in alloy diffusion*

In section 8.3.2 we found that $\tilde{D}_{Zn}$ in α-brass increased by a factor 17 when the zinc content was raised from almost zero to 22%. The thermodynamic factor increases from 1 to 1.8 in this range (R. Hultgren). At a fixed diffusion temperature (785 °C) D_{Zn} increases much more rapidly with

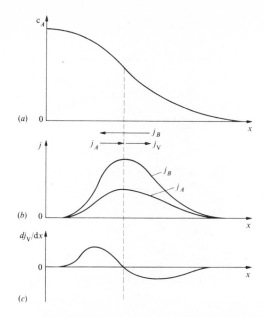

Fig. 8.11. Concentration profile of component A (a) and A and B fluxes for $\tilde{D}_B > \tilde{D}_A$ (b). The vacancy flux $j_V = j_B - j_A$ is not source-free (c): Vacancies are created in the left-hand side of the specimen ($dj_V/dx > 0$) and destroyed on the right-hand side.

v_{Zn} partly because the concentration of thermal vacancies becomes larger as the temperature interval to the solidus temperature decreases (see section 8.2.4: $E_{VD} \propto T_m$). There is also an attractive interaction between zinc atoms and vacancies which, following on from what was said in section 8.2.2, should promote the diffusion of Zn if the concentration is more than a few per cent and the nearest neighbour shells of neighbouring zinc atoms overlap. Th. Hehenkamp *et al.* have measured by the methods of section 10.1 a higher vacancy concentration, a smaller formation energy in alloys which they attribute to the association of a vacancy and B atoms.

The attraction between zinc atoms and vacancies can be understood essentially as an electrostatic interaction. We have already discussed (section 6.3.3) the electronic screening of Zn^{2+} in Cu^+ which leads to mutual repulsion of neighbouring Zn^{2+} ions. The vacancy is (without a Cu^+ ion but with a finite electron density) effectively negatively charged in copper and is thus attracted by the zinc atoms [8.7]. (The excess size would also be an explanation for the attraction between zinc atoms and vacancies.) Lazarus [8.4] has calculated the change in activation energy of D_{Zn} in univalent metals from the screened electrostatic interaction between

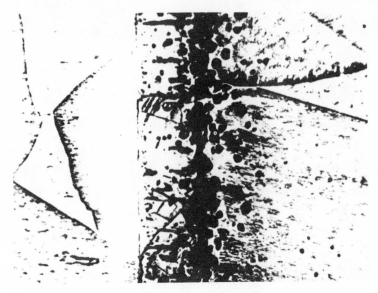

Fig. 8.12. Porous zone to the right of the weld plane in a gold (left)–silver (right) couple after diffusion at 900 °C (W. Seith). 42 ×.

zinc atoms and vacancies. The result agrees qualitatively with experiment. The ratio described in section 8.3.2, $\tilde{D}_{Zn}/\tilde{D}_{Cu} = 2.3$, can be explained in a similar manner (see also the discussion of the correlation factor in section 8.2.2).

The effect of the thermodynamic factor is much more pronounced in systems with a tendency to decompose, see section 5.5.2. In this case a spinodal is defined within the two-phase region by the condition $\partial^2 F/\partial v_B^2 = 0$. Using (5-17), however, this quantity

$$v_A \frac{\partial^2 F}{\partial v_B^2} = \frac{\partial \mu_B}{\partial v_B} = \frac{kT}{v_B}\left(1 + \frac{d \ln \gamma_B}{d \ln v_B}\right) \tag{8-30}$$

is effectively the thermodynamic factor. Below the spinodal (fig. 5.9) $\partial^2 F/\partial v_B^2 < 0$, i.e. the individual diffusion coefficient $\tilde{D}_B$ of (8-28) reverses its sign, leads to 'up-hill' diffusion and creates concentration differences (instead, as in the normal meaning of Fick's first law, of removing them). This phenomenon was first described by U. Dehlinger and R. Becker. 'Spinodal decomposition' will prove to be a vital mechanism in precipitation (see section 9.4).

The significance of the thermodynamic factor becomes obvious in a case of *ternary diffusion* elucidated by L. Darken. γ-Fe with 0.4% carbon is

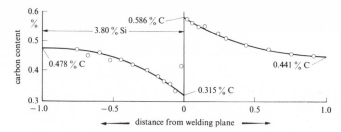

Fig. 8.13. Distribution of carbon after a 13-day anneal at 1050 °C in a diffusion couple iron–silicon/iron (right).

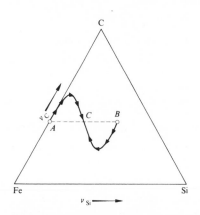

Fig. 8.14. Composition at two points A, B either side of the weld plane as a function of the diffusion time, shown by a succession of arrows. (After L. S. Darken).

bonded to γ-Fe–0.4% C–0.4% Si at 1050 °C. Fig. 8.13 shows that to begin with a carbon concentration gradient *builds up* on both sides of the weld plane which although it is contrary to Fick's first law reduces the μ_C-gradient caused by the difference in Si-concentration. In the long term, after equilibration of the Si content on either side of the weld plane, the carbon level evens out again. Fig. 8.14 shows the concentration–time path for two points to the left and to the right of the weld plane. Finally, if *several phases* arise in a binary diffusion system, for example a simple eutectic system, it should be noted that two-phase regions can never be produced by diffusion. Instead there is a concentration jump from one solubility line to the other, see fig. 8.15 and [8.5]. An isothermal section through a ternary phase diagram can be produced very elegantly by deposition from the vapour phase [5.11]. A, B and C are evaporated from three different sources on to a triangular surface so that the fraction of each in the layer of constant

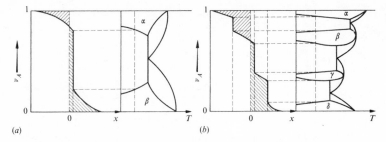

Fig. 8.15. Two phase diagrams and schematic diffusion profiles for the interdiffusion of the pure components.

thickness varies linearly between the corners of the triangle (fig. 5.20). The components are allowed to diffuse to equilibrium and the resulting phases analysed.

8.4   Diffusion in interfaces and along dislocations

In the preceding sections we have assumed that diffusion occurs in a crystal lattice without grain boundaries or dislocations. In practice it is found that at moderate temperatures these lattice defects strongly accelerate diffusion. Based on the models of grain boundaries described in chapter 3, see figs. 3.6 and 3.10, it is quite possible that the packing density of the atoms is lower in the grain boundary than in the perfect lattice so that atoms can *change places more easily*. This will be examined quantitatively in the following.

8.4.1   *Analysis of grain boundary diffusion* [8.6]

We consider a grain boundary perpendicular to a surface coated with a radioactive isotope of concentration c_0. After annealing at constant temperature, the concentration profile of the isotope plotted in fig. 8.16 for $c/c_0 \approx 10^{-5}$ is obtained. The grain boundary obviously acts as an 'irrigation channel' for the diffusing isotope by transporting it rapidly in the y-direction from which it can then diffuse in the x-direction. The situation can be described approximately according to J. C. Fisher as follows: $c = c_0$ for $t \geqslant 0$ and $y = 0$. Let the diffusion coefficient of the isotope in a grain boundary layer of thickness δ ($\approx 10^{-7}$ cm) be D_G^* and in the lattice $D_L^* \ll D_G^*$. The change in the fraction of isotope with time in an element of the grain boundary is

$$\frac{\partial c}{\partial t} = -\frac{\partial j_y}{\partial y} - \frac{2}{\delta} j_x = D_G^* \frac{\partial^2 c}{\partial y^2} + \frac{2D_L^*}{\delta} \frac{\partial c}{\partial x}\bigg|_{x = \pm \delta/2}. \tag{8-31}$$

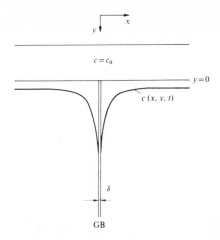

Fig. 8.16. Schematic concentration profile for diffusion from the surface ($y = 0$) preferentially along a grain boundary, GB, (y-direction) of thickness δ.

In the surrounding volume the normal diffusion equation applies with $D_L^* \ll D_G^*$. The one-dimensional equation (8-2) is used because the flow should be perpendicular to and away from the grain boundary. The concentration in the grain boundary $c(x = 0, y) = c_G$ very quickly reaches a quasi-stationary profile from which diffusion can proceed in the x-direction following (8-2). Substituting the thin film solution $c(x)$, (8-4), with an initially unknown multiplying factor $c_G(y)$ in (8-31), c_G can be calculated to obtain the complete solution

$$c(x, y, t) = c_0 \exp\left(\frac{-y}{(\pi D_L^* t)^{1/4} \sqrt{(\delta D_G^*/2 D_L^*)}}\right) \times$$

$$\times \left[1 - \mathrm{erf}\left(\frac{x}{2\sqrt{(D_L^* t)}}\right)\right]. \tag{8-32}$$

Slices of thickness Δy parallel to the surface are analysed for their isotope content, i.e. $\bar{c}(y, t) = \int_{-\infty}^{+\infty} c(x, y, t)\,\mathrm{d}x$ is determined whereby the last factor in (8-32) becomes a constant. A plot of $\ln \bar{c}$ against y gives a straight line with the inverse slope $(\sqrt{(D_L^* t)} \pi^{1/4} \sqrt{\eta})$ with a grain boundary factor $\eta \equiv D_G^* \delta / 2 D_L^* \sqrt{(D_L^* t)}$. (Bulk diffusion from the surface would, according to (8-4), give straight lines in a plot of $\ln \bar{c}$ against y^2!) Thus, knowing D_L^*, η and hence the quantity $D_G^* \delta$ can be obtained. Fig. 8.17 shows concentration profiles for different values of η. Fisher's solution requires conditions

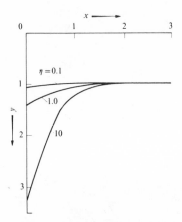

Fig. 8.17. Calculated concentration profiles for $c = 0.2c_0$ show increasing penetration by the isotope in the y-direction near the grain boundary ($x = 0$) with increasing grain boundary (diffusion) factor η. (After R. Whipple.)

corresponding to $\eta \gg 1$. If $t = 10^5$ sec, $D_L^* = 10^{-11}$ cm^2/sec and $\delta = 4 \times 10^{-8}$ cm, D_G^*/D_L^* must be large compared with 5×10^4. Calculations have been made in the meantime with less restrictive conditions, see [8.6].

8.4.2   Results for grain boundaries and dislocations

Experimental results for polycrystalline zinc at 90 °C yield a ratio $D_G^*/D_L^* \approx 10^6$, where the activation energy for grain boundary diffusion $E_{GD} = 59$ kJ/mol $= 14$ kcal/mol is considerably less than for diffusion in the lattice, $E_{LD} = 96$ kJ/mol $= 23$ kcal/mol. (Diffusion in zinc is anisotropic, the values of E_{LD} measured parallel and perpendicular to the c-axis differ by 10%.) In order to be able to give a value for D_{0G}, a value ($\approx$ the lattice parameter) must be assumed for δ. D_{0G} is somewhat smaller than D_{0L}. Diffusion along grain boundaries thus becomes noticeable only at low temperatures where the factor η is much greater than 1, although normally $\sqrt{(D_L^* t)} \gg \delta$. Even in the case of small grains only a tiny fraction X of the atoms are found in the grain boundaries with the result that grain boundary diffusion can predominate only at such low temperatures that $XD_G > D_L$.

The dependence of D_G on the orientation difference between the grains meeting at the grain boundary is of particular interest. Fig. 8.18 shows that the penetration depth $\sqrt{(D_L^* t)}\sqrt{\eta}$ along the grain boundary increases with orientation difference $\theta \leqslant 45°$, in other words that diffusion is more rapid in high-angle than in low-angle boundaries. This seems to result more from the difference in η than from the effect of the orientation difference on the activation energy E_{GD} which is seldom found to be less than $E_{LD}/2$. In the

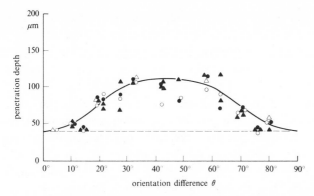

Fig. 8.18. Penetration depth of a Ni isotope in boundaries of nickel bicrystals with different orientation difference θ (7.8 h at 1100 °C). After [8.6].

case of a low-angle grain boundary consisting of edge dislocations with core diameter $r_0 (\approx b)$ separated by h, an effective $\delta = r_0^2/h \approx (r_0^2/b)\theta$ must be defined, see sections 3.2.1 and 11.3.1, which causes η to increase with the orientation difference. As would be expected, D_G is different in tilt and twist boundaries, and in tilt boundaries themselves it is also different parallel and perpendicular to the dislocation direction, see also [8.8], [15.6].

Everything indicates that grain boundary diffusion in metals proceeds by a vacancy mechanism. Separating the terms as in section 8.2.1, $E_{VD} = E_{VF} + E_{VM}$, the energy of formation of a vacancy E_{VF}^G is (several kJ/mol) smaller than E_{VF}^L by an amount equal to the binding energy of the vacancy to the grain boundary or dislocation. The essential part of the difference $(E_{LD} - E_{GD})$ clearly stems from the very small migration energy E_{VM}^G. According to what was said in section 8.2.2 about correlation, a value of $f \approx 0.1$ must be expected for the (virtually) one-dimensional diffusion of the isotope along dislocations.

8.4.3   Surface diffusion

The mathematical analysis in section 8.4.1 applied to the grain boundary shown in fig. 8.16 describes a case of surface diffusion if the left-hand grain is omitted. Since the grain boundary is a plane of symmetry the solution (8-32) remains the same (apart from a factor 2) and gives the spread of the isotope in the y-direction measured from the edge of the specimen over its surface. δ is now the thickness of an affected surface layer; D_S the surface self diffusion coefficient replaces D_G. The evaluation depends on the models used for the surface and for the surface diffusion. Especially important are the roughness and orientation of the surface. Atoms adsorbed on to the surface (ad-atoms) and atom complexes play an important part. (Their observation is possible by the FIM technique, section 2.4.) Ad-atoms are taken from the last lattice plane parallel to the surface by a process similar to evaporation and deposited parallel to the surface by a process similar to evaporation and deposited on the surface. Their activation energy for migration is certainly very small. It thus becomes likely that there will be a connection between the activation energy E_{SD} and the energy of sublimation W_S. The ratio E_{SD}/W_S lies between 0.1 and 0.6. Surface contamination, e.g. by oxygen, disrupts the measurements considerably. D_{0S} is often found to be much bigger than D_{0L} for lattice diffusion indicating long 'jump paths' for the ad-atoms or complexes. Thus surface diffusion is most likely to compete with lattice diffusion at low temperatures (in the case of small particles, see below). The two work together to create or smooth out surface roughness produced artificially or arising where a grain boundary meets the surface [8.10]. The driving force to attain equilibrium by surface diffusion is the surface tension $\tilde{E}_S$. All surface projections or depressions change, according to [8.9], with $(Bt)^{1/4}$ where $B = D_{SD} \delta \tilde{E}_S \Omega / kT$.

8.4.4   Sintering

The diffusion controlled process by which a metal powder aggregate is densified to a solid body is called sintering. It has great technological importance in powder metallurgy for the manufacture of parts with complicated external shapes or made of material which is difficult to deform plastically. The prepressed powder grains adhere as they come into contact forming so-called 'necks', which thicken in the initial stage of sintering. The pores become rounded off, as shown by a cross-section of a wire bundle in fig. 8.19. In the second stage the pores gradually close and the material densifies by transport of vacancies to internal and external surfaces. The vacancies follow the chemical potential gradient μ_p which according to the Gibbs–Thomson equation arises as a result of the 'vapour pressure of small droplets', in this case pores. For a cavity with radius of curvature r_p and (free) surface energy $\tilde{E}_S/\text{cm}^2$, the removal of dn atoms gives

$$\frac{\partial F_p}{\partial n} = \mu_p \approx \tilde{E}_S \frac{\partial(4\pi r_r^2)}{\partial(V_p/\Omega)} = \frac{2\tilde{E}_S\Omega}{r_p}. \tag{8-33}$$

As an approximation for the gradient of $\mu_p(r)$ the maximum value $\mu_p(r_p)/r_p$ can be used. Substituting this in (8-25) a diffusion current is obtained which

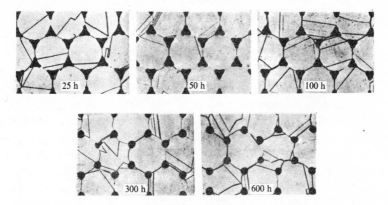

Fig. 8.19. Section through a bundle of copper wires (diameter 30 μm) after annealing at 900 °C for the times given. The wires grow together at necks and the pores become rounded off.

can be related geometrically to the change in shape of the pores or the reduction in pore volume. The result is a time law governing these changes. It is, however, necessary to know (*a*) the diffusion mechanism (surface, bulk or grain boundary diffusion or possibly vapour phase transport) and (*b*) where the vacancies are 'destroyed' (at surfaces, grain boundaries, dislocations, other pores). As was first shown by G. Kuczynski [8.11] different assumptions lead to different time laws. M. Ashby [8.19] has delineated the regions in which different sintering mechanisms are expected to dominate in a 'sintering diagram' (analogous to fig. 12.22). In this, the radius of curvature of the pores is plotted against sintering temperature with the sintering time as parameter. Experimental differentiation is complicated by a whole range of other factors such as the 'activity' of the powder, presence of oxides, gases in the pores, etc. [8.12].

8.5   Electro- and thermomigration [8.13], [8.6], [8.17]

8.5.1   *Electromigration*

If a heavy current is passed through an alloy specimen, electrolytic decomposition is expected similar to that in an ionic crystal, resulting in enrichment at the cathode in the component with the larger (positive) nuclear charge. This is usually but not always the case. The direction of the drift movement in the electric field depends more on whether the alloy conducts by electrons or holes. Evidently there is an 'electron wind' which tends to drag atoms which are changing places towards the anode. By incorporating markings as discussed in section 8.3.2, transport of material in an electric field can be demonstrated in pure metals. The greatest effect is, however, observed with highly mobile interstitial solute atoms in metals, for example, carbon in α-iron, which moves towards the cathode.

The theoretical description starts with (8-25) extended by additional terms $(M_{ie}(\partial\mu_e/\partial x))$ and an additional equation for the electrons. Furthermore the chemical potential is replaced by the *electrochemical potential* $\hat{\mu}_i$; $\hat{\mu}_i \equiv \mu_i + Z_i^* e\varphi$. φ is the electric potential of the field $\mathscr{E} = (\mathrm{d}\varphi/\mathrm{d}x)$, Z_i^* is the effective nuclear charge of an ion i. The electrons (like the vacancies) are considered to be in equilibrium, i.e. $\mu_e = E_{\text{Fermi}} = \text{const.}$ The non-diagonal terms in (8-25a) are again neglected. For a very dilute solution of B in A we can use (8-27) which gives

$$j_B = -D_B \frac{\partial c_B}{\partial x} - \frac{c_B Z_B^* D_B}{kT} e\mathscr{E}. \tag{8-34}$$

The second term is a 'transport term' $c_B v_B$ as in (8-20b), where $v_B = (D_B/kT)(Z_B^* e\mathscr{E})$ the drift velocity in the electric field. (D_B/kT is the mobility = drift velocity per unit force.) The effective charge must take into account the screening of the B ions in the metal. The 'electron wind', however, is included in the non-diagonal term $(B-e)$ and up to now has been described only in terms of a model, e.g. by an expression due to Huntington–Fiks

$$Z_B^* = Z_B - \tfrac{1}{2} Z_A \frac{|\Delta\sigma_{BS}|}{\sigma} \frac{c_B}{c_{BS}} \frac{m^*}{|m^*|}. \tag{8-35}$$

$\Delta\sigma_{BS}/\sigma$ is the relative decrease in electrical conductivity of the alloy due to the fact that a fraction c_{BS}/c_B of the B atoms are in the saddle point position during a diffusive jump. The conduction electrons (effective mass $m^* > 0$) or holes ($m^* < 0$) have a maximum effect at this point; they are scattered most strongly by saddle point atoms. The product $(\Delta\sigma_{BS}/\sigma)(c_B/c_{BS})$ thus represents the relative difference in the effective cross-section for scattering of the charge carriers in the lattice position and the saddle point position. In an experiment to demonstrate this, direct currents of several 10^4 amp/cm^2 are passed isothermally through a layer of Sb placed between two copper rods. Fig. 8.20 shows a concentration profile measured with the microprobe after passage of the current. This profile corresponds to the solution in fig. 8.1 depending on $(x \mp v_B t)^2/4D_B t$. Equations (8-34) and (8-35) are confirmed satisfactorily in as far as the quantities in these equations are known. Surprisingly enough the direct field force on the solute ion seems to be determined by the nominal valency Z_B without taking screening into account.

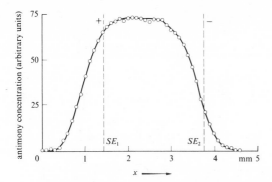

Fig. 8.20. Concentration profile of Sb after electromigration in copper at 866 °C (Cu–0.8 wt.% Sb alloy between the weld planes SE, copper on the outside). After T. Hehenkamp, Göttingen.

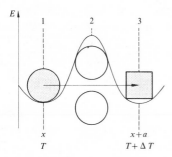

Fig. 8.21. Passage of an atom (lattice plane 1) through a saddle point position (2) into a vacancy (3) on the hotter side.

8.5.2 Thermomigration

If the electric potential gradient across the diffusion specimen is replaced by a temperature gradient, the force $Z^*e\mathscr{E}$ in the transport term in (8-34) has to be replaced by $(Q^*/T)\,\mathrm{d}T/\mathrm{d}x$, i.e. Z_B^* by the 'heat of transport' Q_B^*. In a system of currents of the type given by (8-25) for the components A, B, vacancies and the heat flow j_Q

$$Q_B^* = \frac{M_{BQ}}{M_{BB}} = \left(\frac{j_Q}{j_B}\right)_{(\mathrm{d}T/\mathrm{d}x)\to 0}$$

is the heat flow per unit material flow of B in the limiting case of a vanishing temperature gradient. An atomistic model due to K. Wirtz for a pure metal B considers three lattice planes, 1, 2, 3 in the direction of the T-gradient (fig. 8.21). The flow of B atoms to the right is $\vec{j}_B = c_B a v_V(1) c_V(3)$ and to the left

$\bar{j}_B = c_B a v_V(3) c_V(1)$. With (3-1) and (8-8)

$$\bar{j}_B - \bar{j}_B = c_B a v'_0 c_\infty \left\{ \exp\left[-\frac{E_{VM}}{kT} \right] \exp\left[-\frac{E_{VF}}{k(T+\Delta T)} \right] \right.$$

$$\left. - \exp\left[\frac{-E_{VM}}{k(T+\Delta T)} \right] \exp\left[-\frac{E_{VF}}{kT} \right] \right\} \qquad (8\text{-}36)$$

$$\approx c_B a \bar{v}_V \bar{c}_V \left\{ \left(1 + \frac{E_{VF}}{kT} \frac{\Delta T}{T} \right) - \left(1 + \frac{E_{VM}}{kT} \frac{\Delta T}{T} \right) \right\}$$

$$= c_B D_B \frac{(E_{VF} - E_{VM})}{kT} \frac{\Delta T}{a} \frac{1}{T}; \qquad (8\text{-}37)$$

$\bar{v}_V(T), \bar{c}_V(T)$ are average values for the jump frequency and density of the vacancies respectively.

Comparison with (8-34) making the substitutions given above for $Z^* e \mathscr{E}$ yields $Q_B^* = E_{VF} - E_{VM}$. If the migration energy is not fully supplied in the initial position but part is supplied in the saddle point position, the first term is multiplied by a factor $\leqslant 1$. It reduces the information that can be gained from Q_B^* quite apart from the fact that macroscopic concepts have been used without justification in the microscopic range in the derivation of (8-36). In practice large negative values of Q_B^* are often obtained for noble metals. The temperature gradient also gives rise to an electron current which again is responsible for electromigration of B ions, see [8.17] and [8.18].

8.6   Oxidation of metals [8.5]

The oxidation of metals which in certain cases is known as 'tarnishing' or 'scaling' is of great practical interest because it converts metals to thermodynamically more stable compounds, often those from which they were originally obtained by smelting (cf. also sulphides and chlorides). As for non-metals in general, only an outline of the problem can be given here. Special monographs [8.5], [8.14] have been published on the subject. We shall restrict ourselves here to explaining some aspects linked with diffusion which should not be overlooked by the metal physicist.

We assume that firmly adhering coating layers are formed by the reaction of a metal (Me) with oxygen

$$Me|_{solid} + \frac{v}{2} O_2|_{gas} \; \rightleftharpoons \; MeO_v|_{solid}. \qquad (8\text{-}38)$$

We do not consider microstructural defects, pores in the layers which form

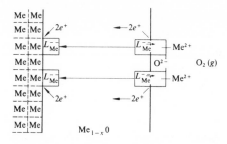

Fig. 8.22. Formation of metal ion vacancies L_{Me}^{2-} and electron holes e^+ which then migrate through the oxide.

especially if there is a large difference in specific volume between metal and oxide ('Pilling–Bedworth rule') and which are partially compensated by plastic deformation of the oxide due to internal stresses, providing they do not actually prevent the formation of a (protective) oxide layer. A layer of thickness ξ grows mainly by *transport of metal ions through the oxide* to the oxygen at the surface. The concentration gradient of Me decreases proportional to ξ^{-1} where ξ is the thickness of the layer. The current, i.e. the increase in thickness with time is therefore

$$\frac{d\xi}{dt} = \frac{k_z}{\xi}; \qquad \xi^2 = 2k_z t. \tag{8-39}$$

This *parabolic growth law* was first discovered by G. Tammann, k_z is the scaling constant which is proportional to a mean diffusion coefficient $\overline{D}_{Me}$ in the oxide and hence exponentially dependent on temperature. According to the theory of C. Wagner which describes the growth of thick oxide layers and following (8-26)

$$\frac{d\xi}{dt} \propto \frac{k_z}{\xi} \propto \frac{\overline{D}_{Me}}{kT} \frac{|\Delta F_{MeO_v}|}{\xi}, \tag{8-40}$$

where ΔF_{MeO_v} is the free energy of formation of MeO_v from metal and gaseous oxygen at the existing partial pressure. There are numerous other growth equations for different special conditions, especially diffusion mechanisms. It is particularly important to maintain neutrality in the movement of metal ions in the oxide. This is often accomplished by electronic or hole conduction in the oxide, see fig. 8.22 for the case of divalent negative Me vacancies moving towards the Me/MeO phase boundary accompanied by (positive) holes e^+ in the valency band of the oxide. It is clear that the electrical conduction mechanism of the oxide is important here. It is assumed that the incorporation of Me^{2+} vacancies (designated L^{2-}_{Me}) is relatively easy on the metal side. The concentration c_V of L^{2-}_{Me}, i.e. of vacancies in the metallic sublattice of the oxide, is determined by the supply of oxygen p_{O_2} at the MeO/O_2 interface which is striving to form new oxide lattice cells. The law of mass action for the reaction

$$Me^{2+}O^{2-} + \tfrac{1}{2}O_2 \rightleftharpoons 2Me^{2+}O^{2-} + L^{2-}_{Me} + 2e^+ \tag{8-41}$$

for a large proportion of MeO (i.e. $c_{MeO} = const$) but infinite dilution is

$$\frac{c_V c^2_{e^+}}{p^{1/2}_{O_2}} \propto \exp\left(-\frac{\Delta F_{MeO}}{kT}\right) \equiv K.$$

Since $2c_{e^+} \approx c_V$ for reasons of electrical neutrality, $c_V = (4K)^{1/3} p^{1/6}_{O_2}$. The concentration of metal ion vacancies in the oxide permits the diffusion of metal through the oxide, i.e. $\bar{D}_{Me} \propto k_z \propto p^{1/6}_{O_2}$ in (8-40). Assuming electronic conduction in place of hole conduction the scaling constant k_z is found to be relatively independent of the partial pressure of oxygen p_{O_2} because $c_{e^+} \propto c_V$ is very small at the MeO/O_2 interface. The essential part of the Wagner scaling theory is thus the association of ion diffusion with electronic charge transport in the coating layer. If the velocity of either of these two processes is very small, the base metal is resistant to oxidation (Al_2O_3 with vanishingly small electronic conductivity).

The situation is more complicated in the oxidation of alloys (AB) depending on whether both components oxidize or not and whether the oxides form solid solutions with one another, etc. If oxygen is soluble in the base metal A, and B forms a very stable oxide, we speak of internal oxidation. A dispersion of BO_y forms in A which hardens the base metal (see section 14.3; example $Cu–SiO_2$). All this leads on naturally to a consideration of the reaction between a metal and a liquid electrolyte ('corrosion'), but this would take us into the field of electrochemistry.

9
Precipitation

An *AB* alloy decomposes into two phases of different composition on cooling if like atom pairs possess a smaller energy than unlike atom pairs, i.e. if the pair exchange parameter is positive $\varepsilon > 0$, see section 5.5. The thermodynamics of the phase diagram provides no information about the distribution and morphology of the two phases nor how these change with time. These questions will be considered in the following. According to W. Gibbs there are two different possibilities for so-called *continuous precipitation*. Firstly, thermal fluctuations can in some places produce 'nuclei' of the second phase, the composition of which *deviates strongly from that of the matrix* and which under certain circumstances may correspond to that of the equilibrium phase shown in the phase diagram. These nuclei then grow by normal diffusion in the concentration gradient of the depletion zone resulting from an enrichment of the nucleus in *B* atoms (fig. 9.1(*a*)) (mechanism of nucleation and growth, sections 9.1 and 9.2). The second possibility involves *small variations in concentration throughout* the specimen, which gradually increase in amplitude by up-hill diffusion (fig. 9.1(*b*)). In section 8.3.4 we saw that this is possible inside the spinodal where the thermodynamic factor of the diffusion coefficient $\tilde{D}_B$ becomes negative. We shall discuss the mechanism of spinodal decomposition in section 9.4.

In addition to these continuous processes, *discontinuous precipitation* is observed in some systems but this occurs only along a reaction front moving into supersaturated material. (Eutectoid decomposition analogous to eutectic crystallization, section 4.6, will be described in section 9.5.) These precipitation processes often yield a whole range of particle sizes. The larger precipitates then grow and the smaller ones dissolve thus reducing the interfacial energy in a given volume (Ostwald ripening, section 9.3). The morphology of the precipitated particles (particularly in the case of so-called Widmanstätten plates) is determined by the anisotropy of the interfacial energy if the precipitates are *incoherent* (see section 3.4). In the case of *coherent* particles, the shape is determined by the elastic distortion

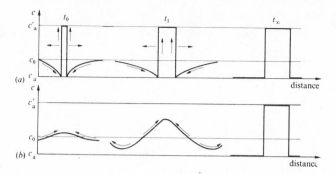

Fig. 9.1. Two possible decomposition processes $(t_0 \to t_1 \to t_\infty)$ (a) by nucleation and growth ('down-hill diffusion') and (b) spinodally ('uphill diffusion'). c_a, c_a' are the equilibrium concentrations of the final phases.

energy. Classic examples of these precipitation reactions take place in supersaturated Al 2 at.% Cu solid solutions. In 1906, A. Wilm observed that the hardness of these alloys increased with time and thereby discovered the technologically important Duralumin (see chapter 14).

9.1   Nucleation of precipitates [9.1]

9.1.1   *Energy contributions*

Compared with the nucleation of a crystal from a pure melt discussed in section 4.1, the composition of the nucleus is an additional variable in the nucleation of precipitates from a supersaturated solid solution. Furthermore nucleation in the solid state is generally accompanied by a *distortion* (δ) which makes the change in free energy $\Delta F_{\text{total}}(r)$ on forming a nucleus (phase β, sphere of radius r) in a solid solution (α) more positive. Equation (4-1) for homogeneous nucleation thus becomes

$$\Delta F_{\text{total}}(r) = (-\Delta f + \Delta f_{\text{el}})\tfrac{4}{3}\pi r^3 + \tilde{E}_{\alpha\beta}4\pi r^2. \tag{9-1}$$

The three coefficients must be explained in more detail:

(a) Δf. The saving in free energy resulting from the formation of a precipitate containing a fixed number $N_{k'}$ atoms within $N_{m'}$ matrix atoms can be derived from fig. 9.2. If F_m is the free energy per atom at the composition $v_B = m$ then the free energy of the system (volume V) changes on decomposition $m \to m' + k'$ by

$$V \cdot \Delta f = (N_{k'} + N_{m'})F_m - N_{m'}F_{m'} - N_{k'}F_{k'}$$

$$= N_{m'}(F_m - F_{m'}) + N_{k'}(F_m - F_{k'}), \tag{9-2}$$

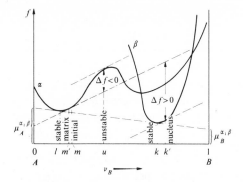

Fig. 9.2. Free energies of the phases α and β in the AB system. The starting alloy m decomposes into $(m' + k')$ with a saving of energy $\Delta f > 0$, or into $(m' + u)$ only if energy is supplied $(\Delta f < 0)$. Both k' and u have the same slope $\partial F / \partial v_B$ as m' and are thus in metastable equilibrium with m'. A stable equilibrium exists between $\alpha(l)$ and $\beta(k)$.

or using the lever rule and for $N_{k'} \ll N_{m'}$

$$V \cdot \Delta f = N_{k'} \left[\frac{k' - m}{m - m'} (F_m - F_{m'}) + (F_m - F_{k'}) \right]$$

$$\approx N_{k'} \left[F_m - F_{k'} + (k' - m) \frac{dF}{dv_B} \bigg|_m \right]. \tag{9-3}$$

According to the tangent construction, the nucleus has a composition (k') in exchange with α rather than (k) of phase β in equilibrium with $\alpha(l)$ fig. 9.2, for an initial composition (m). At the composition k', the decrease in energy Δf is a maximum for a given m. A nucleus with a composition u would necessitate expenditure of energy $(\Delta f < 0)$ and is therefore unstable. Such unstable intermediate stages may be traversed in the process of forming the nucleus (k') from m, unless m approaches the spinodal composition (the point of inflection $f'' = 0$). In this case the classic concept of nucleation fails and spontaneous (periodic) decomposition occurs, section 9.4 [9.2].

(b) $\tilde{E}_{\alpha\beta}$. R. Becker assumes an atomically sharp transition from the composition of the matrix $(v_B = m')$ to that of the nucleus (k') at the interface. In this case $\tilde{E}_{\alpha\beta}$ can be expressed by the pair exchange energy ε of section 5.2.2. Imagine two rods $\frac{1}{2}$ cm^2 in cross-section with the compositions m' and k' each cut through and then rejoined to the wrong half. Assuming that the probability that an A atom has a B neighbour is $P^{AB} = v_B$, the difference in energy of all AA, AB and BB bonds across the interface after and before the change is with (5-15) given by

$$\tilde{E}_{\alpha\beta} = \frac{\varepsilon}{a^2}(k'-m')^2, \qquad (9\text{-}4)$$

(where a is the size of the primitive cubic unit cell). J. W. Cahn assumes a steep but not atomically sharp concentration gradient between nucleus and matrix and describes it in terms of the so-called *gradient energy* obtained by expansion of the free energy f in the variables $dv_B/dx \ldots d^2v_B/dx^2$ up to the second order, see [9.2]. According to this (for an interface perpendicular to the x-axis)

$$\tilde{E}_{\alpha\beta} = \int_{-\infty}^{+\infty}\left[\Delta f(v_B(x)) + \kappa\left(\frac{dv_B}{dx}\right)^2\right]dx. \qquad (9\text{-}5)$$

$\Delta f = f(v_B(x)) - \frac{1}{2}[f(v_B(+\infty)) - f(v_B(-\infty))]$ is the free energy density at a point x in the interface relative to that at a distance outside it. $\kappa > 0$ is an expansion coefficient obtained by differentiating f twice [9.2]. The actual variation $v_B(x)$ between m' and k' in the interface minimizes this energy. According to measurements with the atom probe (section 2.4.2) the interface between Cu-rich precipitates and the iron matrix is in fact atomically sharp, fig. 9.3, if the ageing temperature is smaller than half the critical temperature of the miscibility gap. This is to be expected from lattice theory of the interface [9.12].

(c) Δf_{el}. If the lattice parameter depends on the concentration, described approximately by an atomic size factor $\delta = (1/a)\,da/dv_B$, a (hard) β precipitate in a (soft) α matrix has an associated distortion energy Δf_{el} (per unit volume of precipitate)

$$\Delta f_{el} = \frac{\tilde{E}_\alpha \delta^2}{1-v}(k'-m')^2\varphi\left(\frac{c}{b}\right). \qquad (9\text{-}6)$$

$\hat{E}_\alpha$ is the isotropic elastic modulus, v is Poisson's ratio, $\varphi(c/b)$ is a shape factor depending on the axial ratio of the precipitate considered as an ellipsoid of rotation (c is the semi-axis in the direction of the axis of rotation and b that at right angles to it). Fig. 9.4(a) shows $\varphi(c/b)$ after F. R. N. Nabarro. According to this, a disc possesses the smallest distortion energy, while that of a sphere is greater than that of a cigar-shaped precipitate of the same volume. This is in agreement with the observed morphology of small coherent precipitates, e.g. of Cu (Θ'') in Al (see section 9.1.2) which with $\delta = -12\%$ forms plate-like particles, whereas Zn in Al ($\delta = -2\%$) and Co in Cu ($\delta = -1.6\%$) precipitate as spheres. If the modulus $\hat{E}_\alpha$ is anisotropic, the precipitate platelets line up on elastically 'soft' planes, usually $\{100\}$.

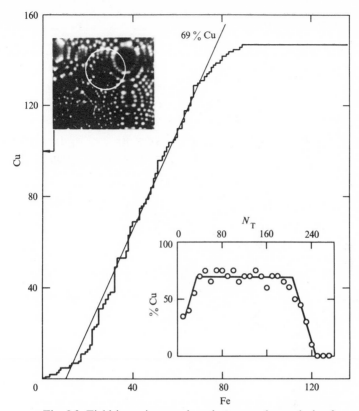

Fig. 9.3. Field ion micrograph and atom probe analysis of a copper precipitate (in the circle of diameter 10 nm) in an Fe–1.4% Cu alloy after 3 h at 500 °C. On removal of atoms from the precipitate at the tip, the probe counts 150 Cu atoms to 80 Fe atoms, in other words a composition $\sim Cu_2Fe$. The total number $N_T = 230$ of atoms removed gives the thickness of the precipitate as ~ 5 nm and an interface thickness of ~ 1 nm. (S. Brenner, S. R. Goodman, J. R. Low, U.S. Steel Corp., Monroeville.)

If the distortion energy is high, the incorporation of dislocations into the interface can be advantageous (semi-coherent boundary) because their extra half-planes accommodate the mismatch δ, e.g. around the periphery of the platelet. The dislocation energy of a low-angle grain boundary increases only linearly with δ (section 3.3) whereas according to (9-6) the energy of the coherent distortions increases as the square of δ. Above a critical δ_c ($\approx 5\%$ in the case of a precipitate three atomic layers thick [9.1]) a partially coherent boundary is obviously more favourable energetically than a fully coherent boundary. This explains why small precipitates of a second phase with a large δ form at *existing* dislocations, in other words by

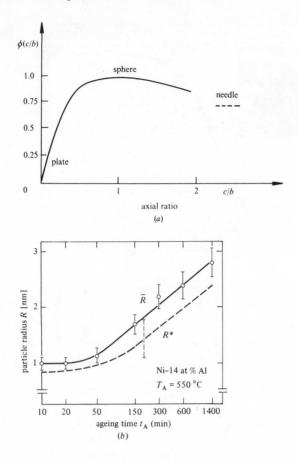

axial ratio

(a)

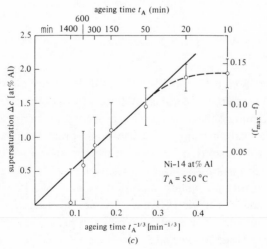

(b)

(c)

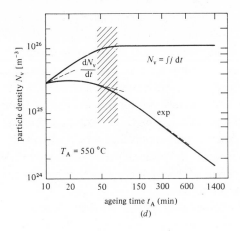

(d)

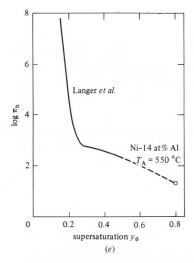

(e)

Fig. 9.4. (a) Shape factor of the distortion energy as a function of the axial ratio of an ellipsoidal precipitate. (b) Measured mean particle radius $\bar{R}$ and calculated radius R^* of critical nucleus vs aging time (H. Wendt). (c) Decrease of supersaturation Δc (and increase of precipitated volume fraction f) vs aging time (running from left to right). (d) Experimental particle density N_v and integrated density of nuclei $\int j \, dt$ vs aging time with Ni 14% Al. (e) Calculated half-time of decomposition vs relative supersaturation [9.14] compared with the measured value by H. Wendt on Ni 14% Al.

heterogeneous nucleation. This will be considered in more detail in the discussion of the Al–Cu system. Incoherent precipitates cause much less distortion than coherent precipitates but have a larger $\tilde{E}_{\alpha\beta}$. The nucleation of incoherent precipitates is promoted at existing grain boundaries.

The critical nucleus size and energy can now be calculated from (9-1) as in section 4.1 using the energy contributions described above. From this the rate of nucleation can be obtained as a function of the undercooling or supersaturation (relative to the solubility line). In practice nucleation is always accompanied by growth of precipitates which reduces the supersaturation and forces the first, smallest precipitates to dissolve again, see [9.3]. It is therefore difficult to test nucleation theory experimentally. D. Turnbull [9.4] attempted an experimental verification on Cu with $(1–3)\%$ Co, which precipitates homogeneously as spherical particles, and obtained satisfactory agreement with classical theory for a value $\tilde{E}_{\alpha\beta} = 0.2\,\mathrm{J/m^2}$, corresponding to Becker's sharp interface.

While the measurements of the electrical resistivity monitor the disappearance of solute from the matrix, the atom probe FIM directly observes the first precipitated particles [9.13]. A corresponding test of the nucleation theory in Cu 2.7% Co showed some discrepancy with the assumption that equilibrium composition particles of 90% Co form the nuclei [9.21]. Better agreement with the classical theory was obtained in Ni 36% Cu 9% Al [9.15] and particularly in Ni 14% Al [9.13]. In the latter alloy spherical, almost misfit-free particles of the Ni_3Al (γ') phase form right from the beginning, even at 1 nm radius. Since in this system the heats of formation of the two phases are known, the driving free energy Δf_v of decomposition could be calculated as a function of supersaturation. The interfacial energy $\tilde{E}_{\gamma\gamma'} \approx 14\,\mathrm{mJ/m^2}$ is also known from the Ostwald ripening kinetics of the system (section 9.3) and is in agreement with Becker's theory of the sharp interface.

From these two quantities the radius R^* of the critical nucleus can be obtained by equations (4-1a) and (9-3) and can be compared with the average radius $\bar{R}$ of the observed particles (fig. 9.4(b)). The ageing time at $550\,°\mathrm{C}$ determines the remaining supersaturation of the matrix which is also measured (fig. 9.4(c)). The agreement is quite striking. Finally the density of precipitated particles can be measured and compared with the total number density of particles formed, $N_v = \int j\,\mathrm{d}t$. One sees in fig. 9.4($d$) that the observed N_v becomes already at small ageing times smaller than the integrated density of nucleated particles because of Ostwald ripening! There is no pure nucleation period but the first nuclei compete right away by their size according to section 9.3. On the other hand does the plateau in $\bar{R}(t)$ at small times show that the initial average particles are nuclei formed in an identical non-depleted matrix. The term 'initial' is not to be understood literally as the measurement needs a certain ageing time and the nucleation rate j has to reach the stationary value. During this incubation time Al atoms diffuse together and disperse again several times before enough of them have clustered to form a critical nucleus. Also during this time the extra vacancies quenched-in from the homogenization temperature of the alloy (1150 °C) anneal out (section 10.2). Recently a combined theory of the nucleation – and Ostwald ripening kinetics according to sections 9.1.1 and 9.3 – has been formulated [9.13], [9.14], which well describes the time law of decomposition shown in figs 9.4(*b*–*d*) without additional free parameters.

A characteristic feature of the influence of Ostwald ripening on the early stages of decomposition of a highly supersaturated alloy is the large half-time τ_h needed for the decay of the supersaturation to one-half of its initial value. This time is in fact much larger than that for classical nucleation. Fig. 9.4(*e*) shows the steep decrease of τ_h with small normalized supersaturations $y_0 \approx 3(\Delta c/c_0)$ according to equations (4-1*a*), (4-2) compared with the much slower one according to the combined theory at larger y_0, in agreement with the results of Ni 14% Al.

9.1.2   *The aluminium–copper system* [9.5]

The precipitation processes in supersaturated α-Al–Cu solid solutions ($v_{Cu} \leqslant 2.5$ at.%) are extraordinarily diverse, technologically important and instructive in illustrating nucleation mechanisms. The stable precipitate phase in equilibrium with α is known as Θ ($CuAl_2$). It is tetragonal and completely incoherent with α (fig. 9.5(*b*)). On account of the high interfacial energy $\tilde{E}_{\alpha\beta} > 1$ J/m^2 the critical nucleation energy ΔF^*_{total} is also very large. This permits the formation of a series of metastable phases, also rich in copper but more similar in structure to α. They are called GP I, GP II (or Θ'') and Θ' and appear in this order. Fig. 9.5(*d*) shows the Θ'' phase also called Guinier–Preston zone of type II ('GP II') which is coherent with α because it has the same lattice dimensions as α in the *a*-direction although the *c*-axis is 5% shorter. {100} planes in Al occupied solely by Cu are called GP I zones and are associated with considerable lattice distortion (fig. 9.6) due to the small atomic radius of Cu relative to Al. GP II thus represents a superlattice of GP I zones. The tetragonal basal plane of the Θ' phase (fluorite lattice, fig. 9.5(*c*)) forms a coherent boundary with α, but perpendicular to this, dislocations (with Burgers vector $(a/2)[001]$, see chapter 11) are needed to compensate the mismatch between the *c* parameters of Θ' and α.

Fig. 9.7 shows the (metastable) solubility limits corresponding to the formation of Θ, Θ' and Θ'' from α. (That of GP I lies below that of Θ''.) If a

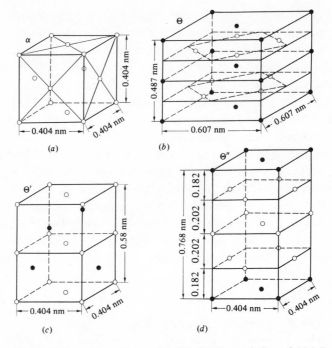

Fig. 9.5. Crystal structures of α, Θ″, Θ′, Θ consisting of Al (○) and Cu (●). Θ′ and Θ″ are coherent with α along (001).

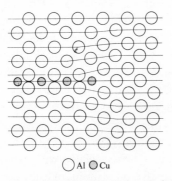

○ Al ○ Cu

Fig. 9.6. Section through a Guinier–Preston zone (GP I) parallel to the (200) plane. (After V. Gerold.)

solid solution containing about 2 wt.% Cu in Al is quenched to different temperatures $T < 420\,^{\circ}C$ and held there, the formation of the different phases can be observed either in the electron microscope or by X-ray diffraction. Θ″ and GP I form homogeneously, Θ′ preferentially at dislocations and Θ at grain boundaries (fig. 9.8). Fig. 9.9 shows that two

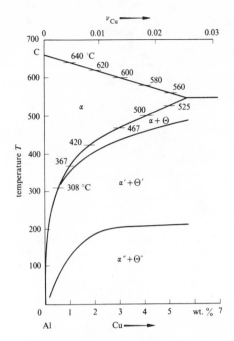

Fig. 9.7. Solubility of Cu in Al in the presence of various phases, some metastable, as a function of temperature.

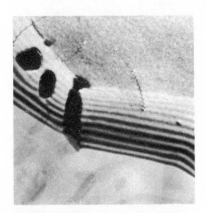

Fig. 9.8. Nucleation of Θ at grain boundaries in Al–3 wt.% Cu after quenching and 3 min at 300 °C. 40 000 ×. After E. Hornbogen, [9.5].

differently oriented semi-coherent Θ' platelets have formed at the same dislocation. As will be seen in chapter 11, a slip dislocation in the α phase has a Burgers vector $a/2[110]$ which can in fact dissociate according to $a/2[110] = a/2[100] + a/2[010]$ and hence assist in the nucleation of two of

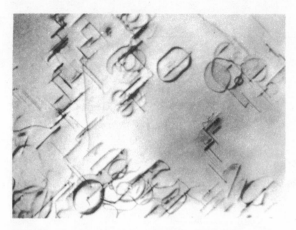

Fig. 9.9. Nucleation of two Θ' orientations at one type of glide dislocation in Al–3 wt.% Cu after a 10 min anneal at 300 °C. 21 000×. (E. Hornbogen.)

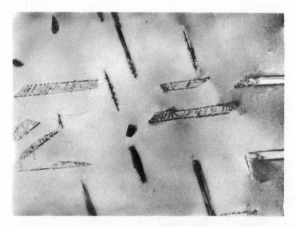

Fig. 9.10. Dislocations in the Θ'/α interface after 1000 min at 300 °C. 22 000×. (E. Hornbogen.)

the three possible Θ' platelet orientations. The dislocations in the Θ'/α interface are visible in the case of large Θ' particles (fig. 9.10). If an alloy containing GP I, Θ" or Θ' is aged further, particles of the next most stable phase are formed (in accordance with Ostwald's sequence rule) by resolution (retrogression) of the existing particles and reprecipitation. This 'sequence hypothesis', which postulates *in-situ* nucleation of Θ' at Θ" particles (possibly with the help of a dislocation) or of Θ in the surface of a Θ' particle, is confirmed by electron microscopy, fig. 9.11. The dispersion of

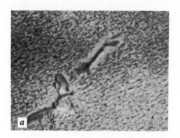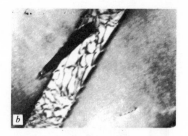

Fig. 9.11. Precipitation sequence $\Theta'' \to \Theta' \to \Theta$ (E. Hornbogen).
(*a*) Al–5 wt.% Cu, 100 min at 200 °C, formation of Θ' from Θ'' at a
dislocation, subsequent deficiency of Θ'' around the dislocation,
45 000 × ; (*b*) Al–3 wt.% Cu, 1000 min at 300 °C, nucleation of Θ in the
interface of Θ'. 45 000 × .

particles becomes increasingly coarser during this process, and this has a
deleterious effect on the hardness of the alloy in its technological
applications (Duralumin 'Overaging') (see chapter 14). Another
particularly interesting feature is the extremely rapid formation of GP I and
Θ'' by means of quenched-in vacancies (see chapter 10) in the course of 'cold
ageing', whereas the formation of Θ' and Θ during 'hot ageing) is described
by the normal diffusion constant of Cu in Al.

9.2   Rate equations for the growth of precipitates [9.1], [8.1], [9.6]

We assume that at time $t = 0$ precipitate nuclei capable of growth
are already present in the supersaturated matrix. For *short times* the growth
zones of the different precipitates do not overlap. The initial concentration
of B in the matrix is c_0, the concentration in equilibrium with the precipitate
is c_B', and that of the precipitate itself c_k. We want to know the average
concentration in the matrix $\bar{c}_B$ after a diffusion time t, or the precipitated
fraction

$$(c_0 - \bar{c}_B)/(c_0 - c_B') \equiv X(t).$$

The quantity $\bar{c}_B$ can be measured directly, for example for the precipitation
of carbon from α-iron from the relaxation strength of the anelasticity due to
dissolved carbon ('Snoek effect', see section 8.2.3). The precipitation of
interstitial atoms is easier to treat than that of substitutional atoms, in
which one atom must make room for its precipitating partner. We first
assume a spherical precipitate growing by diffusion for which distortional
and interfacial energy effects can be neglected. The flow of material through
the interface brings about a change in the concentration of the matrix with

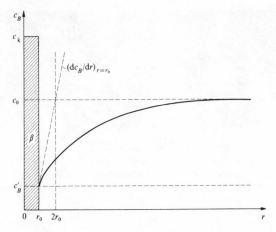

Fig. 9.12. Concentration profile near a precipitate (β).

time according to

$$\tfrac{4}{3}\pi R^3 \frac{d\bar{c}_B}{dt} = j_B(r_0)4\pi r_0^2, \tag{9-7}$$

where R characterizes the radius of influence of a particle (half the distance between particles) and $r_0(t)$ the particle radius. We obtain the current j_B from the concentration profile in fig. 9.12. We can assume that the profile moves with the interface, i.e. $r_0(t)$, in a stationary manner. $c_B(r)$ must then satisfy the time independent diffusion equation (8-3), the solution

$$c_B(r) = c_0 - (c_0 - c'_B)r_0/r \tag{9-8}$$

of which satisfies the boundary conditions shown in fig. 9.12. According to section 8.1 we thus obtain

$$j_B(r_0) = -\tilde{D}_B \frac{\partial c_B}{\partial r}\bigg|_{r_0} = -\tilde{D}_B \frac{c_0 - c'_B}{r_0}. \tag{9-9}$$

Since the number of B atoms must remain constant

$$\tfrac{4}{3}\pi R^3(c_0 - \bar{c}_B) = \tfrac{4}{3}\pi r_0^3 c_k. \tag{9-10}$$

r_0 can be eliminated from (9-7) and (9-9) to give

$$\frac{d\bar{c}_B}{dt} = -(c_0 - \bar{c}_B)^{1/3}\left(\frac{c_0 - c'_B}{\tau}\right)^{2/3};$$

$$\frac{1}{\tau^{2/3}} \equiv \frac{3\tilde{D}_B(c_0 - c'_B)^{1/3}}{c_k^{1/3}R^2}. \tag{9-11}$$

The growth rate for short times (for which $X < 0.2$) is obtained by integration as

$$(c_0 - \bar{c}_B) = \left(\frac{2t}{3\tau}\right)^{3/2}(c_0 - c'_B); \quad \text{i.e.} \quad X(t) = \left(\frac{2t}{3\tau}\right)^{3/2}. \tag{9-12}$$

For long times ($X > 0.9$) neighbouring precipitates compete for the last remaining B atoms between them. According to C. Wert and C. Zener and also F. Ham

$$X = 1 - 2\exp\left(-\frac{t}{\tau}\right). \tag{9-13}$$

C. Zener has justified a growth rate with the same form as (9-12) by totally different reasoning. He assumes that with time the B atoms displacing the interface must cover increasingly long distances $\sqrt{(\tilde{D}_B t)}$ as the immediate vicinity of the particle becomes depleted in B with the result that r_0 can only grow proportional to $\sqrt{(\tilde{D}_B t)}$, i.e. $(c_0 - \bar{c}_B) \propto r_0^3 \propto (\tilde{D}_B t)^{3/2}$ for a spherical precipitate. Zener considers that an ellipsoid of rotation would become elongated in one dimension in a B-rich region so that long diffusion paths would be unnecessary. The volume of the precipitate would then increase linearly with time, $V_k \propto t$. In the case of a plate, however, the radius would increase proportional to t and the volume initially proportional to t^2, $V_k \propto t^2$.[†] The kinetics would thus indicate the morphology of the precipitate. These arguments did not stand up to later theoretical investigations by F. Ham. He finds that an ellipsoid of rotation grows with a constant axial ratio (c/b) and that V_k increases proportional to $t^{3/2}$ and independently of $(c/b)_0$ (when t is small).

W. Mullins and R. Sekerka have, however, pointed out that considering the surface tension, very small precipitates should grow spherically whereas very large particles should show instabilities of the type discussed in dendrite formation in section 4.3. A projection outwards from the interface increases the concentration gradient and hence the growth rate at this point, a bulge inwards remains as growth proceeds unless the undulation is so fine that it is smoothed out by the action of the interfacial energy. Finally, the distortion energy can promote the growth of plate-like particles, see section 9.1.1(c).

Fig. 9.13 shows a comparison with experiment in the case of Fe_3C growing in α-iron. The diffusion constant for C in α-Fe calculated from this is in reasonable agreement with values measured by other methods. This

---

† We encountered a special case of such (three-dimensional) growth kinetics without long-range diffusion in the formation of ordered domains in section 7.4(b).

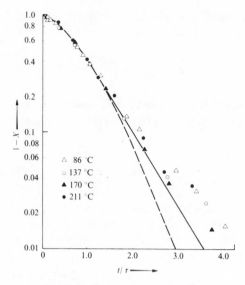

Fig. 9.13. Precipitation of carbon from α-iron. $(1-X)$ is the fraction
not yet precipitated as a function of time (full line: Ham's theory,
dotted line: continuation of original curve as
$(1-X)\propto\exp(-(2t/3\tau)^{3/2})$).

case deviates from the model on which (9-12) is based, however, in that there
is a region of distortion surrounding the particle because (Fe_3C) needs more
space than $(3Fe)$. In analogy to (8-34), the diffusion current contains a drift
term $(D_B c_B/kT)$ grad $U(\mathbf{r})$ in which $U(\mathbf{r})$ is the interaction potential between
the B atoms and the distortion. A similar situation exists for the formation
of precipitates on edge dislocations (with $U(r,\theta)=-(A/r)\sin\theta$, see chapter
11, where r,θ are polar coordinates about the dislocation line). The
diffusion equation with drift term is very unwieldy and was not solved
generally until 1970 by A. Seeger. For precipitation on edge dislocations he
obtains $X(t)\propto\sqrt{(D_B t)}$ instead of the $X\propto t^{2/3}$ law derived earlier by A. H.
Cottrell and B. A. Bilby *using only the drift term*. In all the cases discussed
above we have assumed that the rate of precipitation is limited by the rate of
diffusion in the surrounding matrix. The rate of incorporation of atoms into
the interface can also be decisive, in which case completely different
concentration profiles and growth rates are obtained dependent on the
structure of the interface, see [9.1], [9.6].

9.3   Ostwald ripening [1.3]

Even when as a result of precipitation the concentration of B atoms
in the matrix (α) has reached the value of the solubility $\check{c}_B$, the

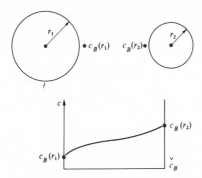

Fig. 9.14. Schematic curve for the concentration between two
precipitates with different radii.

microstructure is still not in equilibrium. A considerable quantity of energy is
associated with the $\alpha\beta$ interface which can be dissipated by the conversion
of numerous small precipitates to a few large ones. There is always a range
of precipitate sizes and hence a variation in the 'vapour pressure' of B
created by the particles in their immediate vicinity according to the Gibbs–
Thomson equation (8-33). We consider the model of two spherical particles
of radii r_1, r_2; these differ in chemical potential due to surface curvature by

$$\Delta\mu_p = 2\tilde{E}_{\alpha\beta}\Omega\left(\frac{1}{r_1} - \frac{1}{r_2}\right) = kT\frac{\Delta c_B}{\check{c}_B}, \qquad (9\text{-}14)$$

(using (5-17) for dilute solutions). A concentration difference

$$\Delta c_B = c_B(r_1) - c_B(r_2)$$

(fig. 9.14) is thus set up in the matrix between the particles which permits
large particles to grow at the expense of smaller ones until $c_B(r \to \infty) = \check{c}_B$ is
reached. Again, the concentration gradient at the particle surface (1) is
approximated by $\Delta c_B/r_1$. The growth rate for this particle is thus obtained
from the diffusion current into its interface

$$\frac{1}{\Omega}\frac{dV_1}{dt} = \frac{4\pi r_1^2}{\Omega}\frac{dr_1}{dt} = -D_B\frac{\Delta c_B}{r_1}4\pi r_1^2, \qquad (9\text{-}15)$$

i.e.

$$\frac{dr_1}{dt} = \frac{D_B\Omega^2\check{c}_B 2\tilde{E}_{\alpha\beta}}{kT}\left(\frac{1}{r_1\bar{r}} - \frac{1}{r_1^2}\right) \qquad (9\text{-}16)$$

if we let particle 2 represent the 'average particle' (radius $\bar{r}$) of the
distribution. If $r_1 > \bar{r}$, particle 1 grows, in the opposite case it shrinks.

Integration of (9-16) with a distribution function $N(r_1)$ first calculated by C. Wagner gives

$$\frac{d\bar{r}}{dt} \propto \left(\frac{\overline{(r^{-1})}}{\bar{r}} - \overline{(r^{-2})}\right) \propto \frac{1}{\bar{r}^2}$$

whereby the numerical factor depends on N. This then yields the Lifshitz–Wagner law (see [9.6])

$$\bar{r}^3 - r_0^3 \approx \frac{D_B \tilde{E}_{\alpha\beta} \check{c}_B \Omega^2}{kT} t \tag{9-17}$$

in which r_0 is the initial radius. This law, which we have already encountered in two-dimensional form in the growth of ordered domains (section 7.4(c)) has been confirmed in numerous cases for the ripening process, e.g. for the Ni 14% Al and CuCo alloys mentioned in section 9.1.1 in which the size of the precipitates can be followed by means of their magnetic properties [9.7]. Surprisingly equation (9-17) holds even for a still finite supersaturation δc_B of the matrix. In the derivation of this equation one assumes

$$\delta c_B = (\kappa t)^{-1/3} \ll \check{c}_B \tag{9-17a}$$

with $\kappa = D_B (kT)^2 / 9\tilde{E}_{\alpha\beta}^2 \check{c}_B^2 \Omega$.

Fig. 9.4(b) shows that equation (9-17a) describes the experiments in a wide range of supersaturations while the decrease of the particle density $N_V \sim t^{-1}$, which also follows from equation (9-17), is observed only for $\delta c_B \ll \check{c}_B$ (fig. 9.4(c)). One notices that the parameters D_B and $\tilde{E}_{\alpha\beta}$ enter equations (9-17) and (9-17a) in different combinations and can thus be determined separately from the data. We have not taken distortion into account in (9-14) and have limited ourselves to spherical precipitates. If the distortion is very large the precipitates are plate-like, see section 9.1.1(c), and the interaction between the distortions can stabilize the precipitates against further Ostwald ripening [9.10].

9.4   Spinodal decomposition [9.2]

As mentioned at the start of this chapter, there is a second precipitation mechanism in addition to nucleation and growth which is possible within the region of the phase diagram bounded by the spinodal ($\tilde{c}_B(T)$, (5-24)). The process consists of creating one-dimensional periodic variations in the concentration. Since an understanding of this process depends basically on solving a one-dimensional diffusion equation, it is in its early stages more susceptible to quantitative description than the first mechanism based on fluctuations. (For the later stages the opposite is true.) The theory was developed in 1961 by John Cahn after preliminary work by M. Hillert and represents a fundamental advance in physical metallurgy. The presentation given here is that of J. Cahn [9.8] and J. Hilliard [9.2]. Consider an AB alloy cooled to the spinodal region $\alpha\alpha'$ in fig. 5.9 in which $(\partial^2 F/\partial v_B^2) < 0$. We want to study the behaviour with time of a small deviation from the homogeneous composition which we consider expanded into Fourier components in the x-direction. One of these components describes a deviation from the homogeneous concentration at time $t = 0$ as

$$c_B(x, 0) - c_0 = C_\beta\, e^{i\beta x}. \tag{9-18}$$

Its variation with time is governed by the diffusion equation (8-2)

$$\frac{\partial c_B}{\partial t} = \tilde{D}_B \frac{\partial^2 c_B}{\partial x^2} = D_B\left(1 + \frac{d \ln \gamma_B}{d \ln v_B}\right)\frac{\partial^2 c_B}{\partial x^2}$$

$$= D_B \frac{v_A v_B}{kT} \frac{\partial^2 F}{\partial v_B^2} \frac{\partial^2 c_B}{\partial x^2} \tag{9-19}$$

incorporating the thermodynamic factor given by (8-28) and (8-30). This equation can be solved for finite times in agreement with (9-18) using

$$c_B(x, t) - c_0 = C_\beta\, e^{i\beta x + Rt}, \tag{9-20}$$

where $R = -\tilde{D}_B \beta^2$. For normal diffusion, $\tilde{D}_B > 0$, the fluctuation dies away with time ($R < 0$). Within the spinodal, $\tilde{D}_B \propto (\partial^2 F/\partial v_B^2) < 0$, it grows exponentially, $R > 0$; indeed, the shorter the wavelength $\lambda = 2\pi/\beta$ of the fluctuation, the faster it grows. This is certainly unrealistic, even in our continuum model, and results from the fact that we have not taken into account the energy of the steep concentration gradients that would arise. J. Cahn achieves this by incorporating a *gradient energy*, (9-5), in the chemical potential, the gradient of which, according to (8-26), generally controls the diffusion current.

The free energy of the alloy can then be written

$$F = Q \int\left[f(v_B) + f_{el}(v_B) + \kappa\left(\frac{dv_B}{dx}\right)^2\right] dx. \tag{9-21}$$

In this Q is the cross-section of the specimen perpendicular to the x-direction and $f(v_B)$ is the free energy density of a distortion-free alloy containing small concentration gradients. These are accounted for in the term $\kappa(dv_B/dx)^2$ obtained by expanding f to the second order in $v_B' = (dv_B/dx)$, $v_B'' \ldots$, (9-5). The coefficient $\kappa > 0$ represents an interface energy between regions of different concentration α and α' as described in section 9.1.1(b). We have also included in (9-21) the density of the (free) *elastic distortional enthalpy* f_{el} which arises from the change in the lattice parameter a with local concentration and is given according to (9-6) by

$$f_{el} = \frac{\hat{E}}{1-v}\left(\frac{a(v_B)-a_0}{a_0}\right)^2 = \frac{\hat{E}\delta^2}{1-v}(v_B - v_{B0})^2. \tag{9-22}$$

($\hat{E}$ is the isotropic modulus, v Poisson's ratio, $v_{B0} \equiv c_0/N_V$, $\delta = \mathrm{d}\ln a/\mathrm{d}v_B$.) The concentration profile $v_B(x)$ should make the free energy (9-21) a minimum given the condition $\int (v_B - v_{B0})\,\mathrm{d}x = 0$, that the mean concentration remains constant. This variational problem is equivalent to solving an Euler–Lagrange differential equation

$$\frac{\mathrm{d}}{\mathrm{d}x}\left(\frac{\partial I}{\partial v'_B}\right) - \frac{\partial I}{\partial v_B} = 0 \tag{9-23}$$

in which $I = f(v_B) + f_{el}(v_B) + \kappa(v'_B)^2 - \tilde{\mu}_B(v_B - v_{B0})$ and $\tilde{\mu}_B$ is a constant, the Lagrange multiplier. Equation (9-23) then gives

$$\tilde{\mu}_B = f' + f'_{el} - 2\kappa v''_B = \text{constant}, \tag{9-24}$$

instead of the earlier equilibrium condition *without gradient energy* $\mu_B = f' + f'_{el} = \text{const}$. Generalizing (8-26) we can write for the diffusion current

$$j_B = -\frac{D_B}{kT}v_B\frac{\mathrm{d}\tilde{\mu}_B}{\mathrm{d}x} = -\frac{D_B}{kT}v_A v_B[(f'' + f''_{el})v'_B - 2\kappa v'''_B], \tag{9-25}$$

obtaining under the assumption of constant coefficients (like f'') which do not depend on concentration in place of (9-19) the diffusion equation

$$N_V\frac{\partial v_B}{\partial t} = D_B\frac{v_A v_B}{kT}\left[(f'' + f''_{el})\frac{\mathrm{d}^2 v_B}{\mathrm{d}x^2} - 2\kappa\frac{\mathrm{d}^4 v_B}{\mathrm{d}x^4}\right] \tag{9-26}$$

which now takes into account the effect of elastic distortion and gradient energy on the diffusion process. The solution for the initial perturbation (9-18) is again (9-20) with

$$R(\beta) = -\tilde{D}_B\beta^2\left(1 + \frac{2\delta^2\hat{E}}{f''(1-v)} + \frac{2\kappa}{f''}\beta^2\right). \tag{9-27}$$

The first term corresponds to the earlier result. The second term is negative and constant. It has the effect that up-hill diffusion ($R > 0$) does not occur at the chemical spinodal $f'' \lesssim 0$ but is delayed to larger negative values of f'' such that $(f'' + 2\delta^2\hat{E}/(1-v)) \lesssim 0$. The equation characterizes the *coherent*

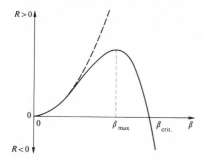

Fig. 9.15. Dependence of the exponent R of the development of concentration fluctuations with time upon the wave number, equation (9-27) (dashed curve is the approximation of equation (9-20)).

spinodal which, depending on the distortion parameter δ, lies from 20 °C (AlZn) to 600 °C (AuNi) lower than the chemical spinodal (fig. 5.9). The third term, similarly negative or zero, prevents short wavelength concentration profiles by making R negative for large β (levelling out of the concentration, fig. 9.15). There is thus a wavelength $\lambda = 2\pi/\beta_{max}$ for which the Fourier component grows most rapidly (with $\exp(R_{max}t)$) with the result that every initial perturbation leads to a periodic decomposition structure with approximately this λ. The elastic anisotropy ensures the preferential occurrence of products of decomposition with plane normals corresponding to a minimum $\hat{E}$. The structure is presumed to be coherent.

A vital approximation made in this calculation is, however, the assumption that the coefficients in the diffusion equation (9-26) are independent of concentration, in other words that it is linear, which for f'' at least, according to (5-24), is permissible only for small fluctuations in concentration. Later stages in the decomposition can be treated only numerically. Langer [9.16] predicts that the concentration amplitude rises more slowly with the time (to saturation) than in Cahn's theory (exponential law, equation (9-20)); also the width of the B-rich areas and the dominant wavelength now increase slowly with ageing time. Fig. 9.16 shows an example. It is difficult to define the point at which the alloy ceases to be single phase. Since f'' vanishes at the spinodal the critical wavelength $2\pi/\beta_{max}$ increases strongly with rising temperature. At the spinodal itself both this theory and the nucleation theory break down, see [9.11].

The Cahn diffusion equation (9-26) has been confirmed directly for vapour deposited sandwich structures with periodically alternating composition (Cu–Pd) and wavelengths of 1 to 3 nm [9.2]. Based on the equilibrium diagram, the levelling out of the concentration with time, in

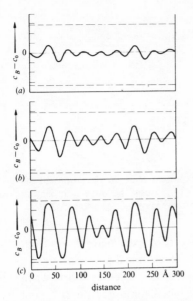

Fig. 9.16. Numerically calculated concentration profiles of the non-linearized diffusion equation for Al–37% Zn after ageing at 100 °C for (a) 8 min (b) 15 min (c) 23 min. The dotted lines mark the equilibrium concentrations for the decomposition. After de Fontaine and Hilliard [9.2].

other words the disappearance of a periodic microstructure, was followed using X-ray diffraction. Periodic decomposition with a long wavelength (period $\lambda \gg a =$ lattice constant) gives rise to satellite reflections or *side bands* in addition to the lattice reflections as already discussed in section 7.1 for superlattices with long periods. In particular the small angle X-ray scattering (the side bands of the $\{000\}$ reflection which are not found with superlattices, see fig. 7.6) can be investigated in alloys which have undergone spinodal decomposition because this is affected only by the different atomic scattering amplitudes f_A, f_B of the components, cf. [2.12] and not by lattice distortion. There is a direct relationship between the increase in the Laue-scattering in the side bands (section 2.3.2) and in the concentration amplitudes. If the Fourier expansion of the latter, according to (9-20)

$$c_B(x, t) - c_0 = \int C_\beta \exp[i\beta x + Rt] \, d\beta \qquad (9\text{-}28)$$

is compared with the amplitude of the Laue-scattering (see (2-7) and section

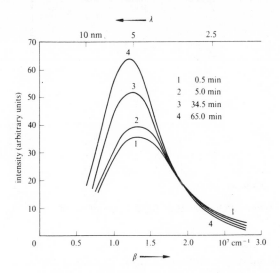

Fig. 9.17. Small angle X-ray scattering by an Al–22% Zn alloy after quenching from 425 °C and annealing at 65 °C for the times given [9.2].

2.3.2)

$$A_{\text{diff}}^{\text{Laue}}(k) = \text{constant} \int (c_B(x) - c_0) \exp[-2\pi i k x] \, dx, \qquad (9\text{-}29)$$

where the constant equals the difference in scattering amplitudes of B and A atoms, then by a Fourier transformation

$$C_\beta \exp[R(\beta)t] = \frac{2\pi}{\text{constant}} A_{\text{diff}}^{\text{Laue}}\left(k = \frac{\beta}{2\pi}\right) \qquad (9\text{-}30)$$

and for an X-ray intensity at small angles $\theta = k\lambda^{\text{X-ray}}/2$

$$I_{\text{diff}}^{\text{Laue}}\left(k = \frac{\beta}{2\pi}, t\right) = I\left(k = \frac{\beta}{2\pi}, t = 0\right) \exp(2R(\beta)t). \qquad (9\text{-}31)$$

These expressions together with (9-27) have been surprisingly confirmed for AlZn alloys, as shown in figs. 9.17 and 9.18. According to experiment $\ln(I_{\text{diff}}^{\text{Laue}})$ in fact increases linearly with time if $\beta < \beta_{\text{crit}}$ (defined in fig. 9.15) but decreases linearly for $\beta < \beta_{\text{crit}}$. The slope of the straight line $2R(\beta)$ according to (9-31) is divided by β^2 and then plotted against it. If f'', $\hat{E}$ and δ are known, this gives the parameters κ and $\tilde{D}_B$ (9-27), see fig. 9.18. The period of the spinodal microstructure of Al 22% Zn is 5 nm at 65 °C. A fine lamellar composite like this has very advantageous mechanical properties, see

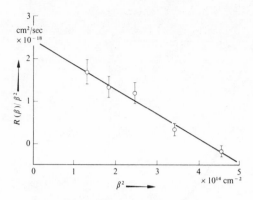

Fig. 9.18. Evaluation of the results in fig. 9.17 using equations (9-31) and (9-27).

chapter 14; the corresponding ferromagnetic alloy 'Alnico' is an excellent permanent magnet.

For the permanent magnet alloys 'Alnico 5' (Fe 26% Co 14% Ni 8% Al) and 'Chromindur' (Fe 28% Cr 15% Co) the spinodal character of the decomposition and the compositions of the resulting phases have been confirmed recently by atom probing (F. Zhu, A. Hütten (Göttingen, [9.17])). Surprisingly in the latter case the decomposition followed the tie line. The time laws of ageing did not observe Cahn's linearized theory of the spinodal decomposition though. (Neither did it in a FIM study of the alloy Cu 2.7% Ti by Biehl and Wagner [9.18].) The dominant wavelength of the composition modulation was not constant but increased with time from the very beginning of ageing, as predicted by Langer's theory [9.16]. Slightly different alloy compositions (Cu 1.9% Ti [9.19], Fe 29% Cr 24% Co (F. Zhu)) decompose according to the classical nucleation mechanism as evidenced by the final composition of the first FIM-observed particles. There must therefore be a well-defined spinodal separating the two mechanisms of decomposition in the inner and outer regions of the miscibility gap.

9.5   Discontinuous precipitation and eutectoid decomposition [1.3], [9.1]

Many precipitation processes occur at a *reaction front* which advances into the supersaturated material. Ahead of the reaction front lies the supersaturated matrix and behind it a microstructure consisting of equilibrium phases. These processes include: (*a*) the *eutectoid decomposition* of an alloy, e.g. the pearlite reaction at 723 °C in the

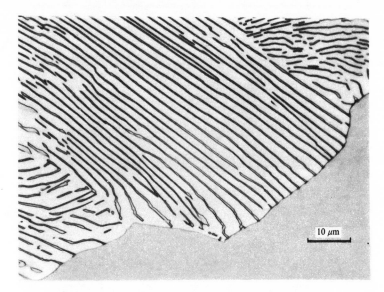

Fig. 9.19. Reaction front for the decomposition of pearlite
(J. R. Vilella, U.S. Steel Corp.) (Fe–0.89% C; 0.29% Mn; 0.19% Si;
0.08% Cr; 700 °C/water).

iron–carbon system $\gamma \rightarrow \alpha + Fe_3C$ (see section 6.1.2 and fig. 6.4); (*b*)
discontinuous precipitation, which similarly leads to a cellular or lamellar
structure of the two equilibrium phases behind the reaction front (*duplex
structure*). In contrast to (*a*), one of the phases in (*b*) has the same structure
as the supersaturated parent phase albeit a different composition and
orientation. Nucleation occurs at a grain boundary in both cases but it is
still a matter of dispute which of the two reaction products, if not both,
determines the rate of nucleation in each case. The cell or lamella 'colony'
subsequently grows at a constant rate into the supersaturated matrix in the
longitudinal direction (fig. 9.19). The process is very similar to eutectic
solidification discussed earlier. We can apply the same formalism as we
used in section 4.6 in describing these processes. In the case of *pearlite
growth* austenite (γ) takes the place of the melt; the eutectoid reaction
products are ferrite (α) and cementite (Fe_3C). According to section 4.6 the
periodicity S of the lamellae is again determined by competition between
the length of the diffusion path and the length of the (projection on to the
plane perpendicular to the lamellae of the) α/Fe_3C interface behind the
reaction front. Equation (4-14) describes the levelling off of the
concentration brought about by bulk diffusion of C in γ ahead of the

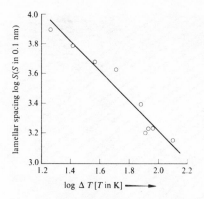

Fig. 9.20. Distance S between the pearlite lamellae dependent on the undercooling $= \Delta T$, after J. C. Fisher.

reaction front, which is advancing at a velocity R

$$8D_\gamma \frac{\Delta c}{S} = R(c^\gamma - c^\alpha).$$ (9-32)

According to fig. 4.23, $\Delta c = c^{\gamma\alpha} - c^\alpha$ where $c^{\gamma\alpha}$ is the metastable equilibrium line below the eutectoid temperature. In a substitutional alloy (Fe–Zn) diffusion (of Zn) is expected to occur in the reaction front (layer thickness δ). It is then necessary, according to section 8.4.1 to multiply D_γ by a grain boundary diffusion factor $\eta \approx D_G \delta / D_\gamma S$ which takes into account the limited space in the interface available for diffusion. In the first case (Fe–C) we expect $RS \propto D_\gamma$, in the second case (Fe–Zn) $RS^2 \propto D_G \delta$. According to section 4.6, $S \propto \tilde{E}_{\alpha/Fe_3C}/\Delta T_1$, where ΔT_1 is the undercooling of γ (fig. 4.23). This is in fact observed (fig. 9.20). The temperature dependence of R for $\Delta c \propto \Delta T_1$ is then given by

$$R \propto (\Delta T_1)^2 \exp\left(-\frac{E_{\gamma D}}{kT}\right) \text{ for bulk diffusion (case 1)}$$

or   (9-33)

$$R \propto (\Delta T_1)^3 \exp\left(-\frac{E_{GD}}{kT}\right) \text{ for grain boundary diffusion (case 2)}$$

in qualitative agreement with experiment, see fig. 9.21. J. Cahn [9.9] has criticized the above description by C. Zener and D. Turnbull because it does not take into account, even qualitatively, a whole series of observations:

(*a*) It is well known that substitutional alloying elements such as Cr, Ni, Mo retard pearlite growth without noticeably affecting the diffusion of carbon.

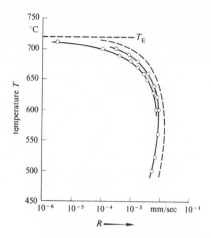

Fig. 9.21. Growth rate of pearlite in Fe–0.78 wt.% C after J. C. Fisher dependent on the undercooling below the eutectoid temperature T_E. Dashed curve according to equation (9-33), case 1.

(*b*) It cannot be assumed that eutectoid decomposition proceeds *completely* to equilibrium at a *finite* velocity. On the contrary, a second slow reaction proceeding by bulk diffusion is often observed in discontinuous precipitation following the rapid reaction described above. (A *continuous* precipitation reaction as described in section 9.1 occasionally occurs before the first discontinuous reaction. The discontinuous reaction is favoured by a high degree of supersaturation and rapid grain boundary diffusion.)

(*c*) Pearlite colonies often grow out from a grain boundary as *hemispheres* into only *one* of the grains, fig. 9.22(*a*), namely that grain with which α and Fe_3C are incoherent (whereas the boundaries between them and the other γ grain are semi-coherent [4.6]).

All this indicates that R is really determined by an independently determined mobility of the reaction front in the sense of (7-11) and not by the rate of diffusion of C in γ which gives only an *upper limit* for R. The distance S between the lamellae must therefore be derived independently of R, taking into account the fact that only a fraction $P < 1$ of the free energy change ΔF for the reaction $\gamma \rightarrow \alpha + Fe_3C$ is available. According to Cahn therefore there are three unknowns, R, S and P. Furthermore, S is such that $R \cdot \Delta F$ is a maximum (instead of the rather suspect principle of maximum R used in section 4.6). The experimental results are, however, not sufficiently detailed to test the theory quantitatively.

In conclusion, two limiting cases of the discontinuous reaction should be

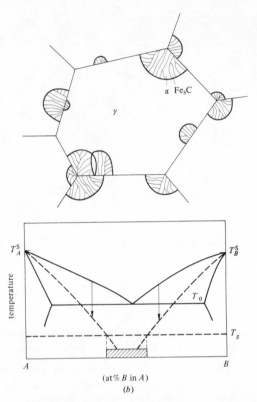

Fig. 9.22. (*a*) Nucleation and growth of pearlite colonies at grain boundaries. (*b*) Schematic phase diagram with T_0 curve (of equal free energies, liquid/solid). The intersection with the glass temperature T_g defines a glass-forming range (hatched).

mentioned: (*a*) the *bainite transformation* which replaces the pearlite transformation at high rates of cooling of Fe–C; (*b*) the *massive transformation* observed on rapidly cooling brass alloys capable of decomposition. In both cases the reaction is most probably short-range decomposition (within atomic dimensions) which permits a change in the crystal structure and which proceeds at a reaction front. In the case of bainite the change in structure takes place by a martensitic reaction, i.e. by a co-operative shear process (see chapter 13). The mechanism of both these processes has yet to be understood completely, see [9.2], [1.3].

At the intersection of the free energy-curves of the α- and β-phases in fig. 9.2 a phase change is possible without change of composition. At this particular composition and temperature (T_0) the driving energy for decomposition is zero though; only for $T < T_0$ it becomes finite as the $f_\beta(v_B)$

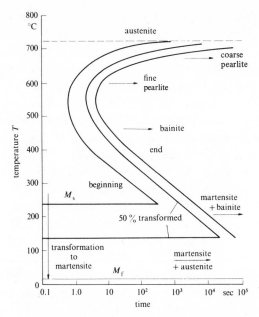

Fig. 9.23. TTT diagram of a eutectoid steel showing curves of equal fractions transformed $X(t, T)$.

curve moves down with respect to $f_\alpha(v_B)$. The temperature $T_0(v_B)$ thus characterizes the upper limit for a polymorphous transformation of the alloy of composition v_B. Therefore T_0 must be below the glass temperature T_g for a system $\alpha=$melt, $\beta=$crystal in order to obtain a glass of the composition v_B, see fig. 9.22(b) and [9.20].

9.6   TTT diagrams [1.2], [1.3], [4.6]

The course of the precipitation reactions described above can be summarized in a so-called time–temperature-transformation diagram, a sort of kinetic phase diagram in which (log t) replaces the concentration as the abscissa. (Instead of having temperature as the ordinate, $1/T$ is sometimes plotted on the negative axis.) The time t required to reach a specified state of precipitation is determined in an isothermal experiment and plotted in a diagram of the type shown in fig. 9.23. The curves for the onset and completion of pearlite (and bainite) formation are in the shape of a letter C. It takes longer for precipitation to begin (nucleation curve) and to reach completion at both high and low temperatures than at intermediate temperatures. This is to be expected from (9-33) because at first the increasing undercooling ΔT_1 accelerates the transformation, while on further cooling diffusion ceases. This principle holds for almost all diffusion controlled reactions (whereas martensitic transformations proceed almost independently of time and *depend only on the undercooling*). Not quite correctly one adds to these TTT-curves those for the beginning and end of the martensitic transformation: Martensite forms in this temperature range *during cooling*. A *heat treatment* is shown in TTT diagram as a $T(t)$ curve. If we assume (without any justification) that the isothermally determined curves for a constant degree of transformation $X(t, T)$ can also be used to describe specimens cooling continuously, the intersection of $T(t)$ and $X(t, T)$ indicates the microstructure of the specimen cooled according to this 'schedule'. (In reality the curves for beginning and end of diffusion-controlled transformations differ in 'continuous cooling diagrams' (CCT) from those in isothermal TTT diagrams; the necessary times are longer in the former.) This is the practical significance of the TTT diagram! If martensite is to be produced, the cooling curve must avoid the 'pearlite knee' in fig. 9.23, i.e. the 550 °C range must be traversed very quickly. This is impracticable for thick specimens and leads to undesirable quenching stresses. The obtained microstructure varies with the distance from the surface. Such an effect can be avoided by alloying which changes the hardenability of the steel, i.e. its ability to form martensite. Substitutional alloying (e.g. by molybdenum) often slows down the decomposition of the austenite and shifts the transformation curves towards larger times. If it leads to carbide formation, the pearlite and bainite reactions might be changed differently up to complete suppression of the latter, see fig. 13.16. Carbon by itself moves the curves towards shorter times, so that the pearlite reaction cannot be quenched even by rapid cooling.

Alloying additions usually lower the martensite temperatures M_s and M_f 20 times stronger (per wt.% carbon) than substitutional elements. In addition to all other effects of alloying on properties, they permit heat treatments to be performed at convenient rates, so that the microstructure (and hence the strength) can be chosen almost at will by suitable heat treatment and this explains the technological importance of steel (see section 13.6).

Similar C curves can be drawn for the nucleation and growth of precipitates in the Al–Cu system described in section 9.1.2, a different curve for each of the homogeneously or heterogeneously (at certain lattice defects), nucleated, metastable or stable phases. A further application concerns crystallization competing with glass formation. As in the case of phase diagrams, certain points on the TTT diagram can be characterized by metallographic microsections. Clearly it is vital for a metallurgist to be reasonably well acquainted with these diagrams.

10

Point defects especially those created by quenching and irradiation

10.1 Measurement of the equilibrium vacancy concentration

In section 8.2.1 the vacancy was recognized as being the predominant means of diffusion in metals. Vacancies are present as part of a structure in *thermal equilibrium*; the equilibrium concentration $c_V(T)$ at a temperature T can be calculated from (3-1). Vacancies can be detected directly but tediously by counting the occupied lattice sites at a quenched metal tip using a field ion microscope [10.5]. In principle, it is also possible to demonstrate their existence by any physical property which varies proportional to c_V, for example the electrical resistivity. Nevertheless, since c_V decreases exponentially with decreasing temperature from 10^{-4} at the melting point, high temperature measurements are necessary to study vacancies in equilibrium. The electrical resistance of metals at these temperatures is, however, largely determined by lattice vibrations and scarcely influenced by lattice defects. There are very few remaining methods, only two according to A. Seeger [10.1], which are sufficiently sensitive to vacancies at high temperatures to permit their effects to be separated from those of the 'background'. These are (differential) thermal expansion and the life time of positrons.†

The first method, originally proposed by C. Wagner, compares the macroscopic thermal expansion $\Delta l/l$ with the microscopic thermal expansion, given by the temperature dependence of the lattice parameter $a(T)$. The formation of a vacancy increases the volume of the crystal by V_{VF}. If the surrounding lattice were rigid, V_{VF} would equal Ω. Since, however, the atoms surrounding the vacancy relax inwards, the volume created is smaller than the atomic volume Ω. As shown by J. D. Eshelby [10.10], the lattice parameter is able to register this difference through its influence on the

---

† Unfortunately the specific heat at temperatures near T_m not only contains the energy of formation of the vacancies which increase in number with increasing temperature (see section 10.2) but also involves other less well known anharmonic terms and is thus not quite as suitable for the determination of c_V.

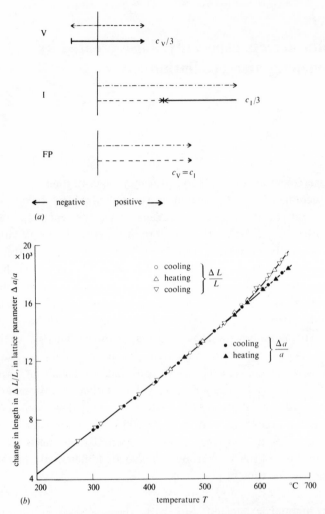

(a)

(b)

Fig. 10.1. (a) Macroscopic change in length (dashed) and change in lattice parameter (dash–dotted) together give information only about the *number* of vacancies or interstitial atoms (full line). In the case of Frenkel pairs the resultant is zero; (b) experimental results of R. O. Simmons and R. W. Balluffi for the specimen length and lattice parameter of aluminium as functions of temperature.

macroscopic change in length, fig. 10.1(a). Eshelby starts with the displacement field $\mathbf{u}$ (2-6) of a spherical cavity in an infinitely large, elastic isotropic medium. In a finite body $\mathbf{u}$ must include a contribution $\mathbf{u}_i$ from the image forces, which are necessary to compensate surface stresses. Eshelby calculates the volume change $(3\Delta l/l)'$ caused by $(\mathbf{u} + \mathbf{u}_i)$ and the associated change in the reciprocal lattice dimensions, which are expressed by $(3\Delta a/a)$ for a statistical distribution of distortion centres. The two quantities are found to be identical. In vacancy formation, the displaced atom must be accommodated at the surface, which means an increase in volume of Ω. Thus

$$\frac{\Delta l}{l} - \frac{\Delta a}{a} = \frac{1}{3}\left\{\frac{V_{VF}}{\Omega} - \left(\frac{V_{VF}}{\Omega} - 1\right)\right\}c_V = \frac{1}{3}c_V, \tag{10-1}$$

is obtained for cubic crystals containing vacancies but no (self) interstitial atoms. Otherwise, the term on the right-hand side of (10-1) would be $(c_V - c_I)/3$. As has already been stated in section 8.2.1 the equilibrium concentration of interstitial atoms c_I is usually negligibly small compared with $c_V(T)$ on account of their high energy of formation.

Fig. 10.1(b) shows measurements by R. O. Simmons and R. W. Balluffi on aluminium. It must be taken into account in evaluating these that divacancies are also present in equilibrium and therefore included in the right-hand side of (10-1). The formation of a divacancy requires less energy than the formation of two single vacancies $E_{DF} = 2E_{VF} - E_{DA}$ because the association saves two 'broken bonds'. For Al, Seeger gives $E_{VF} = 0.65$ eV, $E_{DA} = 0.25$ eV, i.e. $E_{DF} = 1.05$ eV compared with $E_{VF} = 0.65$ eV. According to (3-1), therefore, $c_D \ll c_V$ but the ratio c_D/c_V nevertheless increases with increasing temperature, see (10-3b). Multiple vacancies are virtually never present in equilibrium.

The second method, that of positron annihilation, is fortunately most sensitive in the medium temperature range $(T \approx 0.6T_m)$ and is thus not affected by the problem of divacancies. A positron produced in a nuclear reaction has a life time of about 2×10^{-10} sec in the metal before it is annihilated by an electron. During this period, which is relatively long compared with the period of vibration of the atoms $(\approx 10^{-13}$ sec), it becomes 'thermalized' and attracted by a vacancy. (This has a negative charge, see section 8.3.4.) If the positron is captured by the vacancy a second characteristic life time is observed for the positron which influences the measured mean life time by an amount proportional to $c_V(T)$ (for not too large values of c_V). On the other hand, interstitial atoms repel positrons and hence cannot form a bound state. Although measurement of the positron annihilation does not yield absolute values for c_V, the temperature dependence of the life time provides very accurate values of E_{VF} (unaffected by E_{DF}!). Fig. 10.2 shows the vacancy concentration for Al calculated using the first method together with a line of slope E_{VF} obtained from measurements of the positron life time. The influence of divacancies is

232   *Point defects*

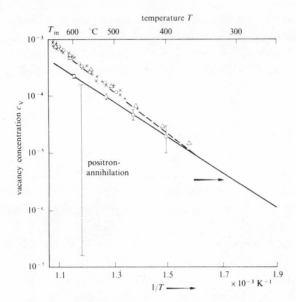

Fig. 10.2. Experimental results of various authors for the
T-dependence of $c_V = 3(\Delta l/l - \Delta a/a)$ in aluminium. The experimental
errors give rise to an uncertainty $\Delta c_V = \pm 10^{-5}$ shown as four error
bars. Measurements of the positron lifetime at vacancies give the slope
of the full line in the range shown accurate to $\pm 10^{-7}$. The maximum
sensitivity of this method is at $c_V \approx 5 \times 10^{-6}$ (arrow). The open
triangles are derived from measurements of the specific heat. After
A. Seeger [10.1].

apparent from the curvature of the plot near the melting point and
corresponds to $c_D(T_m)/c_V(T_m) \approx 0.28$. The volume of formation of the
vacancy is found to be $V_{VF} = 0.65\,\Omega$, i.e. the lattice parameter relaxes
inwards near a vacancy in Al, on average by about 12%. The relaxation of
neighbouring ions around a vacancy can be obtained using electron theory
from the screening effect, section 6.3.3. The negative charge on the vacancy
extends beyond its nearest neighbours in a univalent metal (fig. 6.24). They
therefore relax inwards, but the next nearest neighbours relax outwards on
account of the Friedel oscillations. In multivalent metals these relaxations
can occur in the opposite direction, as shown by fig. 6.24. Similar arguments
can be applied to the interaction between two vacancies.

The sum $(V_{VF} + V_{VM})$ is the determining factor in the dependence of the self
diffusion coefficient on hydrostatic pressure. V_{VM} is the activation volume
for vacancy migration and is determined by the additional lattice dilation
caused by an atom in the saddle point position. In the case of aluminium,
$V_{VM} = 0.19\,\Omega$.

10.2 Quenching and annealing of excess vacancies

If a metal wire is cooled rapidly from a high 'quenching' temperature T_Q to a low temperature, the vacancies in equilibrium at T_Q are 'quenched-in' and can thus be observed at the lower temperature. Here they make a significant contribution to, for example, the electrical resistivity $n\rho_{el}$. If the specimen is then heated to a medium ('recovery') temperature T_R and ρ_{el} is observed as a function of the time t, a decrease in the resistivity $(-\Delta\rho_{el})$ indicates that the vacancies are annealing out, and this continues until the equilibrium concentration $c_V(T_R) \ll c_V(T_Q)$ is reached. In addition to this isothermal annealing out, a study can be made of the isochronal recovery of the specimen, in which the specimen is heated for constant time intervals and $\Delta\rho_{el}(T)$ noted. A *recovery stage* is then expected at temperatures at which the vacancies become mobile. The aim of such measurements is to obtain E_{VF} and E_{VM}, particularly for those metals for which they cannot be measured by more direct methods. For example, the positron is bound less strongly to the vacancy in a univalent metal than in a polyvalent metal. Also, a large inwards relaxation around a vacancy decreases its effective negative charge. Isotopes suitable for diffusion measurements are often unavailable (Al for example), with the result that $(E_{VF} + E_{VM})$ is not known. A knowledge of the quenching and annealing process is important in metallurgy because it is used to produce certain precipitates, as explained in section 9.1.2. As proved by experience, however [10.2], the process and its analysis are extremely complicated.

Fig. 10.3 shows the additional electrical resistance produced in gold by quenching from various temperatures T_Q. If $\Delta\rho_{el} \propto c_V$, E_{VF} would be ≈ 0.96 eV. In fact, this value is increased by a contribution from divacancies, which are present at T_Q or which form from single vacancies during the quench, even at quenching rates of several 10^4 K/sec. Larger values of dT/dt produce quenching stresses and strains, the dislocations so produced act as vacancy sinks, and the quenched-in vacancy concentration is reduced. The quenched specimens are then annealed isothermally (at T_R). Fig. 10.4 shows that on increasing T_R the rate of annealing-out (which is proportional to the diffusion constant for vacancies) increases in the ratio $\exp(-E_{VM}/kT_1):\exp(-E_{VM}/kT_2)$. This gives the activation energy $E_{VM} \approx 0.83$ eV. In fact divacancies are produced on annealing which, *in the fcc lattice, can migrate faster than single vacancies* (for gold $E_{DM} \lesssim 0.79$ eV)! In contrast to the fcc lattice the nearest neighbours of a vacancy in the bcc lattice are not the nearest neighbours of one another. This means that a divacancy becomes dissociated every time one of its neighbouring atoms changes position. The simplest form of annealing kinetics is determined by

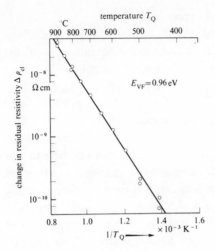

Fig. 10.3. Change in the residual resistance $\Delta\rho_{el}$ on quenching gold wires from different temperatures T_Q (after J. S. Koehler).

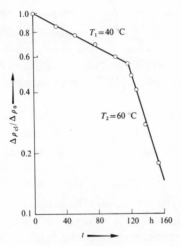

Fig. 10.4. The fraction $\Delta\rho_{el}/\Delta\rho_0$ of the change in residual resistance remaining in gold wires quenched from 700 °C after tempering for t hours at 40 °C or 60 °C.

the reactions

$$V + V \underset{K_2}{\overset{K_1}{\rightleftharpoons}} D, \qquad V \overset{K_3}{\longrightarrow} \text{sinks}, \qquad D \overset{K_4}{\longrightarrow} \text{sinks}.$$

Their time dependence can be described by the theory of chemical reactions as

$$\frac{dc_V}{dt} = 2K_2 c_D - 2K_1 c_V^2 - K_3 c_V, \tag{10-2a}$$

$$\frac{dc_D}{dt} = K_1 c_V^2 - K_2 c_D - K_4 c_D. \tag{10-2b}$$

For interaction between nearest neighbours only, in the fcc lattice, we obtain

$$K_1 = 84v_0 \exp\left(-\frac{E_{VM}}{kT}\right) \equiv 84v_V,$$

$$K_2 = 14v_0 \exp\left(-\frac{E_{VM}+E_{DA}}{kT}\right) = 14v_V \exp\left(-\frac{E_{DA}}{kT}\right),$$

$$K_3 \approx Na^2 v_V,$$

$$K_4 \approx Na^2 v_D \equiv Na^2 v_0 \exp\left(-\frac{E_{DM}}{kT}\right),$$

where N is the dislocation density. It is assumed that the dislocations act as sinks and as such absorb each vacancy which arrives. The numbers 14 and 84 are explained by counting the random steps which lead to the dissociation or association of a divacancy. For example, 5 nearest neighbour sites of a half divacancy are also nearest neighbours of the other half, only 7 (out of 12 in the fcc lattice) lead to dissociation. The ratio 84/14 is understandable from (10-3b).

Equation (10-2) cannot be solved in closed form except in special cases, such as when the divacancies exist in a dynamic equilibrium between formation, dissociation and annihilation. This implies $dc_D/dt = 0$, i.e. from (10-2b)

$$c_D = \frac{K_1 c_V^2}{K_2 + K_4}. \tag{10-3a}$$

(In the special case $K_4 \ll K_2$ single and divacancies are in equilibrium with each other so that

$$c_D = \frac{K_1}{K_2} c_V^2 = 6 \exp\left(\frac{E_{DA}}{kT}\right) c_V^2 \tag{10-3b}$$

where $6 = n/2$ the number of divacancy orientations in the fcc lattice, $n = 12$ nearest neighbours.)

Substitution of (10-3a) gives (10-2a) with $c_V(t=0) \equiv c_{V0}$

$$\frac{1}{c_V} = \left[\left(\beta + \frac{1}{c_{V0}} \right) \exp(K_3 t) - \beta \right]$$

with

$$\beta = \frac{2K_1}{K_3} \left(1 - \frac{K_2}{K_2 + K_4} \right).$$

For $K_3 t \ll 1$ expansion yields

$$\frac{1}{c_V} - \frac{1}{c_{V0}} = \left(\beta + \frac{1}{c_{V0}} \right) K_3 t, \tag{10-4a}$$

which corresponds to a second-order reaction,

$$\frac{dc_V}{dt} \propto (-c_V^2); \qquad \frac{1}{c_V} \propto t.$$

On the other hand if $K_3 t \gg 1$ a first-order reaction is obtained,

$$-\frac{dc_V}{dt} \propto c_V,$$

i.e.

$$\frac{c_V}{c_{V0}} = (1 + \beta c_{V0})^{-1} \exp(-K_3 t). \tag{10-4b}$$

Experimentally, the effective energy of migration $E_{M,eff}$ can be determined by measuring the change in slope $- \Delta(\ln dc_V/dt)$ for a change in annealing temperature $\Delta(1/kT)$. $E_{M,eff}$ does not generally correspond to either of the two characteristic migration energies E_{VM}, E_{DM}.

Similar reactions occur during the quenching process. $c_V(T)$ and $c_D(T)$ are obtained from (10-2) by dividing both sides by the rate of quenching dT/dt. Fig. 10.5 shows the result of a numerical integration with assumed values for T_Q, N, (dT/dt), E_{VM}, E_{DM}, E_{DA} and the association energy E_{TA} for trivacancies. It can be seen that the ratio c_V/c_D becomes much smaller during the quench.

In order to describe the recovery process realistically we must take into account the reactions by which triple, quadruple, etc., vacancies are formed as well as the association with solute atoms to form complexes with characteristic mobility. The observed kinetics measure a complicated average of the reaction single to double to triple to quadruple vacancies which in addition depends on the distribution of immobile sinks such as dislocations, large vacancy agglomerates, solute atoms and surfaces [10.3]. According to a careful evaluation of recovery measurements it appears that

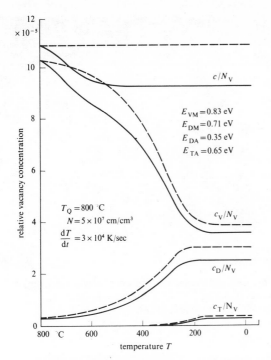

Fig. 10.5. Calculated changes in the concentration of single, double and triple vacancies c_V, c_D, c_T and in the total vacancy concentration $c = c_V + 2c_D + 3c_T$ in Au during quenching from 800°C. In the case of the dashed curves dislocation sinks are assumed to be absent ($N = 0$) (after Furukawa). (The c_i represent volume concentrations of the different sorts of vacancies, N_V is the atomic density.)

single vacancies in gold move with $E_{VM} = 0.89$ eV; the energy of formation after correcting for the influence of divacancies is $E_{VF} = 0.87$ eV. The sum $E_{VM} + E_{VF} = E_{VD} = 1.76$ eV is almost identical with the experimentally measured activation energy for self diffusion in gold. (Here again a slight divacancy influence can be observed [10.2].)

Vacancy agglomerates can be observed using field ion microscopy and transmission electron microscopy [10.2]. Apparently, small (planar) arrangements of vacancies transform into prismatic dislocation loops, fig. 4.8. A stacking fault can be created within these loops which, if it has a low energy, closes up to form a stacking fault tetrahedron in three dimensions, as observed in gold (fig. 10.6, see chapter 11). As well as these homogeneously produced agglomerates, dislocations serve as preferential vacancy sinks, as will be explained in chapter 11. Screw dislocations then rotate into the form of a helix, see fig. 11.6.

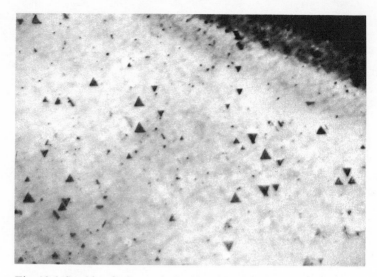

Fig. 10.6. Stacking fault tetrahedra in gold after quenching from
1000 °C and tempering 1 h at 200 °C. Transmission electron
micrograph by S. Mader, Stuttgart. 51 000 × .

By virtue of their different atomic size and their (screened) extra charge,
solute atoms often exert an attractive force on vacancies of the order of
0.1 eV. E_{VF} would be reduced by this amount in the neighbourhood of a
solute atom, i.e. c_V would be increased [10.2a]. The formation of GP zones
in Al–Cu and Al–Ag occurs more rapidly for higher quenching
temperatures T_Q corresponding to an increase in $c_V(T_Q)$. The kinetics for the
combined diffusion of solute atom-vacancy complexes is complicated, as
indicated in section 8.2.2, see [10.3]. The rate constant for the formation of
GP zones has an activation energy in the region of $E_{VM} = 0.62$ eV in Al. The
vacancies do not precipitate in the GP zones but must diffuse back into the
matrix, where they again form complexes with solute atoms thus allowing
these to migrate to GP zones. L. Girifalco [10.4] has called this mechanism
the 'vacancy pump'. Each vacancy transports on average 1000 solute atoms
to the GP zones before it itself arrives at a sink. This gives us a starting point
for the description of substitutional solute precipitation, which was largely
ignored in chapter 9. Similar effects due to quenched-in vacancies are to be
expected in the kinetics of isothermal ordering. In this case E_{VF} and E_{VM}
depend on the degree of order, since wrong bonds are more easily broken
than right ones [8.7].

10.3   Effects of irradiation with high-energy particles [10.7], [10.8]

The use of metals in nuclear reactors has extended the field of metallurgy to include the effects of radiation and lattice defects created by high energy particles. Neutrons with energies of about 1 MeV are sufficient to produce very complicated lattice defects in simple metals. For this reason elementary effects of irradiation are studied after electron bombardment of thin metal specimens at low temperatures.

To describe these effects we use the mechanics of elastic collisions between incident particles of mass M_1 and kinetic energy E_K (assumed not to be in the relativistic range) and stationary lattice atoms of mass M_2, see [10.6]. Accordingly a maximum energy of

$$U_{max} = E_k \frac{4M_1 M_2}{(M_1 + M_2)^2} \tag{10-5}$$

is transferred to M_2. If U_{max} is larger than the threshold or 'Wigner energy' E_d (in most metals about 25 eV dependent on lattice direction), the atom is ejected from its lattice site and can itself act as a bombarding particle ('primary knock-on atom'), etc. If $U_{max} \gg E_d$ a *displacement cascade* is obtained. The total number of displaced atoms in a cascade (not counting the recombinations of neighbouring interstitial atoms and vacancies which will be discussed later) is about $(\bar{U}/2E_d)$. For 1 MeV electrons in copper $U_{max} = 68$ eV and the mean transferred energy is $\bar{U} = 33$ eV. Thus only one *Frenkel pair* (vacancy + interstitial atom) is produced per impact. Before the next impact can take place the electron has been retarded by Coulomb interaction to such an extent that $U_{max} < E_d$. Consequently only one Frenkel pair is produced by each incident electron. On the other hand for 1 MeV neutrons $\bar{U} = 20$ keV and $\bar{U}/2E_d \approx 400$! Reaction radiation thus produces large displacement cascades.

Closer examination of the impact process reveals a series of *lattice structure effects*. If the direction of impact of the primary knock-on atom coincides approximately with a close packed direction ($\langle 110 \rangle$ for fcc) the energy transfer from one atom to the next at low energies is focussed along this lattice direction for a distance of about 100 atomic spacings. If a static lattice defect lies in the path of this *focusson*, a permanent displacement can result. At higher energies (~ 35 eV for copper) matter is transported by the focussed impact. A vacancy remains behind while a configuration of $(n + 1)$ atoms on n lattice sites propagates in the $\langle 110 \rangle$ direction. This configuration is known as a *crowdion*. It moves by exchange collisions which can destroy the order in an alloy along its path. If the colliding particles have high energies the effective diameter of the lattice atoms

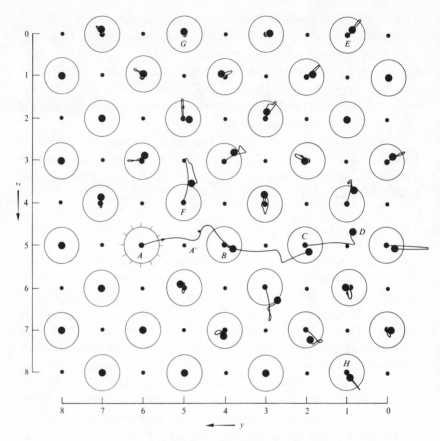

Fig. 10.7. Calculated atomic movements following the impact of an atom *A* at 15° to the *y*-axis with 40 eV. The large circles mark the initial positions of the atoms in the fcc (100) plane and the small dots those in the plane below (G. Vineyard, Brookhaven Nat. Lab.).

appears to be reduced and the lattice seems to be empty in certain directions ('channels'), e.g. $\langle 110 \rangle$. The primary knock-on atom is focussed or 'channelled' along these directions. These phenomena were predicted in computer simulations. Fig. 10.7 shows the dynamics of the formation of a Frenkel pair *AD* together with the focusson propagation along *AE*, *BH*, etc., as a result of the impact of a 40 eV primary particle. Focussed impacts can be rendered visible experimentally by bombarding copper and gold single crystals with (1–5) keV argon ions. Agglomerates of interstitial atoms can then be observed in the TEM at the end of the focussed impact at depths of about 10 and 20 nm below the $\{110\}$ copper surface. This corresponds to the mean free path of crowdions along different $\langle 110 \rangle$ directions and is

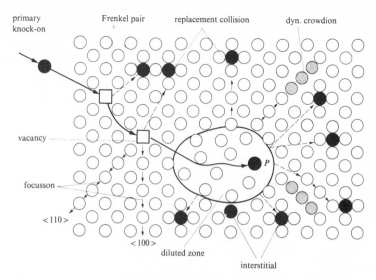

Fig. 10.8. Schematic view of the lattice defects created by a primary knock-on atom (after A. Seeger).

much greater than the penetration depth of the argon ions themselves (3.5 nm) [10.11].

This extensive focussed transport of matter occurs only for interstitial atoms and not for vacancies. The end of a displacement cascade is therefore a *zone of reduced density*. Fig. 10.8 shows this and the other defects resulting from an impact in a schematic representation due to A. Seeger. In copper, these zones have a diameter of about 10 lattice constants and thus can be observed only in the FIM, see [10.2]. Larger agglomerates of such zones can be observed in the TEM after n-irradiation.

10.4   Recovery stages after irradiation [10.9]

In view of the effects of irradiation with high-energy particles discussed in 10.3, Frenkel defects can be expected in specimens which have been irradiated at very low temperatures. During bombardment, high kinetic energies are converted to heat but this is largely dissipated within a few 10^{-12} sec. Nevertheless interstitial-vacancy pairs will recombine during the irradiation. The remaining Frenkel pairs recombine in copper on heating to 50 K, without any difference between the microscopic ($\Delta a/a$) and macroscopic ($\Delta l/l$) expansion, i.e. c_V and c_I change in the same way during irradiation and recovery in this range. The existence of interstitial atoms in fcc metals after irradiation can be demonstrated directly by

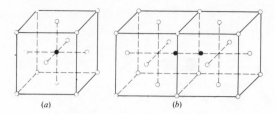

Fig. 10.9. Interstitial atoms in the fcc lattice (a) at the centre of the cube (b) dumb-bell configuration.

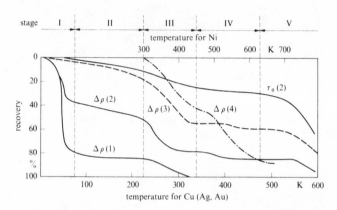

Fig. 10.10. Schematic recovery curves for the electrical resistance $\Delta\rho$ and the critical shear stress τ_0 of fcc metals after (1) electron irradiation (2) neutron irradiation (3) plastic deformation and (4) quenching [10.9].

anelasticity measurements (see section 2.7) now that it has been ascertained theoretically that the dumb-bell is the stable configuration rather than the symmetrical position in the centre of the cube, see fig. 10.9.

Recently the structure of interstitial atoms in electron irradiated aluminium has been measured using diffuse X-ray scattering ('Huang scattering', section 2.3.1) [10.12]. The distortion field u_p must be expressed in terms of anisotropic elasticity theory and the diffuse X-ray intensity must be measured in the neighbourhood of different reflections.

Frenkel pairs and interstitial atoms are not produced to any appreciable extent by quenching or by plastic deformation (see chapter 11). This is shown by a comparison of the isochronal *recovery behaviour* after these treatments with that following n-irradiation, fig. 10.10. If copper is bombarded with electrons (or Al with neutrons) *only* Frenkel pairs are observed, most of which anneal out in stage I of the isochronal residual resistivity–temperature curve. A good many of the lattice defects produced by n-irradiation of copper and similar metals anneal out only at later stages at higher temperatures. The subdivision in fig. 10.10 was made by H. G. van Bueren and A. Seeger. Each stage is associated with the mobility and annealing out of a characteristic type of lattice defect. Stage III is probably explained by the diffusion of interstitial atoms to vacancies (not necessarily their own partners in the Frenkel pair as in stage I) and also, in view of the quenching and deformation results, the movement of divacancies. In stage IV, individual vacancies should become mobile, although this is open to question [10.2]. In stage V the remaining low temperature radiation damage anneals out, in particular the dilute zones and vacancy agglomerates. This takes place by recrystallization as a result of bulk diffusion, see chapter 15. The higher stages are of greater metallurgical interest because they influence the mechanical properties of reactor materials, which are impaired by n-irradiation. Normal reactor radiation above room temperature leaves only defect agglomerates in Cu, Ni and similar metals. Defects of atomic dimensions are produced but anneal out again, although this does involve considerable movement of atoms. (For a radiation dose of 10^{19} n/cm^2 each atom in the metal has on average changed its lattice position more than once!)

This 'stirring' of the atoms during irradiation acts as *microdiffusion* in many metallurgical reactions. Short-range order in Cu–30% Zn can be established in a neutron flux of 5×10^{12} n/cm^2 even at temperatures at which thermal diffusion is negligible. Fig. 10.11 shows that below 150 °C the time constant for this reaction becomes independent of temperature and is thus determined only by the vacancies produced by irradiation [10.3]. The same is true for precipitation, e.g. the formation of GP zones. It appears possible to extend the temperature range in which phase diagrams of alloys can be considered to be equilibrium diagrams to much lower temperatures by irradiation-induced diffusion. In addition, the vacancy flux towards sinks and surfaces produces an inverse Kirkendall-effect for solute atoms (fig. 8.11), i.e. a difference $(j_A - j_B)$. Such a concentration change can lead to a phase change [10.3a] under irradiation.

10.5   Radiation damage to reactor materials

The adverse effects of radiation on metals are particularly apparent in the mechanical properties of structural materials in the nuclear reactor, for example embrittlement by precipitation processes, see chapter 14. Fission products of uranium and products of nuclear transmutations, particularly inert gas atoms, cause heavy damage, not only because they are

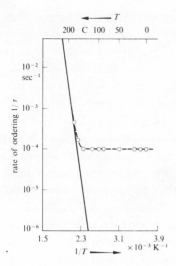

Fig. 10.11. The rate at which order is established in α-brass as a function of temperature: The dashed curve shows the additional effect of irradiation with neutrons. After [10.3].

formed with a high kinetic energy but also because they are usually insoluble in the matrix metal and coagulate to bubbles of gas. This causes the fuel elements to swell when irradiated. This dimensional instability is particularly unwelcome in anisotropic materials. Crystallographically oriented precipitates can form, for example interstitial atoms on the basal planes of graphite or α-uranium, which cause the material to expand perpendicular to this plane and shrink parallel to it. The porosity of the fuel elements and structural steel parts which result from irradiation is troublesome in high temperature nuclear reactors. Point defects thus develop into major technological problems [10.7], [10.8].

In recent years the *voids* which form after irradiation in the reactor with more than 10^{21} n/cm^2 at about one-third of the melting temperature have been more closely examined in various pure metals (Nb, Ta, Ni). TEM studies have shown them to be spherical or polyhedral cavities which grow with increasing irradiation and often arrange themselves on a cubic 'void lattice', fig. 10.12. The total volume of the voids can be greater than 10% of the specimen volume at a pore radius of about 30 nm. Since the irradiation produces interstitial atoms and vacancies in equal numbers one is forced to ask how and why only vacancy agglomerates occur. The answer [10.14], [10.13] is that because of the strong distortion associated with interstitial atoms, these are attracted more strongly to edge dislocations than are vacancies and are thus annealed out preferentially (see section 11.1.4). There

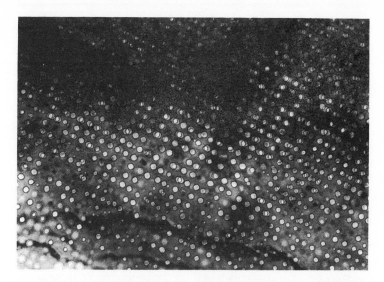

Fig. 10.12. bcc void lattice in ion-irradiated Nb–1% Zr (with 570 ppm oxygen, 3×10^{16} nickel ions/cm^2 at 900 °C) 106 000 ×. (B. Loomis, P. Okamoto, H. C. Freyhardt, Argonne Nat. Lab.)

is a slight interaction between interstitial atoms and voids whereas (on account of the low stationary density of both vacancies and interstitials) that between interstitial atoms and vacancies at moderate temperatures is insignificant. Under these circumstances vacancies can form agglomerates, as shown by an analysis of the diffusion kinetics, which take the form of voids rather than of prismatic dislocation loops (fig. 4.8). A. Seeger [10.13] has shown that small agglomerates of n vacancies in the form of spherical cavities have a lower energy than dislocation loops of radius r: the energy of a cavity like its surface area is proportional to $r^2 \propto n^{2/3}$, that of a loop to $(r \ln r) \propto n^{1/2} \ln n$ (section 11.2.2). Thus the ratio of the energies proportional to $n^{1/6}/\ln n$ is favourable for voids only at low values of n. If larger voids are observed this is probably due to stabilization by gas atoms. The activation threshold for the transformation of voids to loops is relatively high even for $n = 5$. At higher temperatures the voids lose vacancies and disappear so that there is only a relatively narrow temperature range in which voids can form. Unfortunately it is an important range for reactor technology, see [10.14].

11
Line defects – Dislocations

In the preceding chapters we have already employed the concept of the dislocation, which is an essential constituent of the metallic microstructure. So far we have been able to refer to a chapter in Kittel [1.1] in which dislocations were introduced to explain plastic deformation and were described in terms of geometry and elasticity theory. Since the following chapters develop concepts based on the properties of these lattice defects, the present chapter presents the necessary elements of dislocation theory and verifies them by experimental results of a fundamental character. There is no lack of more advanced texts on dislocations, see for example [11.1] to [11.5].

11.1   Topological properties of dislocations
11.1.1   *Definition*

As was stated in [1.1], dislocations are needed to explain why, in principle, metallic crystals are very easy to deform. The fundamental process in plastic deformation consists of shear on a definite crystallographic plane (slip plane) by a unit translation vector **b** (Burgers vector) in a definite lattice direction (slip direction), fig. 11.1. It can be explained in terms of *two possible intermediate steps*, the edge dislocation and the screw dislocation (figs. 11.1(*b*) and (*c*)). In the first case (fig. 11.1(*b*)), slip has started on the right-hand side of the crystal but has not yet reached the left-hand side. In between, therefore, there is a distorted region (edge dislocation) with five half planes above the slip plane and four half planes below it (or three above two) which extends from the front to the back of the crystal at right angles to **b**. It moves on the slip plane, shown here as a dotted line, and marks the boundary between the slipped and unslipped regions of the crystal. In the case of the screw dislocation, fig. 11.1(*c*), the same slip process starts on the front face of the crystal and propagates on the same plane but towards the back. In this case, therefore, the slip vector **b** and the line direction d**s** are parallel to each other and not perpendicular as

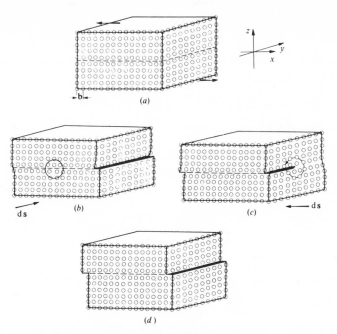

Fig. 11.1. The two possible intermediate steps in the elementary slip process in crystals (*a*) to (*d*), namely (*b*) the edge dislocation and (*c*) the screw dislocation.

in the case of the 'edge' (dislocation). The distorted atomic configuration around the 'screw' (dislocation) has the form of a spiral staircase. Both types of dislocation are stable distortion states of the crystal lattice, which apparently extend over large distances. They are described in a continuum model (section 11.2.1) using linear elasticity theory.

There are only these two limiting cases of dislocation orientation in the slip plane, which continuously change one into the other in the curved dislocation as shown in fig. 11.2(*a*) looking down onto the slip plane. The definition of a dislocation by operations with reference to the crystal surface as in fig. 11.1 is unsatisfactory. Slip can start anywhere within the crystal on the slip plane and can propagate, e.g. up to the circular region shown in fig. 11.2(*b*). It is then bounded by a dislocation loop consisting of positive and negative edges and screws and all intermediate orientations with the arbitrary angle (**b**, d**s**). The arrangement of five over four half planes is characteristic of the positive edge and four over five half planes of the negative edge. The sense of rotation of the staircase for the positive screw is simply the reverse of that for the negative screw. The atomic arrangement within the loop is no different from that outside the loop since a complete

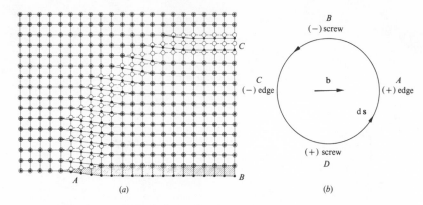

Fig. 11.2. (*a*) View of the slip plane of a partially slipped crystal. *AB* is a slip step. The surface is intersected by a screw dislocation at *A* and by an edge dislocation at *C*. The dots represent the atoms below the slip plane and the circles those above it (W. T. Read, *Dislocations in Crystals*, McGraw-Hill Book Co.); (*b*) dislocation loop in the slip plane showing the different character and sign of the dislocation.

lattice translation has occurred. Consequently the *slip plane* is unimportant in a definition of a dislocation. (It can be replaced by an arbitrary *slip surface* which is bounded by a dislocation line as shown in [1.1]. The volume changes resulting from translation, section 11.1.3, must then be eliminated.) Essential in this (synthetic) definition of a dislocation in a perfect crystal are the line element of the boundary d**s** and the slip vector **b** which states the relative displacement of the sides of the slip surface.

11.1.2 *Burgers circuit*

In the majority of cases we are concerned less with creating a dislocation in a perfect crystal than with locating and investigating it in a real crystal *K*. For this we can use the 'Burgers circuit' (named after J. M. Burgers who discovered the screw dislocation: the edge dislocation was discovered independently in 1934 by G. I. Taylor, E. Orowan and M. Polanyi). A perfect crystal region *K'* is imagined adjacent to the real crystal *K*. A similar path, with the same number of individual steps, is traced in *K'* as in *K* about the strongly distorted core of the supposed dislocation (fig. 11.3). If the path *C* in *K* leads back to the starting position, then the circuit *C'* in the perfect crystal, which is equivalent as regards the number of steps and the direction of steps taken, does not close if *C* surrounds a dislocation. The closing vector **b** is the Burgers vector of the enclosed dislocation. (According to the accepted sign convention, d**s** points

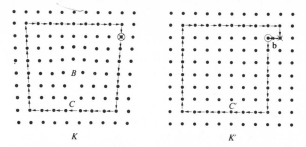

Fig. 11.3. Burgers circuit C round a 'bad' crystal region B in the crystal K and C' in the perfect crystal K' with closing vector $\mathbf{b}$.

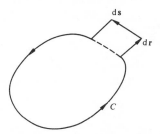

Fig. 11.4. Expansion of a dislocation loop C by an element $[\mathbf{ds} \times \mathbf{dr}]$.

downwards out of the plane of the diagram for a circuit C and vector $\mathbf{b}$ in the directions shown in fig. 11.3.)

A number of dislocation properties follow from this (analytical) definition. If the circuit C is moved continuously along the dislocation line, $\mathbf{b}$ cannot change. Hence a dislocation cannot end in the middle of the crystal (just as the boundary of the slip surface, in the synthetic definition, cannot end in the crystal). $\mathbf{b}_1 = \mathbf{b}_2 + \mathbf{b}_3$ must apply at dislocation junctions. A chain of vacancies (or interstitial atoms) has a closing vector zero. This corresponds to a narrow *dislocation dipole*, i.e. two edge dislocations of opposite sign lying above one another in two neighbouring slip planes.

11.1.3   *Glide and climb*

If the dislocation loop is expanded by the element $[\mathbf{ds} \times \mathbf{dr}]$ (fig. 11.4), then in terms of the synthetic definition this corresponds to a relative displacement of the two surfaces of the element by $\mathbf{b}$. If $\mathbf{dr}$ does not lie in the slip plane $[\mathbf{b} \times \mathbf{ds}]$, a volume element

$$\mathrm{d}V = \mathbf{b} \cdot [\mathbf{ds} \times \mathbf{dr}] = \mathbf{dr} \cdot [\mathbf{b} \times \mathbf{ds}]$$

is created by the displacement, i.e. $\mathrm{d}V/\Omega$ vacancies are created or $\mathrm{d}V$ is

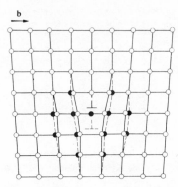

Fig. 11.5. Edge dislocation climbing out of the dashed position by incorporation of vacancies.

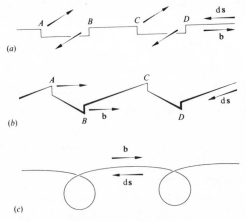

Fig. 11.6. A screw climbing by means of jobs A to D (a) leads to the formation of a helix dislocation (c).

doubly occupied and interstitials are formed. Displacements of the dislocation for which $dV \neq 0$ are called *non-conservative* or *climb* and those for which $dV = 0$ *conservative* or *glide*. In this sense a screw dislocation has no prescribed slip plane, $[\mathbf{b} \times d\mathbf{s}] = 0$ and thus can glide in any (equivalent lattice) plane. As illustrated in fig. 11.5, an edge dislocation climbs by absorbing vacancies or interstitial atoms on the extra half plane. If a screw dislocation rotates locally to an edge orientation, as for example at jogs (section 11.1.4), and then interacts with a non-equilibrium concentration of vacancies such as exists after quenching, a dislocation '*helix*' is formed (fig. 11.6). A planar accumulation of vacancies, on which the lattice collapses, leads to the formation of 'prismatic' dislocation loops. In this case **b** is

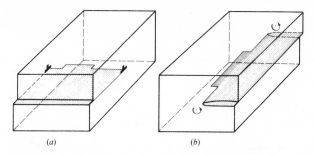

Fig. 11.7. Kinks in an edge (*a*) and in a screw (*b*).

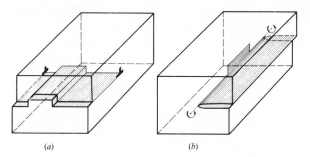

Fig. 11.8. Jogs in an edge (*a*) and in a screw (*b*).

perpendicular to the plane of the loop which in consequence can glide only on a cylindrical surface defined by $[\mathbf{b} \times d\mathbf{s}]$, see figs. 4.8 and 11.19.

11.1.4   *Kinks and jogs*

Glide and climb of straight dislocations often start locally at *kinks* and *jogs* as shown in figs. 11.7 and 11.8. A kink moves the dislocation forward by a unit step in the slip plane, a jog displaces it locally from *one* slip plane to the next. The movement of kinks and jogs as elements $d\mathbf{s}$ of a dislocation with Burgers vector $\mathbf{b}$ is governed by the rules in section 11.1.3 characterized by dV. In particular a jog in a screw dislocation can *glide* (conservatively) only *along* the dislocation and it can move *with* the dislocation only by *climb*. This impedes the movement of the screw dislocation and is the principal means of producing vacancies and interstitial atoms during plastic deformation. Since the energy of formation of interstitial atoms is four times larger (in Cu) than that of vacancies, the latter predominate. Jogs in screw dislocations which produce interstitial atoms therefore avoid moving perpendicularly to the dislocation line preferring to move along it forming 'superjogs', which extend over several

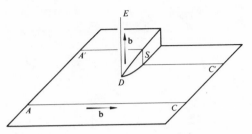

Fig. 11.9. Intersection of two screws: dislocation AC moves across dislocation DE to position $A'C'$ and then includes a jog S.

lattice planes (unless they were able to annihilate with a vacancy jog). Edge dipoles are drawn out at these jogs, see fig. 12.4. Jogs are produced mainly by the intersection of screw dislocations lying in different slip planes, see fig. 11.9.

11.1.5   *The force on a dislocation*

The dislocation movements described above result from the action of internal and externally applied stresses. These forces, which can be described by a stress tensor

$$\boldsymbol{\sigma} = \begin{pmatrix} \sigma_{xx} & \sigma_{xy} & \sigma_{xz} \\ \sigma_{xy} & \sigma_{yy} & \sigma_{yz} \\ \sigma_{xz} & \sigma_{yz} & \sigma_{zz} \end{pmatrix}$$

as a function of position, perform work (force times distance) on displacement of the element, $d\mathbf{a} = [d\mathbf{s} \times d\mathbf{r}]$.

$$dW = (d\mathbf{a} \cdot \boldsymbol{\sigma}) \cdot \mathbf{b}$$

$$\equiv \begin{pmatrix} da_x \sigma_{xx} + da_y \sigma_{xy} + da_z \sigma_{xz} \\ da_x \sigma_{xy} + da_y \sigma_{yy} + da_z \sigma_{yz} \\ da_x \sigma_{xz} + da_y \sigma_{yz} + da_z \sigma_{zz} \end{pmatrix} \begin{pmatrix} b_x \\ b_y \\ b_z \end{pmatrix}.$$

Employing the rules of vector analysis this can be rewritten

$$dW = ([d\mathbf{s} \times d\mathbf{r}] \cdot \boldsymbol{\sigma}) \cdot \mathbf{b} = [d\mathbf{s} \times d\mathbf{r}](\mathbf{b} \cdot \boldsymbol{\sigma}) = d\mathbf{r}[\mathbf{b}\boldsymbol{\sigma} \times d\mathbf{s}]. \qquad (11\text{-}1)$$

On the other hand, defining the force on the dislocation element $d\mathbf{s}$ as $d\mathbf{K}$, it follows that for a displacement $d\mathbf{r}$, $dW = d\mathbf{r} \cdot d\mathbf{K}$. Taken together with (11-1), the force on the dislocation (Peach–Koehler force) can then be expressed as

$$d\mathbf{K} = [\mathbf{b}\boldsymbol{\sigma} \times d\mathbf{s}]. \qquad (11\text{-}2)$$

Special cases

(a) *Edge*: $\mathbf{b} = (b, 0, 0)$, $\mathbf{ds} = (0, -ds, 0)$, slip plane normal in the z-axis, arbitrary $\boldsymbol{\sigma}$. Equation (11-2) then yields $\mathbf{dK} = b \cdot ds(\sigma_{xz}, 0, -\sigma_{xx})$. The x-component dK_x/ds is the *glide force* per unit length, given by the product of the magnitude of the Burgers vector and the shear stress acting in the slip direction in the slip plane at the position of the dislocation line. dK_z/ds is a *climb force* per unit length, which results from the attempt of the normal stress $-|\sigma_{xx}|$ to squeeze out the extra half plane of the edge dislocation in the z-direction. This means that according to fig. 11.5 the dislocation absorbs vacancies (or expels interstitial atoms). Their concentration (c_V) deviates from the equilibrium value (c_0) until the corresponding change in the chemical potential (section 5.2.1) $\Delta\mu_V = kT \ln c_V/c_0$ builds up a 'chemical force' $(\Delta\mu_V b/\Omega)$ per unit length such that the climb force dK_z/ds is just compensated. Conversely an excess of vacancies always gives rise to a climb force on the dislocation. A hydrostatic pressure p reduces the equilibrium concentration of vacancies in the crystal. The associated chemical force causes a dislocation to climb. On the other hand p acts directly on the edge dislocation as did the normal stress $-|\sigma_{xx}|$ discussed above. The excess vacancies can thus be absorbed. There is, however, no macroscopic dislocation movement under the influence of a hydrostatic pressure [11.1].

(b) *Screw*: $\mathbf{b} = (b, 0, 0)$, $\mathbf{ds} = (+ds, 0, 0)$. Equation (11-2) yields $\mathbf{dK} = b \cdot ds(0, +\sigma_{xz}, -\sigma_{xy})$. The screw thus experiences the same slip force in the xy-plane perpendicular to its line as does the edge. There is no climb force on the screw (although there are couples which turn it into the edge orientation). The screw also glides under the action of a shear stress in the xz-plane (and in every plane containing $\mathbf{b}$).

11.2   Elasticity theory of dislocations

11.2.1   *Stress fields of dislocations*

A dislocation is created in an elastic continuum with the isotropic shear modulus G and Poisson's ratio v. Fig. 2.10 shows the elastic displacements $\mathbf{u}(x, y, z)$ around a *screw*. It can be seen directly from fig. 11.10(a) that the displacement field should have the form $\mathbf{u} = (0, b(\alpha/2\pi), 0)$. This corresponds to (2-5). The validity of this expression is tested by two requirements. The proof that it is a screw dislocation is provided by the Burgers circuit (path element dw)

$$\int \frac{\partial u_y}{\partial w} \, dw = \int_0^{2\pi} \frac{\partial u_y}{\partial \alpha} \, d\alpha = b. \tag{11-3}$$

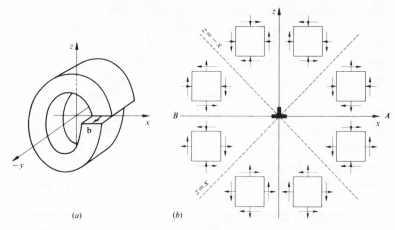

Fig. 11.10. (*a*) Distortion round a screw dislocation. (*b*) Stresses round an edge dislocation.

In addition, (2-5) must fulfil the elastic equilibrium conditions, in this case div grad $\mathbf{u} = \Delta u_y = 0$. The following shear stresses are obtained from $\mathbf{u}(x, y, z)$

$$\sigma_{xy} = G \frac{\partial u_y}{\partial x} = -\frac{Gb}{2\pi} \frac{z}{x^2 + z^2};$$

$$\sigma_{yz} = \frac{Gb}{2\pi} \frac{x}{x^2 + z^2} \tag{11-4}$$

or $\sigma_{\alpha y} = A_s/r$ in cylindrical polar coordinates (r, α) with $A_s = Gb/2\pi$. The stresses diverge at the origin. We therefore remove a cylinder of atomic diameter (r_0) from around the dislocation axis. In the case of a body with finite dimensions as shown in fig. 11.10(*a*), we must ensure that the surface is made stress-free by additional (Burgers vector-free) displacements. These are of no significance in macroscopic specimens.

The corresponding elastic solution for the *edge* is more difficult to obtain because an edge is not cylindrically symmetrical but represents a state of plane strain ($u_y = 0$ for $ds = ds_y$, $b = b_x$). With $A_e = Gb/2\pi(1 - v)$ we obtain

$$\frac{\sigma_{xz}}{A_e} = \frac{x(x^2 - z^2)}{(x^2 + z^2)^2}; \qquad \frac{\sigma_{xx}}{A_e} = -\frac{z(3x^2 + z^2)}{(x^2 + z^2)^2};$$

$$\frac{\sigma_{zz}}{A_e} = \frac{z(x^2 - z^2)}{(x^2 + z^2)^2} \tag{11-5}$$

and furthermore $\sigma_{yy} = v(\sigma_{xx} + \sigma_{zz})$. Here again the terms are neglected which compensate the stresses on the internal and external surfaces of the hollow cylinder surrounding the dislocation.

Equation (11-5) can be discussed with the aid of fig. 11.10(b). In the upper half, $\sigma_{xx} < 0$ represents a compressive force in the direction of **b** and in the lower half it represents a tensile force. Furthermore at $x = \pm z$ both σ_{zz} and σ_{xz} change sign. The shear stress of dislocation (1) can be interpreted according to (11-2) as a glide force $dK_x^{(2)} = b_2\,ds_2\sigma_{xz}^{(1)}$ on a hypothetical second edge dislocation of the same sign on a parallel slip plane. In the forward and backward sectors (A and B respectively) this will be repelled but in the lower and upper sectors it will be attracted to the equilibrium positions at $x = 0$. This explains the tendency to form low angle grain boundaries, section 3.2.1. If the two dislocations have opposite signs, there is a stable equilibrium position at $x = \pm z$, the dislocation dipole, which no longer possesses any long-range stresses. (There is incidentally no stable arrangement for two parallel screws.)

Another interesting difference between the edge and screw dislocation is the finite volume dilatation around the edge dislocation, which depends on the azimuthal angle α in the xz-plane

$$\text{div } \mathbf{u} = \frac{dV}{V} = -\frac{b}{2\pi}\frac{1-2v}{1+v}\frac{\sin\alpha}{r}. \tag{11-6}$$

Thus the zone of maximum compression lies above and the zone of maximum dilatation lies below the slip plane of a positive dislocation. The mean value $\overline{(dV/V)}$ is zero according to general principles of linear elasticity theory (Colonetti's theorem). In second-order elasticity theory $\overline{(dV/V)}$ is proportional to the energy density of the internal stresses of the dislocations, see [11.6]. In this approximation both edge and screw dislocations are associated with an increase in volume. On an atomistic scale this means that the repulsive interaction between the displaced atoms in the dislocation core increases more rapidly with displacement than the attractive forces. The dilatations around the dislocation are, according to A. H. Cottrell and B. A. Bilby, one of the main reasons for its interaction with solute atoms different in size from the matrix atoms ($\delta \neq 0$).

Internal stresses around dislocations have been experimentally confirmed by observation of the stress birefringence in the infrared and by X-ray topography using the double crystal technique, using germanium in both cases, see [11.7] and section 2.2.2. On account of the factor $1/(1-v)$ in A_e, the interaction between edge dislocations is about 50% stronger than that between screws in isotropic media. The maximum shear stress $\sigma_{xz} = \tau_p$, with which two parallel edges on parallel slip planes z_0 apart interact with one another is obtained from (11-5) as $\tau_p \approx A_e/4z_0$. This is the 'passing stress' which must be applied externally to push the dislocations past one another, see [1.1].

11.2.2 *Dislocation energy and line tension*

On integrating the energy densities given by the squares of the stresses over the hollow cylinder (R, r_0) containing the dislocation, the energy per unit cylinder length is found in the case of the screw, (11-4), to be

$$E_L^s = \int_{r_0}^{R} \frac{\sigma_{\alpha y}^2}{2G} 2\pi r \, dr = \frac{Gb^2}{4\pi} \left(\ln \frac{R}{r_0} - 1 \right). \qquad (11\text{-}7a)$$

The (-1) is due to the additional stresses required to make the surface stress-free. The corresponding expression for the edge (with $R \gg r_0$), (11-5) is

$$E_L^e = \frac{Gb^2}{4\pi(1-v)} \left(\ln \frac{R}{r_0} - 1 \right). \qquad (11\text{-}7b)$$

These energies E_L diverge for $R \to \infty$, i.e. an infinitely large body, and in the dislocation core ($r_0 \to 0$). The difficulties are, however, more of a formal nature. The energy of the core has to be calculated using an atomistic model (see section 11.3.1) and added to E_L. Surprisingly enough, the result obtained from elasticity theory (11-7), $E_L \propto \ln R$, is found to be correct down to R values of atomic dimensions. At large R values there are always other dislocations to compensate the stress of the dislocation under consideration, so that it is possible to put $R \propto N^{-1/2}$, where N is the dislocation density (cm^{-2}) of the crystal. (The dislocation density can be obtained either by measuring the total dislocation length in unit volume by transmission electron microscopy or by counting the number of intersection points with unit area of the surface after etching. According to G. Schöck, the two results differ only by a numerical factor.) Following the above principle it is possible to estimate the energy per unit length of a dislocation loop of radius R_1 as $E_1 = (b/4)(A_s + A_e)[\ln(R_1/r_0) - 1]$. The energy per unit length is thus smaller for sharply curved dislocations than for straight ones because it is mostly long-range. For $N = 10^8$ cm^{-2} and $r_0 = 10^{-7}$ cm, $bE_L = \frac{1}{2}Gb^2 \approx 2$ eV for Al but for $R_1 = 10^{-6}$, $bE_1 = 0.7$ eV. In the case of kinks or jogs bE_L is approximately equal to $Gb^3/10$. These elements can be created by thermal fluctuations, unlike macroscopic lengths of dislocation which are constituent parts of the microstructure, see section 3.1. The configurational entropy of a dislocation as a

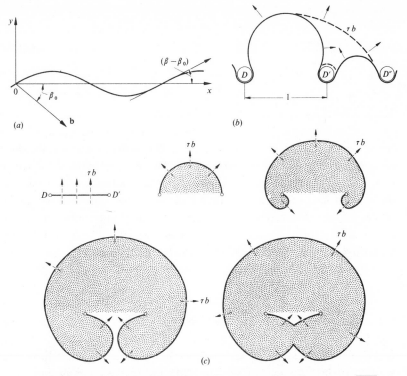

Fig. 11.11. (*a*) Curved dislocation line; (*b*) dislocation segment $\overline{DD'}$ becomes unstable under the force τb and joins with $\overline{D'D''}$ leaving a loop round D'; (*c*) Frank–Read source of length $l = DD'$ produces a dislocation loop and reproduces itself.

macroscopically correlated structure is small, with the result that the free energy of the dislocation is virtually equal to its energy.

The energy per unit length of a dislocation can be imagined as a line tension, which straightens a curved dislocation taking into account certain orientation factors. When a screw dislocation bows out, segments of edge orientation are produced creating dilatational energy in addition to the shear energy. If the energy of a slightly curved dislocation having a tangent at any point $dy/dx = \tan(\beta - \beta_0)$ in fig. 11.11(*a*) is compared with that of a straight dislocation fixed at both ends, the following expression is obtained for the line tension

$$E_T(\beta_0) = \left(E_L(\beta) + \frac{d^2 E_L}{d\beta^2} \right)_{\beta = \beta_0}. \tag{11-7c}$$

The line energy of a mixed dislocation (angle β_0) is obtained by linear superposition using (11-7)

$$E_L(\beta_0) \approx \frac{A_s b}{2}\left(\ln\frac{R}{r_0}\right)\left(1 + \frac{v}{1-v}\sin^2\beta_0\right) \tag{11-7d}$$

which yields the line tension

$$E_T(\beta_0) = \frac{A_s b}{2}\left(\ln\frac{R}{r_0}\right)\left(1 + \frac{2v}{1-v}\cos 2\beta_0 + \frac{v}{1-v}\sin^2\beta_0\right). \tag{11-7e}$$

In the case of a screw $E_T(0°) = [(1+v)/(1-2v)]E_T(90°)$. For $v = \frac{1}{3}$, therefore, its line tension is four times larger than that of an edge (cf. [11.2]).

A dislocation, fixed at both ends, bows out to a radius $r_K = E_L/\tau b$ under a stress τ. If the radius is smaller than half the distance between the anchor points $l/2$, the dislocation is unstable and bulges out round the anchor points until it forms a complete dislocation loop (fig. 11.11(c)). The stress initiating this instability is the *Orowan stress*

$$\tau_l = \frac{2E_L}{bl} \approx \alpha\frac{Gb}{l} \tag{11-8}$$

containing a numerical factor α, which depends on the configuration of the neighbouring dislocations. In general, the dislocation segment *DD'* (fig. 11.11(b)) is not isolated in the slip plane but connected to other dislocation segments which lie against similar obstacles. In these circumstances the breakthrough occurs at τ_l (fig. 11.11(b)). The neighbouring dislocation arcs then join up by annihilating their connections to the anchor points to form a straight dislocation front on the other side of the obstacles and small dislocation loops round each obstacle. The Orowan mechanism is important in alloy hardening (chapter 14). The *Frank–Read source* permits dislocation multiplication as illustrated in fig. 11.11(c). Dislocation configurations like this have occasionally been observed [1.1] but it is not yet known whether they represent the prevailing multiplication mechanism. Other multiplication mechanisms are based on the same principle and prove that many dislocations can be generated on one slip plane, as demonstrated by surface slip steps [11.7].

11.2.3 *Dislocation interactions*

The discovery of dislocations and dislocation multiplication mechanisms not only solved the problem of easy plastic deformation in crystals but actually posed the reverse question, namely why crystals do not shear indefinitely at the stress τ_l corresponding to the observed separation $l \approx 10^{-4}$ cm of grown-in dislocations. In fact interactions between the

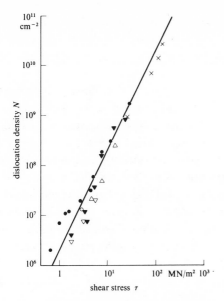

Fig. 11.12. Dislocation density N measured from etch pits on Cu single crystals deformed to the flow stress τ (after J. D. Livingston; the crosses are values obtained by TEM on polycrystals).

multiplying dislocations cause *work hardening*. We have already met two cases of dislocation interaction:

(*a*) the interaction between *parallel edge dislocations* (section 11.2.1) resulting in a passing stress $\tau_p = Gb/8\pi(1-v)z_0$. The separation z_0 of neighbouring slip planes should decrease with increasing dislocation density N according to $z_0 \propto N^{-1/2}$, in other words the *flow stress* required for further plastic deformation should increase according to

$$\tau_p = \alpha_1 Gb\sqrt{N}. \tag{11-9}$$

This expression, first derived by G. I. Taylor, has since been confirmed many times, see fig. 11.12 for $\alpha_1 \approx \frac{1}{3}$;

(*b*) as deformation proceeds, the interaction between perpendicular dislocations, i.e. *cutting of 'forest' dislocations*, section 11.1.4, requires more energy to create kinks and jogs as well as for the production of vacancies associated with the movement of jogs in screws. The separation l of the 'trees' decreases with increasing dislocation density N_F of the forest according to $l \propto N_F^{-1/2}$. Thus the Orowan stress, the upper limit of the flow stress necessary to cut through the forest is given by

$$\tau_l = \alpha_2 Gb\sqrt{N_F}. \tag{11-10}$$

The only difference between this expression and (11-9) is in the orientation of the dislocations determining the flow stress. Normally N_F increases proportional to N on deformation (section 12.3) so that unfortunately the results in fig. 11.12 do not tell us much about the interaction mechanism.

The interaction between arbitrarily oriented dislocations with Burgers vectors $\mathbf{b}_1, \mathbf{b}_2$ can be estimated from a hypothetical *association reaction* $\mathbf{b}_1 + \mathbf{b}_2 \rightarrow (\mathbf{b}_1 + \mathbf{b}_2)$ in which the energy increases or decreases proportional to $(\mathbf{b}_1 + \mathbf{b}_2)^2 - \mathbf{b}_1^2 - \mathbf{b}_2^2 = 2\mathbf{b}_1\mathbf{b}_2$ depending on whether repulsion $(\mathbf{b}_1\mathbf{b}_2) > 0$ or attraction $(\mathbf{b}_1\mathbf{b}_2) < 0$ is observed. Association may, however, be prevented by a lack of opportunities for glide for the participating dislocations. The $(\mathbf{b}_1\mathbf{b}_2)$ criterion shows that dislocations decompose into other dislocations with the smallest possible Burgers vector. They then move apart to a distance r_{12} (see dislocation dissociation in section 11.3.3.1). The interaction energy in the case of parallel screws is [11.3]

$$E_{12} = \frac{G(\mathbf{b}_1\mathbf{b}_2)}{2\pi} \ln \frac{r_{12}}{r_0}. \tag{11-11}$$

A screw dislocation at a distance r from a *free surface* is attracted by image forces which seem to originate from a virtual dislocation of the opposite sign at r on the other side of the surface. Its energy decreases as it approaches the surface according to $(Gb^2/4\pi) \ln(2r/r_0)$ with the result that it leaves the crystal. A dislocation held inside the crystal thus rotates into an orientation perpendicular to the surface. This causes serious problems in the thinning of specimens for transmission electron microscopy (section 2.2). If the surrounding medium is not a vacuum but a substance with modulus $G' \neq G$, the energy of interaction with the surface contains an additional factor $(G - G')/(G + G')$. If $G' > G$, for example an oxide layer of thickness δ, the dislocation is repelled as long as $r < \delta$, whereas if $r > \delta$, it is attracted by the external medium. The dislocation thus remains at a distance δ below the surface. C. S. Barrett has shown that if an anodized Al crystal is deformed and then etched, the crystal undergoes further spontaneous deformation as dislocations leave the crystal.

Although we have already discussed the interaction between dislocations in *different* (and also parallel) slip planes, it is the interactions between dislocations on the same plane which are of particular interest because of their ability to multiply. Dislocations of like sign repel one another and will consequently distribute themselves over a limited area of the slip plane (e.g. between two grain boundaries) such that the density at the boundaries is greater than at the centre. Fig. 11.13(*a*) shows the distribution function $D(x)$ for the stress-free case. ($D(x) \, dx$ is the number of dislocations considered

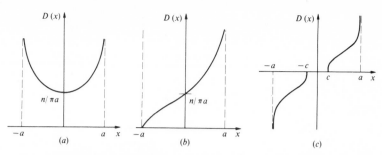

Fig. 11.13. Distribution function for dislocations in the interval $(-a, a)$ for (a) zero stress; (b) constant stress; (c) dislocations of both signs from a source at the centre under a stress τ. Critical shear stress of the source is $\tau c/a$.

smeared out between x and $x + dx$.) If a stress τ_a acts in the slip plane, fig. 11.13(b) applies for dislocations of like sign and fig. 11.13(c) for those of either sign which emerge as sections of loops from a centrally situated Frank–Read source. The distributions $D(x)$ have been derived by G. Leibfried, using continuum theory, from the condition that the total effective stress at any point on the slip plane occupied by a dislocation is zero, i.e. for fig. 11.13(b)

$$\int_{-a}^{+a} D(\xi)\frac{A_e}{x-\xi}\,d\xi + \tau_a = 0, \tag{11-12}$$

where $\int_{-a}^{+a} D(\xi)\,d\xi = n$ equals the total number of dislocations in the interval. The solution of the integral equation in the case of fig. 11.13(b), in which τ_a was chosen to make $D(-a) = 0$, is

$$D(x) = \frac{n}{\pi a}\sqrt{\left(\frac{a+x}{a-x}\right)}, \qquad |x| < a. \tag{11-13}$$

The total stress of the dislocation pile-up outside the interval can be derived mathematically by an analytical continuation of the function $D(x)$ outside the interval in which it is defined (i.e. where it is purely imaginary)

$$\tau(x) = i\pi A_e D(x) \approx \tau_a\sqrt{\left(\frac{2a}{x-a}\right)}, \qquad |x| > a. \tag{11-14}$$

A stress concentration is generated ahead of the right-hand grain boundary, which increases with increasing grain size. This leads (in chapter 12) to the Petch relationship for the flow stress of polycrystals. The total effective stress on the right-hand obstacle is shown to be $\tau(a) = n\tau_a$, i.e. the external applied stress multiplied by the number of piled-up dislocations.

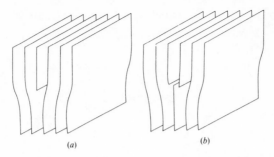

Fig. 11.14. Two symmetry positions for an edge dislocation in its slip plane.

Dislocation pile-ups of this type have frequently been observed in etching experiments and transmission electron microscopy (see [11.3]). The method of continuous dislocation distribution is especially useful in formulating an elasticity theory for a deformed continuum in which infinitesimal dislocations arise at points at which the elastic compatibility conditions are violated (E. Kröner [11.8]).

11.3   Dislocations in crystals

Hitherto, we have described the plasticity of crystals by dislocations in a continuum (which, however, have a lattice translation vector as Burgers vector). In order to describe the dislocation core, which largely determines the dynamics of real dislocations, we must look at the dislocation in the crystal. Finally the stability of Burgers vectors in typical crystal structures will be investigated, particularly the dissociation into partial dislocations with the formation of stacking faults (section 3.1).

11.3.1   *The Peierls potential*

The atomic arrangement in the core of a dislocation varies periodically (period b) for reasons of symmetry as it moves along the slip plane, as shown in fig. 11.14 for an edge dislocation. The associated periodic change in the dislocation energy with position is called the 'Peierls potential'. In order to calculate this, it is necessary to know the atomic interaction forces over the slip plane. Lattice planes lying further away can be treated by continuum theory. For the sake of simplification let us assume that an atom at position x in the lattice plane directly above the (non-material) slip plane reacts to a displacement $2u(x)$ in the direction **b** with a restoring force $K_1 = (Gb/2\pi c)\sin(4\pi u(x)/b)$. ($c$ is the distance between the lattice planes perpendicular to the slip plane; this expression reduces to

Hooke's Law for small $(2u/b)$.) The restoring force is just balanced at each x by the displacing force K_2 of the adjoining continuum containing the extra half-plane. The force $K_2(x)$ is represented as the effect of a distribution of infinitesimal dislocations along the slip plane

$$K_2 = \frac{G}{2\pi(1-v)} \int_{-\infty}^{+\infty} \frac{b'(\xi)}{x-\xi} \, d\xi. \tag{11-15a}$$

Using the definition of the dislocation by the Burgers circuit

$$b = \int_{-\infty}^{+\infty} b'(\xi) \, d\xi = 2 \int_{-\infty}^{+\infty} \frac{du}{d\xi} \, d\xi$$

$b'(\xi)$ becomes $(du/d\xi)$. (In reality the displacements $2u(x)$ are distributed antisymmetrically over the atoms in the two planes immediately above and below the slip plane.) Equating the forces $K_1(x) = K_2(x)$ for each x yields Peierls integral equation

$$\frac{1-v}{2} \frac{b}{c} \sin\left(\frac{4\pi u(x)}{b}\right) = \int_{-\infty}^{+\infty} \frac{du}{d\xi} \frac{d\xi}{x-\xi}. \tag{11-15b}$$

The equilibrium of forces thus provides an equation from which $2u(x)$ can be determined.

Peierls' solution of the equation for a primitive cubic lattice is

$$u(x) = -\frac{b}{2\pi} \arctan\left(\frac{2(1-v)(x-x_0)}{c}\right) \tag{11-15c}$$

if the centre of the dislocation is at x_0. The displacement field thus defines a dislocation core of width $\zeta = c/2(1-v)$ which is very small indeed. Better descriptions than those using a sinusoidal potential over the slip plane or the continuum approximation for the remainder of the crystal result in more realistic values for ζ. Very small values $\zeta/c \approx 1$ are obtained for *localized* binding, similar to that in germanium, whereas dislocations in close packed metals have much wider cores (see [11.4]).

The Peierls potential (per unit dislocation length) can be obtained by summing the work done in effecting the displacements $u(x)$

$$U_{\text{PN}}(x_0) = E_{\text{PN}} \cos\frac{2\pi x_0}{b} \tag{11-16}$$

and from this the 'Peierls–Nabarro force', which is the minimum force necessary to move a rigid dislocation by one lattice constant

$$\tau_{\text{PN}} = -\frac{1}{b} \frac{dU_{\text{PN}}}{dx_0}\bigg|_{\text{max}} = \frac{2\pi}{b^2} E_{\text{PN}} \equiv \frac{2G}{1-v} \exp\left(-\frac{2\pi\zeta}{b}\right). \tag{11-17}$$

The wider the dislocation, i.e. the larger the value of ζ, the smaller is the minimum force τ_{PN} required to move it. The Peierls force is negligibly small, $\tau_{PN} \approx 10^{-5}G$, for fcc and cph metals with their close packed, 'smeared out' atoms. In the case of Ge and Si, $\tau_{PN} \approx 10^{-2}G$ and $bE_{PN} \approx 0.2$ eV; bcc metals lie somewhere between these limiting values. In reality dislocations in Ge and bcc metals are mobile at $\tau \ll \tau_{PN}$ at finite temperatures due to the thermally assisted formation of kink pairs in the Peierls potential (see figs 11.7(a) and (b)). These can then propagate sidewards. The energy of a kink consists of Peierls and elastic line energy

$$E_k = \frac{4b}{\pi} \sqrt{(E_{PN} E_L)}. \tag{11-18}$$

If the frequency of the thermally activated formation of kink pairs is the same as that of a periodic elastic stress applied to the material, there is a loss of elastic energy due to 'internal friction', see section 2.7. In close packed metals this is observed to be a maximum at an applied frequency of 10^3 Hz at 70 K, corresponding to $E_k \approx 0.1$ eV ('Bordoni maximum', see [11.9]). In the case of Ge, however, dislocation velocity measurements at temperatures between 500–700 °C have shown E_k to be ≈ 0.3 eV, see [11.7] and section 11.4.1.

11.3.2   *Slip systems in important crystal structures*

Based on the energy estimates in section 11.2.3 the shortest translation vectors should be the Burgers vectors of dislocations and thus function as slip directions in crystals. This is indeed the case, namely $\mathbf{b} = (a/2)\langle 110 \rangle$ for the fcc, $(a/2)\langle 111 \rangle$ for the bcc and $(a/3)\langle \bar{2}110 \rangle$ for the cph structure. It is more difficult to justify the choice of slip plane, which together with $\mathbf{b}$ forms the slip system. The most closely packed planes are the most widely separated and should therefore shear most easily, i.e. have the lowest Peierls force. Nevertheless for fcc and cph crystals τ_{PN} is smaller than τ_0, where τ_0 is the observed critical shear stress for macroscopic slip, i.e. the stress to overcome obstacles on the slip plane such as the dislocation interactions described in section 11.2.3. The $\{111\}$ planes are indeed the primary slip planes of the fcc structure, as are the $\{0001\}$ basal planes for the cph with $c/a > c/a|_{id}$. At higher temperatures, however, other planes are observed. In the bcc structure, slip occurs on the $\{112\}$ and, especially at low temperatures, on the $\{110\}$ planes. For a given family of planes the one with the highest ratio m_s of force on the dislocation to externally applied stress (11-2) should operate first (cf. section 12.1). This is Schmid's law of the critical resolved shear stress. It is obeyed at the onset of deformation by the

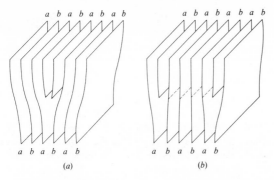

Fig. 11.15. Stacking sequence of fcc (110) double planes (a, b). Edge dislocation (a) and dissociated edge dislocation (b). Where a and b half planes meet between the partial dislocations a stacking fault is created.

fcc and cph slip systems yet is usually broken in the bcc structure, presumably as a result of dislocation dissociation discussed below [11.10].

11.3.3   *Partial dislocations and stacking faults*
11.3.3.1 Shockley partial dislocations and stacking faults

Perfect dislocations in the fcc structure can dissociate into *Shockley partial* dislocations with a consequent saving of energy (section 11.2.3)

$$\mathbf{b} = \frac{a}{2}[101] \Rightarrow \frac{a}{6}[112] + \frac{a}{6}[2\bar{1}1] = \mathbf{b}_{P1} + \mathbf{b}_{P2}.$$

All three vectors lie in $(11\bar{1})$. The $\mathbf{b}_{Pi}$ are, however, no longer translation vectors but create a stacking fault. Stacking faults were introduced in section 6.2.1 and fig. 6.11. In place of the translation CC described by $\mathbf{b}$, two steps CB and BC are observed, which circumvent the hill 'A' via two saddle points. As the $\mathbf{b}_{P1}$ dislocation moves through the slip plane, a cph stacking sequence is created which is eliminated by the following $\mathbf{b}_{P2}$ dislocation, see section 6.2.1 and fig. 11.15. This stacking fault possesses a surface energy γ, which prevents the two partial dislocations from separating too widely as a result of the elastic repulsive forces between them. The equilibrium separation w is given by the equilibrium of forces

$$\gamma = \frac{G(\mathbf{b}_{P1} \cdot \mathbf{b}_{P2})}{2\pi w} = \frac{Ga^2}{24\pi w}. \tag{11-19}$$

For noble metals with $\gamma = 0.02 \, \text{J/m}^2$, $w \approx 6a$. The width of the stacking fault can be measured by transmission electron microscopy, see section 2.2.1.2.

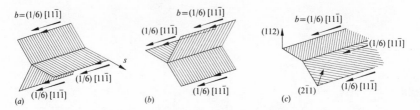

Fig. 11.16. Dissociated screw dislocation in the bcc lattice. The forms (a), (b), (c) have different symmetries and therefore need different forces to move them, e.g. to the right in the horizontal (112) plane.

In the cph structure, stacking faults occur in (0001) as a result of

$$\frac{a}{3}\langle\bar{2}110\rangle \Rightarrow \frac{a}{3}\langle\bar{1}100\rangle + \frac{a}{3}\langle\bar{1}010\rangle$$

with a sequence $BCBCB|ACACAC$ corresponding to an fcc layer. The γ values are very high in the bcc structure so that the dissociation occurs within the dislocation core and consequently cannot really be described by elasticity theory (11-19). There are a series of energetically favourable reactions, for example that of A. Sleeswyk

$$\frac{a}{2}[11\bar{1}] = \frac{a}{6}[11\bar{1}] + \frac{a}{6}[11\bar{1}] + \frac{a}{6}[11\bar{1}]$$

for a screw dislocation along $[11\bar{1}]$, fig. 11.16. The three partial dislocations lie in three (or two) slip planes of type $\{112\}$. The dislocation is thus incapable of glide and is described as 'sessile'. The stacking fault must be constricted before the dislocation can move in one of the three (or two) slip planes. The work required to do this varies depending on whether the force due to the applied stress acts to the right $(+s)$ or to the left $(-s)$. In the configuration shown in fig. 11.16(c) and for slip on (112) the partial dislocation on $(2\bar{1}1)$ must be constricted up to the intersection of the two planes against the applied stress in order to be able to move to the right $(+s)$. It moves to the left assisted by the applied stress. This asymmetry of the flow stress on reversing the slip direction is indeed observed in bcc metals even though the phenomenon and its explanation are in reality more complicated, see [11.10], [11.10a].

11.3.3.2 Consequences of dislocation dissociation in the fcc lattice

As a result of dissociation, a screw dislocation obtains a well defined slip plane, that of the stacking fault. If it wants to change planes (*cross-slip*) and transfer, for example in the fcc lattice, from $(\bar{1}11)$ to $(1\bar{1}1)$,

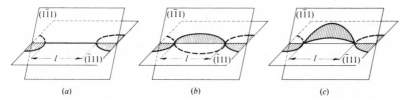

Fig. 11.17. Cross-slip by a dissociated screw dislocation in the fcc lattice after [11.5]. In stage (a) a construction of length l forms in $(1\bar{1}1)$, in (b) and (c) extends on $(\bar{1}11)$.

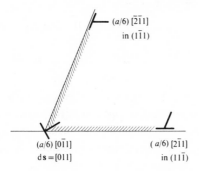

Fig. 11.18. Lomer–Cottrell dislocation in the fcc lattice with a 'stair-rod' dislocation at the line of intersection of two slip planes.

both of which contain the Burgers vector $(a/2)[110]$, the partial dislocations must first recombine over a certain length as shown in fig. 11.17. This process, described by G. Schöck and A. Seeger [11.5] is thermally activated so that the stress to initiate cross-slip depends strongly on temperature as well as on the stacking fault energy (see section 12.4). Similar difficulties are involved in the cutting of dissociated dislocations and in the consequent formation of jogs. Slip on intersecting planes is thus hindered in materials with a low γ and climb is more difficult (chapter 12).

The dislocation reactions described in section 11.2.3 often produce *sessile* dislocations. The following reaction

$$\frac{a}{2}[101] + \frac{a}{2}[\bar{1}\bar{1}0] \Rightarrow \frac{a}{2}[0\bar{1}1]$$

in $(11\bar{1})$ in $(1\bar{1}1)$ in (100)

can occur in the fcc lattice along the intersection $[0\bar{1}1]$ of two slip planes with a resultant saving of energy. The product does not lie in a $\{111\}$ plane and is therefore sessile (Lomer dislocation). If the reacting dislocations are dissociated, fig. 11.18, an even more stable three-dimensional obstacle is

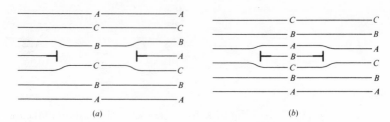

Fig. 11.19. Formation of a prismatic dislocation loop with (a) an intrinsic, (b) an extrinsic stacking fault by condensation of (a) vacancies, (b) interstitial atoms.

formed, the Lomer–Cottrell dislocation. The partial dislocation with $\mathbf{b}_P = (a/6)[0\bar{1}1]$ lying at the intersection of the two stacking faults is called a 'stair rod' dislocation (after the rod which holds a stair carpet in place). The Shockley partial dislocation also plays an important part in twinning and martensite reactions, see chapter 13. Next to the shear modulus, the stacking fault energy is the most important parameter in crystal plasticity [11.5].

11.3.3.3 Frank partial dislocations and prismatic dislocations in the fcc lattice

As described in section 10.2 vacancies condense in discs on a $\{111\}$ plane causing local collapse of the lattice, fig. 4.8. The result is a prismatic dislocation loop with $\mathbf{b}_P = (a/3)\langle 111\rangle$ which is sessile as shown in fig. 11.19 because the stacking fault does not lie in the plane containing $\mathbf{b}_P$. At high γ a Shockley partial dislocation can form within the prismatic loop with a resultant reduction in energy by the reaction

$$\frac{a}{3}[11\bar{1}] + \frac{a}{6}[112] \Rightarrow \frac{a}{2}[110],$$

removing the stacking fault in the loop and creating a perfect prismatic dislocation around it. Prismatic loops with and without faults have been observed in the transmission electron microscope in Al and Au respectively, which have different stacking fault energies. In the case of loops with low energy stacking faults an energetically more favourable form arises, the *stacking fault tetrahedron*, each surface of which contains a stacking fault. Equivalent stacking faults are produced by condensing interstitial atoms, but these are of an *extrinsic* rather than an *intrinsic* nature as shown by fig. 11.19. (The stacking sequence $A|\underline{BC}|B|ABCABC$ contains two

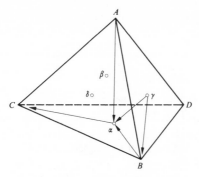

Fig. 11.20. The Thompson tetrahedron showing the types of
dislocation possible in the fcc lattice.

neighbouring intrinsic stacking faults.) The two types of faults can be
differentiated using transmission electron microscopy, section 2.2.1.2.

The overall picture of perfect and partial dislocations in the fcc structure
can be illustrated by the *Thompson tetrahedron*. Its vertices are located at
four nearest neighbour atomic sites, fig. 11.20. Its side faces $\alpha, \beta, \gamma, \delta$ are the
four $\{111\}$ planes, and its edges AC, BC, AB correspond to perfect
dislocations in δ. $A\alpha$ is a Frank partial dislocation, which reacts with a
Shockley partial dislocation αC to form a perfect dislocation AC. The
dissociation of a perfect dislocation into Shockley partial dislocations is
written $BC = B\alpha + \alpha C$. Two Shockley partial dislocations produce a stair
rod partial dislocation by the reaction $\gamma B + B\alpha = \gamma\alpha$. Similar constructions
are helpful for other crystal structures.

11.4   Dislocation dynamics
11.4.1   *Dislocation velocity*

The mean dislocation velocity v varies within wide limits
depending on stress and temperature. In the simplest case in which
localized obstacles on the slip plane can be surmounted with the aid of
thermal fluctuations $v = v_0 \exp(-U(\tau)/kT)$. For some years now it has been
possible to measure v directly by marking the dislocation positions on the
crystal surface before and after a stress pulse, e.g. by etching. The velocity v
is obtained by dividing the distance between the etch pits by the pulse length
if the two are proportional to each other. Fig. 11.21 shows the experimental
results for a series of typical materials at 300 K (see also [11.13]). Three
distinct ranges are visible. At low and high τ relative to the critical resolved
shear stress τ_0, v is similarly small or large and approximately linear in τ. In
the intermediate range v shows a rapid increase at τ_0 which can often be

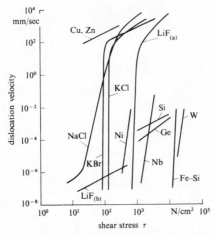

Fig. 11.21. Dislocation velocities in different crystals depending on the shear stress. All values at 20 °C except Ge (450 °C) and Si (850 °C) [11.7], [11.12].

described by a power law $v = B\tau^m$. This region will be treated empirically in section 12.3. In the region of small τ we are dealing with quasi-viscous dislocation movement, which has been treated in terms of a model for the case of germanium [11.12].

The dislocation velocity is limited by the transverse velocity of sound c_t. Near c_t the displacement field of the (screw) dislocation moving at constant v must be treated relativistically, i.e. subjected to a Lorentz transformation in which the coordinate x in the direction of motion (fig. 2.10) becomes $x' = (x - vt)/\sqrt{(1 - v^2/c_t^2)}$. The solution (2-5) then becomes

$$u_y = \frac{b}{2\pi} \arctan\left(\frac{z(1 - v^2/c_t^2)^{1/2}}{x - vt}\right) \tag{11-20}$$

i.e. the displacement field of the screw dislocation transforms from circular cylindrical symmetry to that of an elliptical cylinder flattened in the direction of motion. At the limiting velocity the dislocation energy diverges as

$$E_L(v) = E_L(0)/\sqrt{(1 - v^2/c_t^2)}. \tag{11-21}$$

For $v \ll c_t$ the rest mass m_d of the dislocation (per length b) can be obtained by expanding the above expression together with (11-7a)

$$m_d = \frac{E_L(0)}{c_t^2} = \frac{Gb^3\rho_m}{4\pi G} \ln\frac{R}{r_0} = m\frac{\ln R/r_0}{4\pi} \approx m \tag{11-22}$$

in which $m \approx \rho_m b^3$ is the atomic mass and ρ_m is the bulk density of the material. The dislocation with its extended distortion field has only a very low rest mass: one atom mass per lattice plane perpendicular to the dislocation line. The accelerating times are therefore negligibly small.

11.4.2   *Dislocation damping*

The friction coefficient $B = \tau b/v$ at high stresses is found to be proportional to the temperature over a wide range. The dislocation is damped by collisions with phonons. At low temperatures when the phonon density is small, electron collisions become measurable. In the case of superconductivity these disappear as well and v becomes large with the result that inertia effects become noticeable in dislocation dynamics and crystal plasticity [11.11a]. Furthermore a dislocation moving over the Peierls potential emits sound waves (like a chain moving across a corrugated sheet). The frictional effects and hence v are thus largely determined by the structure of the dislocation core. They also manifest themselves in measurements of the anelasticity in the 10^8 Hz range in which pinned dislocation segments are excited to damped resonance vibration (A. Granato and K. Lücke, see [11.4]).

12
Plastic deformation, work hardening and fracture

12.1 Crystallography of plastic shear

According to what has been said in section 11.1.1 about dislocation movement on slip planes and from the observations of slip traces on deformed crystals reproduced in [1.1], there is no doubt that the plastic deformation of metals occurs essentially by slip on crystallographic planes in crystallographic directions. Plastic deformation is not usually measured in a pure shear test, however, but in a tensile or compression test on a rod, section 2.6.1. Shear on planes inclined to the tensile axis in fact leads to an extension of the rod, as shown in fig. 12.1(a) for the case of 'single' slip. In order to understand the slip process in terms of dislocation theory, the experimental parameters of the tensile test (σ, ε as detailed in section 2.6.1) must be resolved on to the slip system, i.e. to 'crystallographic coordinates', or in other words converted to the shear stress τ and shear strain a in the slip system. In a geometrical method described by Schmid and Boas [12.1] a force K_A acts along the longitudinal axis of a rod of cross-section q. τ is the force per unit area resulting from K_g the component of K_A resolved in the slip direction, acting on area F of the slip plane (fig. 12.1(b))

$$\tau = \frac{K_g}{F} = \frac{K_A \cos \lambda}{q/\cos(90-\chi)}. \tag{12-1a}$$

λ and χ define the 'orientation' of the rod axis with respect to the slip system for a specimen of length l. λ_0 and χ_0 define the initial orientation for a specimen of original length l_0 (A in fig. 12.2 for the fcc lattice). During deformation the crystal lattice rotates relative to the rod axis as shown in fig. 12.1(b) such that

$$\frac{\sin \chi_0}{\sin \chi} = \frac{\sin \lambda_0}{\sin \lambda} = \frac{l}{l_0} = 1 + \varepsilon. \tag{12-1b}$$

Since the volume remains constant during shear deformation, $lq = l_0 q_0$. Utilizing the definitions of stress σ and strain ε given in section 2.6.1,

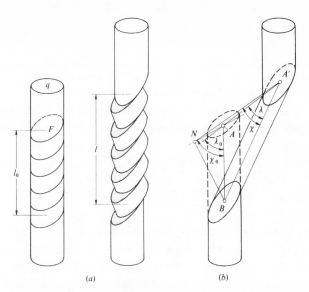

Fig. 12.1. (*a*) Shear in a single crystal leads to (*b*) a change in orientation between the crystal lattice and the axis of the rod ($\chi_0 \to \chi$, $\lambda_0 \to \lambda$). It is assumed here that the ends of the specimen do not deform but can rotate freely in the grips.

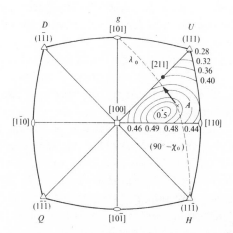

Fig. 12.2. Slip systems in the fcc lattice and path traced by the crystal axis (initial orientation A) in a tensile test shown in the standard triangle of the stereographic projection. Lines with equal Schmid factor are shown for the primary slip system (H, g).

equations (12-1a) and (12-1b) can be solved to give

$$\tau = \sigma \cos \lambda \sin \chi_0. \tag{12-1c}$$

Equation (12-1c) corresponds to (11-2) for the special σ of the uniaxial tensile test, where $K = \tau b$ is the force per unit length of the dislocation in the slip system. The ratio $\tau/\sigma = m_s$ is known as the Schmid factor. The shear a is defined as the crystallographic displacement of two points on parallel slip planes, unit distance apart, i.e. in fig. 12.1(b), $a = AA'/BN$. With the aid of the geometric relationship $AA'/\sin(\lambda_0 - \lambda) = l_0/\sin \lambda$ and (12-1b) we obtain

$$a = (\cot \lambda - \cot \lambda_0) \frac{\sin \lambda_0}{\sin \chi_0}. \tag{12-1d}$$

λ can be calculated by substituting the experimental value of ε in (12-1b). The shear a can then be obtained using (12-1d) and the shear stress τ with the help of (12-1c) and the experimental value of σ. For the most favourable initial orientation ('0.5' in fig. 12.2) with $\lambda_0 = \chi_0 = 45°$, m_s assumes the maximum value one half; $\mathrm{d}a/\mathrm{d}\varepsilon = 2$ is then also a maximum according to (12-1d). For unfavourable orientations ($\chi_0, \lambda_0 \neq 45°$) both values are lower. Fig. 12.2 shows the variation of m_s in the standard triangle for an fcc lattice which contains all the possible orientations of the specimen (see section 2.9). It also shows the path traced by the specimen axis A during deformation according to (12-1b). At the boundary of the standard triangle other slip systems become equivalent and yet the simultaneous operation of several slip systems with the same m_s is seldom observed. The slip system with the greatest dislocation density provides too dense a 'dislocation forest' to be cut easily by dislocations in a second system. Only after the symmetry boundary [100]–[111] has been crossed and the Schmid factor of the second system is significantly higher than that of the first system does the second system take over. The path traced by the specimen axis thus 'overshoots' the symmetry boundary.

The face centred cubic lattice possesses twelve crystallographically equivalent slip systems of the type slip plane $\{111\}$, slip direction $\langle 110 \rangle$ although the resolved stress is different on each system. The four slip planes are named according to their function. The *primary slip plane* $H = (11\bar{1})$ in fig. 12.2, together with the slip direction $g = [101]$, forms the slip system with the largest Schmid factor m_s, operative at the onset of plastic deformation (Schmid's law). At the symmetry boundary [100]–[111] the stress on the *conjugate system* $D = (1\bar{1}1), [110]$ is the same as on the primary system and the two systems operate alternately. The specimen axis oscillates about the symmetry boundary finally converging at $[211]$

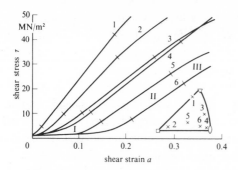

Fig. 12.3. Work hardening curves for Cu single crystals of different orientation at 293 K (J. Diehl, Stuttgart). The lines mark the beginning and end of stage II (stresses τ_{II}, τ_{III}).

irrespective of the initial orientation. $Q = (1\bar{1}\bar{1})$ is the *cross-slip plane*, which contains the same slip direction $g = [101]$ as the primary slip plane. This means that screw dislocations can cross-slip from the primary plane on to the cross-slip plane, see section 11.3.3.2. Finally the *'critical' slip plane* $U = (111)$ with slip direction $[10\bar{1}]$ is sometimes observed in tensile tests where the initial orientation is close to the symmetry boundary $[100]$–$[110]$. (In compression tests, however, where to a first approximation the axis rotates in the opposite direction, it is observed for all orientations.) Similar considerations can be applied to the bcc lattice, see section 11.3.3.1. For $c/a > 1.63$ and at low temperatures, the cph structure possesses only one slip plane $\{0001\}$, which contains three slip directions $\langle \bar{2}110 \rangle$.

As a generalization of Schmid's law we would expect all *work hardening curves* for crystals of different initial orientations but measured at the same shear strain rate $\dot{a}$ to be coincident when converted to shear stress/shear strain. This is not the case, as shown in fig. 12.3 for copper at 293 K. The stress–strain curve shows the initial stage I with a small slope $d\tau/da|_I = \Theta_I$ only in the case of favourable orientations. Stage I ends at a stress τ_{II}, strain a_{II} at which the critical shear stress for secondary systems has been reached. For unfavourable orientations the stress–strain curve starts with the steeper stage II, slope Θ_{II}. Obviously this is due to the simultaneous (minor) operation of other slip systems. At large strains in stage III the slope Θ again decreases. Similar stress–strain curves have been obtained for bcc crystals, although stage II have often degenerated to an inflection point. At low temperatures cph crystals with $c/a > c/a_{ideal}$ show stage I only. Since only easy glide is possible in this structure, this is compatible with the above interpretation.

In the following two sections, we want to interpret stress–strain curves and creep curves $\dot{a}(a)$ obtained in static tests ($\tau = \text{const}$), see section 2.6.1, in terms of dislocation theory. We shall first approach the problem using the general principles of dislocation theory, section 12.2. This is possible only in exceptional cases, such as the deformation of crystals with the diamond structure (Ge, Si), and even then is valid only approximately and for the initial stage of the stress–strain curve. (In [12.26] a phenomenological generalization of the theory to highly strained metals is proposed.) In section 12.3 an alternative theory is presented for stage II of the stress–strain curve based on slip line observations.

12.2   Slip and dislocation movement

According to section 11.1 the movement of dislocations in the slip plane by a mean distance dx results in an increase in shear da. (As in section 11.1 we are considering only straight dislocations here.) If the density of mobile dislocations is N_M, then from the Orowan relationship [11.5]

$$da = bN_M \, dx,$$

or   (12-2)

$$\dot{a} = bN_M v,$$

where v is the mean dislocation velocity. (The equation is easily extended to dislocations which are not straight. In the case of dynamic deformation $\dot{a}$ is the *plastic* shear strain rate to which must be added the elastic deformation rate ($\dot{\tau}/G$)). In (12-2) N_M is dependent on τ and also, by virtue of dislocation multiplication, on v. v is dependent on τ and T, as was described in section 11.4.1, and also, because of work hardening, on N. The total dislocation density N can quite easily be different from the mobile dislocation density because the proportion of sessile dislocations, in dipoles, etc., increases as the dislocation density increases, see chapter 11. The ratio N_M/N is only rarely known, for example for Ge [11.7]. In this case it falls from 1 at the onset of deformation to about $\frac{1}{2}$ at $a = 20\%$. Ge is a suitable material on which to test (12-2) because, due to the high Peierls stress, the dislocations can move only slowly and require considerable thermal activation. Many dislocations must therefore participate to produce a measurable $\dot{a}$. The conditions for fcc metals appear to be exactly the reverse. Dislocation multiplication is described most simply by the empirical relationship

$$dN = N \, dx/x_1, \tag{12-3}$$

with $N_M = N$ [11.7]. $x_1(\tau)$ is the mean distance moved by the dislocations in which their length is increased e-fold. ($x_1 = 70 \, \mu m$ for Ge for q = 10 N/mm^2). The equation has to be integrated with an x_1 which may possibly be

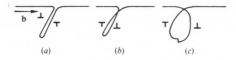

Fig. 12.4. An edge dipole, which has formed on a superjog in a screw dislocation, flips over producing a new dislocation loop, (a) to (c).

dependent on N and for an initial dislocation density N_0. A model for the dislocation multiplication process based on transmission electron microscopy studies on bcc metals and Ge is shown in fig. 12.4. Jogs in screws are pushed together at a point at which an edge dipole is then drawn out. If the stress τ_{eff} on the dipole is larger than the passing stress τ_p (section 11.2.1), the dipole can flip over and produce a dislocation ring. The model predicts $x_1 \propto \text{jog height} \propto (\tau_p)^{-1}$. Specifically $x_1 = 1/(K\tau_{eff})$ where τ_{eff} is the 'effective stress', i.e. the externally applied stress less the mean internal stress (of other dislocations). We assume that the dislocation moves in a periodic internal stress field

$$\tau_i(x) = A(\sqrt{N}) \sin 2\pi x/\Lambda, \tag{12-4}$$

where $A = \alpha_1 Gb$ (11-9), and that the dislocation movement is determined by the sites of greatest resistance where the waiting times are longest. Thus

$$\tau_{eff} = \tau - A\sqrt{N}, \tag{12-5a}$$

and

$$v = B_0 \tau_{eff}^m \exp(-U/kT) \equiv B\tau_{eff}^m. \tag{12-5}$$

v is perhaps better described by averaging the localized velocities over a period Λ of $\tau_i(x)$ as given in (12-4) [11.7]

$$\frac{\Lambda}{v} = \int_0^\Lambda \frac{dx}{B(\tau - \tau_i(x))^m}. \tag{12-5b}$$

Equations (12-2) to (12-5) yield a system of equations which can be used to calculate $\dot{a}(\tau, T)$, the *equation of state for plasticity*

$$\dot{a} = bNB(\tau - A\sqrt{N})^m + \dot{\tau}/G, \tag{12-6a}$$

$$\dot{N} = NKB(\tau - A\sqrt{N})^{m+1}. \tag{12-6b}$$

The validity of (12-6) is limited by putting $N_M = N$ and by the simplifying assumptions of (12-3) to (12-5).

In the creep test ($\tau = \text{const}$) (12-6a) yields the curves $\dot{a}(t)$ and $a(t)$ shown in fig. 12.5. At the beginning $\dot{a}$ is small because there are too few dislocations (12-2) and later on because there are too many (12-4). The maximum creep

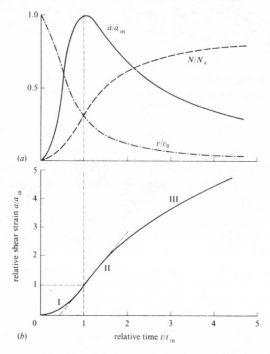

Fig. 12.5. Subdivison of the creep curve (*b*) in its differentiated form $\dot{a}_{in}(t)$ into its factors according to equation (12-2).

rate (with $\ddot{a} = 0$) is obtained from (12-6) as

$$\dot{a}_{in} = \frac{bB_0 C_m}{A^2} \tau^{2+m} \exp\left(-\frac{U}{kT}\right) \tag{12-7}$$

where C_m is a number dependent on m. This relationship between τ and T has been confirmed for Ge and Si in which $m \approx 1$ and U is the energy of formation and migration of a kink pair. The form of the creep curve has been substantiated in independent investigations of (12-3) and (12-5).

In the *dynamic tensile test* ($\dot{a} = $const) equations (12-6) have to be solved for $\tau(\dot{a}, T)$. As shown in fig. 12.6, a pronounced yield point is obtained. At first the stress rises steeply to make the few available dislocations move quickly. Dislocation multiplication then begins and the imposed strain rate $\dot{a} = $const can be maintained at a lower τ, v. Finally as a result of 'work hardening', i.e. the increase in the internal stress with N and a (which is a monotonic function of N), τ again increases. The curve shows an upper and lower yield point (τ_{uy}, τ_{ly}). At τ_{ly} the imposed strain rate $\dot{a}$ is achieved with the lowest stress, analogous to the situation at the point of inflection in the

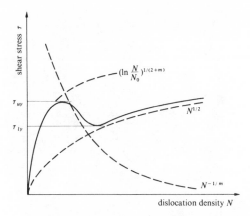

Fig. 12.6. Pronounced yield point resulting from multiplication and interaction of dislocations of density N.

creep curve where, for a given τ, the highest rate $\dot{a} = \dot{a}_{in}$ is obtained. τ_{ly} is evaluated by inversion of (12-7) to $\tau_{ly} \propto \dot{a}^{1/n} \exp(U/nkT)$, $n = m + 2$, in agreement with experiments on Ge and Si. The larger the value of m or of the grown-in dislocation density N_0, the smaller is the yield drop. In the case of metals with the rapid increase in $v(\tau)$ shown in fig. 11.21 and with $N_0 = 10^8/\text{cm}^2$, a pronounced yield drop is no longer observed. In alloys, the dislocations can be pinned by solute atoms, severely reducing N_M (at large N). There is then a 'dislocation multiplication yield drop'. In the case of Fe–C alloys, however, the pronounced yield point would appear to be caused by the unpinning of dislocations from the solute atoms (see chapter 14). In these circumstances the deformation is often not uniformly distributed over the specimen, as was assumed above, but propagates along the specimen as a 'Lüders band'.

12.3   Flow stress and work hardening

In close packed metals the flow stress at a given strain rate is not determined by the lack of mobile dislocations as was described above. N_M may no longer be considered equal to N. The 'flow stress' necessary to cause further deformation at each stage of deformation is no longer determined by dynamic effects as in the case of materials with high Peierls stress, but by the internal stresses of all the dislocations in the crystal. This *static theory of work hardening* must: (1) calculate the flow stress for the given dislocation configuration or, in particular, the dislocation density N, and (2) take into account how this configuration (N) changes for an increase in strain of $\mathrm{d}a$. This is an extremely difficult task considering the complicated *dislocation*

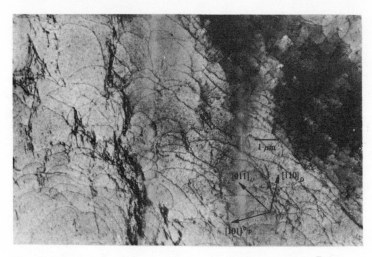

Fig. 12.7. Dislocation arrangement in the main slip plane $(11\bar{1})$ of a copper single crystal deformed into stage II at 78 K. Primary Burgers vector $\mathbf{b}\,\|\,[101]$. Lines of intersection with secondary slip planes U, D are shown. Irradiated under load with $2 \times 10^{18}\,\mathrm{n/cm^2}$ at 4 K after deformation. (H. Mughrabi, *Phil. Mag.*, **23** (1971), 897.)

configurations observed in deformed metals by transmission electron microscopy. Fig. 12.7 shows the dislocation configuration in a Cu single crystal which was deformed to a flow stress $\tau = 12\,\mathrm{N/mm^2}$. It was then irradiated with neutrons to prevent the dislocations from relaxing, a process which occurs especially easily during the thinning of specimens for electron microscopy. The mobile dislocations in the picture are curved. The radius can be used (11-8) to obtain the effective stress. Obviously $N_M \ll N$, most dislocations being fixed in 'braids', dipole or multidipole arrangements. Lomer–Cottrell dislocations have been formed by reaction with the secondary slip system. The greater the value of τ, the greater is the density of such obstacles and the shorter the period Λ of the dislocation configuration. This dislocation configuration is typical of stage II of the stress–strain curve $\tau(a)$ for an fcc single crystal in which slight participation by secondary slip systems has led to maximum work hardening $\Theta_{II} = \mathrm{d}\tau/\mathrm{d}a|_{II} \approx G/300$ (fig. 12.3) although the orientation of the specimen axis is still within the standard triangle. At lower strains deformation takes place on the primary system only. This is the 'easy glide' stage I of fcc metals with $\Theta_I = G/3000$. For cph metals at low temperatures, where only one set of slip planes operates, easy glide continues to $a \approx 1$. At high temperatures and large strains (stage III), thermally activated dislocations can leave their slip planes, whereby Θ again decreases. This is the stage of 'dynamic recovery' which will be discussed in section 12.4. As regards work hardening, bcc metals at medium temperatures behave in a similar manner to fcc except that the critical shear stress is higher and more strongly dependent on temperature on account of the effects described in section 11.3.3.1.

Systematic investigation of the work hardening of metals over the last two decades has yielded a wealth of information not only concerning the correlation between deformation and dislocation configuration using electron microscopy, X-ray topography and etch pit observations but also about the dependence of the stress–strain curve on the orientation of the tensile axis, on the temperature, specimen size, deformation rate, stacking fault energy, etc. [12.2]. The results of *slip step measurements* on deformed crystals are also particularly instructive. Slip steps indicate where and how many edge dislocations have emerged at the crystal surface. The shear strain a is determined by the area of the slip plane traversed by these dislocations, given by the slip step length L. Observations on copper in stage II show that a strain interval da is associated with the activation of dN' new slip planes and sources, each of which emits a fixed number of dislocations $n \approx 20$ over a region of length L, i.e.

$$da = nbL(a)\, dN'. \tag{12-8}$$

The mean free path L which is inversely proportional to the strain in stage II becomes shorter because secondary slip systems 'inject' obstacles on to the primary slip plane, e.g. form Lomer dislocations. Hence $L = \Lambda/(a - a_{\text{II}})$ for a constant $\Lambda (= 5 \times 10^{-4}\,\text{cm})$ and a_{II}. We thus have a model for stage II which describes the change in dislocation configuration in a strain interval da in a very simple manner.

We still need to know the *flow stress* of this array of dislocations for the first part of the theory. The n dislocations which moved the distance L on one slip plane are trapped at the end of the slip line and exert a back stress on the source which assists in suppressing its operation. This pile-up of n dislocations, together with the secondary dislocations which limit L, forms the braids mentioned above, which have been observed in transmission electron microscopy in stage II. These groups are the sources of internal stress in the crystal, which must be overcome by further dislocations moving through the crystal. A. Seeger describes the pile-ups as superdislocations of strength (nb) and, using (11-9), obtains the mean internal stress for the N' groups which he puts equal to the flow stress

$$\tau_{\text{G}} = \alpha G(nb)\sqrt{N'}. \tag{12-9}$$

Using (12-8) and $L(a)$ he obtains the work hardening

$$\frac{1}{G}\frac{\tau_G}{a-a_{II}}=\frac{\Theta_{II}}{G}=\alpha \ \sqrt{\left(\frac{nb}{2\Lambda}\right)}. \tag{12-10}$$

This equation relates the results of slip line measurements on the surface (n, Λ) to work-hardening measurements on the bulk of the crystal. This correlation has been substantiated by experiment. It is more difficult to calculate the absolute value of Θ_{II}, i.e. to relate n and Λ. If the only dislocations contributing to the back stress on a source were those *produced by this source*, we could use the pile-up theory of section 11.2.3 and obtain $nb/\Lambda = $const, and therefore $\Theta_{II}/G = $const, as is observed in the linear stage II of the stress–strain curve. In reality, however, other groups of dislocations act upon the source and the proportion of the back stress resulting from its own dislocations is uncertain theoretically. A whole series of models has been proposed [12.3].

Some of these models emphasize the flow stress contribution by the dislocation *forest*, i.e. the total N_F of all dislocations not lying in the primary slip plane. According to Z. S. Basinski N_F can easily be of the order of N for middle orientations in stage II, although the mean free path and hence the contribution to shear by the forest dislocations is small compared with that of the primary dislocations [12.3]. *Some* of the intersection processes and the resulting jog movements, possibly involving the production of lattice defects, require so little energy that they can be assisted by thermal fluctuations. According to (11-10) the necessary intersection stress τ_S to cut through a forest of density N_F at $T=0$ is obtained by equating the work done by τ_S, $(\tau_S bd/\sqrt{N_F})$, to the energy of jog formation E_J. At *finite temperatures*, thermal fluctuations of frequency

$$v = v_0 \exp(-(E_J - \tau_S bd/\sqrt{N_F})/kT) \tag{12-11a}$$

assist jog formation

i.e. $$\tau_S = \frac{E_J - kT \ln v_0/v}{bd} \cdot \sqrt{N_F} \tag{12-11b}$$

where d is the width of the obstacles (stacking fault width!) and $v_0 \approx 10^{10} \sec^{-1}$ is a frequency factor, fig. 12.8. v determines the mean dislocation velocity in the forest and in favourable cases the shear strain rate $\dot{a}$. Equation (12-11b) is the simplest form of the temperature and strain rate dependent contribution to the flow stress, in particular of the critical shear stress τ_0. As early as 1954, A. Seeger [12.2] showed that (12-11b) explained $\tau_0(T)$ for hexagonal metals at low temperatures very well, see fig.

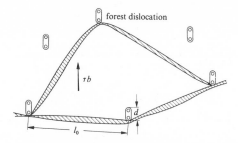

Fig. 12.8. Movement of a dislocation through a 'forest' of dissociated dislocations: width d, distance apart $l_0 = N_F^{-1/2}$.

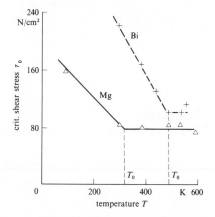

Fig. 12.9. Temperature dependence of the critical shear stress of Mg and Bi [12.1].

12.9. T. Vreeland [12.4] has confirmed the validity of the relationship $\tau_S \propto \sqrt{N_F}$ by direct measurements of the velocity of dislocations on the basal plane of Zn crystals with forest dislocations lying in the pyramidal planes. At larger strains the temperature dependent contribution to the flow stress τ_S is found to be much less than τ_G and Θ_{II} is found to be independent of T and $\dot{a}$.

In the easy glide region, where the dislocation mean free path is of the order of the crystal diameter and the step height is variable $n \propto a$, the rate of work hardening Θ_I appears to be determined by the trapping of edge dislocations in dipole arrays.

In a *creep experiment*, the effective stress on a dislocation in front of an obstacle is steadily reduced by work hardening, i.e. τ_S in (12-11a) should be replaced by $\tau - \tau_G = \tau - \Theta a (\Theta a < \tau)$. For $\dot{a} \equiv sv$, where $s = Nb/\sqrt{N_F}$, we then have

$$\frac{\mathrm{d}a}{\mathrm{d}t} = \dot{a}_0 \exp\left(-\frac{\Theta bd}{kT\sqrt{N_F}} a \right)$$

where

$$\dot{a}_0 \equiv sv_0 \exp\left(-\frac{E_S}{kT} + \frac{\tau bd}{kT\sqrt{N_F}} \right).$$

Integrating:

$$a = \frac{kT\sqrt{(N_F)}}{\Theta bd} \ln\left(\frac{\Theta bd\dot{a}_0}{kT\sqrt{N_F}} t + 1 \right). \tag{12-12}$$

For constant Θ, a logarithmic time law is obtained for creep, which is in fact observed at moderately high temperatures for metals in which the cutting of forest dislocations determines the dislocation velocity.

12.4   Dynamic recovery – cross-slip and climb

Etch pit experiments on deformed specimens show that, starting from relatively few sources, crystals become filled with dislocations throughout their whole volume. Dislocations must thus be able to leave their original slip planes. Screw dislocations achieve this by *cross-slip* (section 11.3.3.2) and edge dislocations by *climb*. The driving force for both processes is the mutual attraction of dislocations of opposite sign. If they meet, they annihilate, resulting in a reduction in *internal stress*. This process is thermally activated and is known as *recovery*. The *applied stress* causing the deformation assists cross-slip, which is hindered by dissociation of the dislocations in one plane, and it also contributes to the climb force (section 11.1.5). Recovery is thus aided by and is more pronounced during deformation: *dynamic recovery*. This determines stage III of the stress–strain curve of cubic crystals as well as creep at high temperatures. Dynamic recovery manifests itself as *work softening*, when a crystal predeformed at low temperatures into stage II is redeformed at a higher temperature (then in stage III), fig. 12.10.

A dissociated dislocation must first be constricted in order to move from one slip plane to a cross-slip plane which also contains the Burgers vector, (section 11.3.3.2, fig. 11.17). According to G. Schöck, the necessary activation energy U_Q decreases logarithmically with the stress and is inversely proportional to the stacking fault energy. The cross-slip frequency v_Q is therefore determined by the following Arrhenius expression

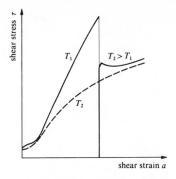

Fig. 12.10. Work softening on raising the temperature.

$$U_Q = -\frac{E_Q^2}{\gamma} \ln \tau/\tau_Q,$$

$$v_Q = v_0 \exp(-U_Q/kT) = v_0(\tau/\tau_Q)^{E_Q^2/\gamma kT}, \qquad (12\text{-}13)$$

(v_0, τ_Q and E_Q are constants). The additional strain rate due to cross-slip, $\dot{a}_Q = \beta v_Q$ attains the order of magnitude of $\dot{a}$ at the start of stage III at the stress τ_{III}. The relationship for τ_{III} is thus [12.5]

$$\ln \frac{\tau_{III}}{\tau_Q} \approx -\frac{kT\gamma}{E_Q^2} \ln \frac{\beta v_0}{\dot{a}}. \qquad (12\text{-}14)$$

The stress for the onset of stage III decreases exponentially with increasing temperature and particularly rapidly for metals with a high stacking fault energy (such as Al, Ni, fig. 12.11). Virtually the whole temperature dependence of the stress–strain curve of fcc metals derives from the cross-slip process. Nevertheless only a slight increase in τ_0 and in τ_{II} (the stress to initiate secondary slip) with decreasing temperature, significantly extends the easy glide region, fig. 12.11. S. Mader was able to identify stage III with cross-slip by the observation of slip steps, fig. 12.12. In stage III, the crystal can be observed in the TEM to 'break down' into 'cells' or subgrains; after annihilation of the screws, the edge dislocations rearrange themselves to form low-angle grain boundaries. In cph metals with $(c/a) < (c/a)_{ideal}$, the dislocations move on to the prism planes at elevated temperatures by thermally activated cross-slip.

At temperatures above half the melting point, the mean diffusion distance for the duration of a deformation test is comparable with the mean distance between dislocations. *Recovery by climb* of edge dislocations then has a noticeable effect on the strain rate. Dynamic equilibrium is established between work hardening and recovery which can be expressed by

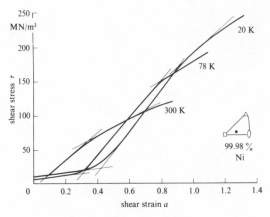

Fig. 12.11. Temperature dependence of the work-hardening curve for nickel single crystals.

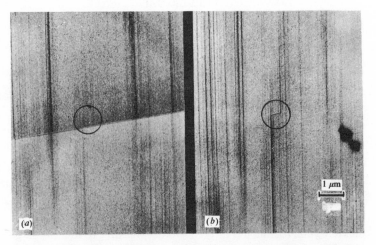

Fig. 12.12. Cross-slip and increase in height of single slip steps on Cu at the beginning of stage III. (b) shows the same region as (a) but after a further 5% deformation (S. Mader, Stuttgart).

$(\partial\tau/\partial a)_{t=\text{const}}$. Thus for $\tau = \text{const}$, stationary creep with a rate $\dot{a}_S$ is observed, according to

$$\mathrm{d}\tau = 0 = \frac{\partial\tau}{\partial t}\bigg|_a \mathrm{d}t + \frac{\partial\tau}{\partial a}\bigg|_t \mathrm{d}a \,; \qquad \dot{a}_S = \frac{-\dfrac{\partial\tau}{\partial t}\bigg|_a}{\Theta}. \qquad (12\text{-}15)$$

The actual calculation of $\dot{a}_S$ using a model due to J. Weertman does not, however, employ (12-15). It describes the stationary creep as shown in fig. 12.13. Dislocations emitted by sources Q_1 and Q_2 under a stress τ just block each other at a passing distance $h = Gb/8\pi\tau(1-v)$. A few of them are rendered sessile by interaction with secondary dislocations. The sources remain exhausted as long as dislocations of the two groups do not annihilate one another by climb. The climb stress per unit length of dislocation exerted by the applied stress is of the same order as the stress for glide, i.e. $K_z = n\tau b$ if there are n mobile dislocations behind the climbing dislocations, see section 11.2.3. A dislocation jog moves by diffusional drift with a velocity $v_J = (D/kT)n\tau b^2$. On average it produces more vacancies than it assimilates and thus the dislocation moves in the sense of the climb stress towards the other group. (D is the bulk diffusion coefficient, see (7-11).) If c_J is the number of jogs per unit length of dislocation, the rate of climb of the dislocation in the z direction is

$$v_z = \frac{D}{kT} n\tau b^3 c_J. \tag{12-16}$$

After a time $(h/2v_z)$ the dislocations annihilate one another and the sources can emit two new dislocations, which after moving a distance L again block one another. If (11-9) is used for the dislocation density N when the stresses are in equilibrium (corresponding to a fixed ratio of τ_{eff} to τ), the strain rate controlled by v_z is

$$\dot{a}_S = \frac{b2L}{h/2v_z} N = \frac{32\pi(1-v)bLn}{G^3kT\alpha_1^2} c_J \tau^4 D(T). \tag{12-17}$$

$\dot{a}_S$, the creep rate, depends on the temperature like the bulk diffusion coefficient, and increases with the fourth power of the stress as long as jogs are available, i.e. as long as c_J is a microstructural constant. $c_J \propto LN_F$ is more likely, however, when the jogs are created by intersection processes in the dislocation forest. $Lc_J \propto L^2 N_F \gg 1$ could then be independent of τ, see section 12.3. These predictions are confirmed experimentally for many metals and so-called type II-alloys (at not too high a value of τ) [12.6]. In the dynamic test, a contribution to $\dot{a}$ from $\dot{a}_S$ can have the same effect as that from $\dot{a}_Q$ for cross-slip described above. Non-crystallographic, wavy slip steps are then observed. When the dislocations are widely dissociated, i.e. for small γ, $\dot{a}_S$ is smaller because the formation of jogs is more difficult.

12.5   Deformation of polycrystals, deformation texture

12.5.1   *Dislocations at grain boundaries*

The plastic deformation of polycrystals differs from that of single

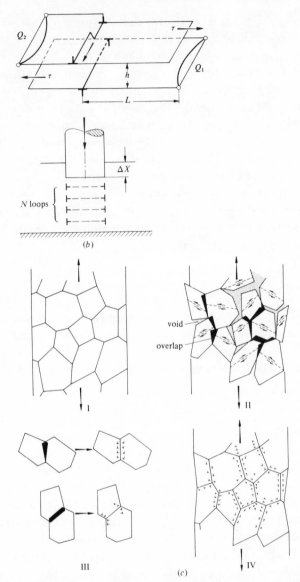

Fig. 12.13. (*a*) Model for the annihilation of two edge dislocations by climb, effected by the movement of jogs along the dislocations. (*b*) Under an indentation of depth Δx there are N prismatic dislocation loops needed geometrically. (*c*) Generation of geometrically necessary dislocations during tensile deformation of a polycrystal: If the grains did not cohere (part III), then open or doubly occupied wedges would appear during deformation. The introduction of suitable dislocations in part III allows rotation of material continuity.

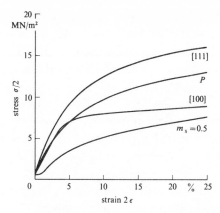

Fig. 12.14. Stress–strain curves ($\sigma/2$ against 2ε) for three aluminium single crystals of different orientation and a polycrystal P (grain size 0.2 mm [12.9]).

crystals in two respects: firstly grain boundaries act as obstacles to dislocation movement. This means, for example, that the easy glide region observed in fcc and bcc metals and which actually dominates in cph metals cannot exist because the mean free path of the dislocations cannot exceed the grain size d and cannot equal the specimen diameter. On the other hand the work hardening mechanisms operative in single crystals in stages II and III should also be observed in polycrystals because here $L < d$. Secondly the individual grains in the polycrystal possess a wide variety of orientations considered here to be a random distribution (although this does not follow necessarily from solidification and in fact a texture is often present, see section 4.3). Even in the case of single crystals the stress–strain curve is strongly dependent on the orientation because of the varying degrees of participation of secondary slip systems (fig. 12.14). In the polycrystal the orientation dependence of the Schmid factor is superimposed on this behaviour with the effect that the individual grains would deform quite differently for a given stress were they not connected to one another. There is obviously no point in considering the polycrystalline stress–strain curve as a single crystal curve averaged over all orientations.

Ashby [12.24] differentiates in the deformation of inhomogeneous materials between *'geometrically necessary'* and *'statistically stored'* dislocations. We have so far used the latter as sources of work hardening. The former term is exemplified in fig. 12.13(b) by the indentation of a cylindrical piston which must produce $N = \Delta x/b$ prismatic dislocations (in addition to an unpredicted number of statistically distributed dislocations which do not contribute to the indentation depth Δx). Fig. 12.13(c)

describes the geometrically necessary dislocations needed for deformation of a polycrystalline aggregate. In part II of this diagram the grains have been deformed according to their orientations and Schmid factors as if they were independent. The introduction of suitable dislocation arrays at the grain boundaries in part III allows them to restore material continuity (part IV). It is plausible to place the geometrically necessary dislocations mainly near the grain boundaries where the incompatibility stresses are strongest while the (not drawn) statistically stored dislocations will harden the crystals more or less homogeneously, depending on their orientations. In the following the inhomogeneity of the total dislocation distribution is no longer taken into account, which will lead to deviations, especially for small grain sizes, from the Hall–Petch and Taylor models to be discussed [12.25].

We consider the two influences mentioned at the beginning of this section consecutively. A dislocation in grain 1 moving towards the grain boundary does not usually find a plane matching its Burger vector in grain 2. Nor can it be absorbed by the high-angle grain boundary without seriously modifying the structure of the latter (section 3.2.2). Thus this dislocation and those following it on the same plane pile up at the grain boundary. Ultimately the stress concentration at the pile up is so large (section 11.2.3) that at a stress τ_0 dislocation sources in the neighbouring grain are activated. If λ is the distance of the source from the pile-up, then from (11-14)

$$(\tau_0 - \tau_1)\sqrt{\frac{d}{\lambda}} = \tau_1 m_{12}$$

or   (12-18)

$$\tau_0 = \tau_1 + m_{12}\tau_1\sqrt{\frac{\lambda}{d}} \equiv \tau_1 + k_y d^{-1/2}.$$

τ_1 is the critical shear stress in grain 1 and τ_1 is the minimum stress for activation in grain 2; m_{12} transforms the shear stress from the slip system in the first grain to that in the second. This relationship, known as the *Hall–Petch relationship*, which describes the dependence of the yield stress of polycrystals on grain size, has been confirmed for many metals. (If the mean free path and the pile-up length of the dislocations are smaller than the grain size then τ_0 should not depend on d.) If the mean distance between dislocations is $\lambda = 10^{-4}$ cm and $m_{12} \approx \frac{1}{2}$, the experimental value of k_y yields a very high value for $\tau_1 = G/120$ in the case of a 0.11 wt.% C steel. The source in the second grain is obviously pinned by carbon atoms.

12.5.2 *Changes in grain shape and work hardening in polycrystals* [12.9], [12.10]

In order to preserve the cohesion of the constituent grains a polycrystal subjected to an externally applied stress can undergo macroscopically homogeneous plastic deformation only if each of these grains is capable of a general plastic shape change. This means that the strain in each grain must conform to five independent components of the strain tensor ε_{iK}. (The sixth component is then fixed by the condition that the volume remains constant.) In 1928 R. von Mises realized that this necessitates the operation of five independent slip systems. (A slip system is independent if the shape change it produces cannot be achieved by a combination of other slip systems [12.7].) At low temperatures cph metals slip only on the basal plane which contains two linearly independent Burgers vectors. These metals should therefore exhibit very little plasticity in the polycrystalline state, as opposed to the single crystal state when they are very ductile. This is confirmed experimentally. Zinc, cadmium and similar polycrystals become ductile only at low temperatures by the activation of non-basal slip (apart from mechanical twinning, chapter 13). Fcc metals possess 12 $\{111\}\langle 110\rangle$ slip systems. Only two of thee three slip directions are, however, independent in each plane. Using the terminology of the Thompson tetrahedron, fig. 11.20, this means that there are four conditions for the Burgers vectors of a plane of the type

$$(DB)_\alpha + (BC)_\alpha + (CD)_\alpha = 0$$

in the slip planes α, β, γ and δ. In addition, shears of the type

$$(CD)_\alpha + (DC)_\beta + (AB)_\delta + (BA)_\gamma = 0$$

lead only to identical rotations of the lattice, in this case about the $\langle 100\rangle$ axis, as can be seen from the Thompson tetrahedron in fig. 11.20. The axis of rotation in the above example is in the line joining the tetrahedron edges AB and CD; there are three such conditions. There thus remain $12-7=5$ independent slip systems, sufficient to provide overall general plasticity of a grain of the fcc lattice, as evidenced by the good ductility of polycrystals. The same holds good for the bcc lattice.

If a polycrystalline tensile specimen is loaded, grains with a large Schmid factor begin to deform and exert stresses on other, less favourably oriented grains, which then themselves influence slip in the favourably oriented grains, section 12.5.1. In this way a general stress state is set up which actually permits five systems to become activated. E. Kröner describes the process in terms of the following model. All grains of one orientation are represented by an average spherical grain and its neighbours by an isotropic continuum. The spherical grain is removed and the specimen subjected to the macroscopic tensile stress $\bar{\sigma}$. Surface stresses must now be applied at the edge of the hole and of the grain (of opposite sign) in such a way that the surrounding material, under the mean stress $\bar{\sigma}$, is unaware that the grain has been removed. The grain and the matrix then deform plastically with the result that a deformation state $\bar{\varepsilon}$ is produced in the hole and ε in the spherical grain. If the grain is to be reintroduced it must first be deformed elastically; according to J. D. Eshelby [12.8] the internal stresses $(\sigma - \bar{\sigma})$ required for this are homogeneous throughout the grain and approximately equal to $G(\varepsilon - \bar{\varepsilon})$. It is precisely this stress $(\sigma - \bar{\sigma})$ which activates the five necessary slip systems in the grain. Its deformation ε is therefore similar to the macroscopic deformation $\bar{\varepsilon}$ (G. I. Taylor 1938) since the internal stresses attain only a fraction of G before deformation commences.

According to Taylor the selection of the five slip systems (out of the available 12 in fcc metals) necessary to produce $\bar{\varepsilon}$ follows the principle of the smallest algebraic total strain $a = \sum_{i=1}^{5} |a_i| = \min$, which corresponds to the principle of the least work of deformation on slip systems of comparable flow stress. The simultaneous operation of several slip systems leads to strong interaction between dislocations and work hardening as described in section 11.2.3. The stress–strain curves of polycrystals should therefore be correlated not with those of favourably oriented single crystals but with those of highly symmetrical orientations especially $\langle 111 \rangle$ and $\langle 100 \rangle$, fig. 12.14. These 'hard' single crystal stress–strain curves $\tau(a)$ should also express the flow stress as a function of the total strain in the polycrystal. This applies for the flow stress σ_x when the polycrystal is stressed in the x-direction as a function of the strain ε_x

$$\sigma_x = \frac{dA}{d\varepsilon_x} = \tau \frac{da}{d\varepsilon_x} \equiv M\tau.$$

The 'Taylor factor' is therefore

$$M \equiv \frac{\sigma_x}{\tau} = \frac{da}{d\varepsilon_x},$$

and the polycrystal work hardening

$$\frac{d\sigma_x}{d\varepsilon_x} = M^2 \frac{d\tau}{da} = M^2 \Theta. \qquad (12\text{-}19)$$

Taylor has calculated $M = 3.06$ for fcc metals for a random distribution of the grain orientations, which is in good agreement with measured work hardening curves, fig. 12.14. (The Taylor factor is thus larger than the mean reciprocal Schmid factor, see section 12.1.) In the absence of an initial linear stage, the critical flow stress of a polycrystal is usually defined as a 0.2% proof stress.

12.5.3   *Deformation textures* [12.11]

In principle, if the operative slip systems of a polycrystal are known, so also are the changes in orientation of the grains with deformation, (see section 12.1). Depending on the type of deformation and largely independent of the initial orientation, only very stable final grain orientations are observed after heavy deformation. The deformation thus imparts to the material a *texture* or preferred orientation, as observed in section 4.3 for solidification. In a tensile test on a favourably oriented fcc single crystal the alternate operation of primary and conjugate slip systems leads to a final orientation of $\langle 211 \rangle$ (section 12.1). In the case of compression the axis rotates towards the slip plane normal, ending in the case of double slip in an fcc metal half way between two $\langle 111 \rangle$ poles, i.e. $\langle 110 \rangle$. If even more systems are operative in tension, $\langle 111 \rangle$ and $\langle 100 \rangle$ are observed as final orientations but in compression $\langle 110 \rangle$ remains the only possibility. This is actually observed in fcc metals. Rolling, a process of technological importance, corresponds roughly to a superposition of the two processes: tension in the rolling direction and compression at right angles to the sheet. Correspondingly a $(011)[21\bar{1}]$ rolling texture is expected for fcc metals, i.e. an $[011]$ orientation of the normal to the sheet plane and a $[21\bar{1}]$ orientation in the rolling direction, if the tension activates only those slip systems which also reduce the sheet thickness.

We want now to compare these predictions with actual practice. First we must briefly discuss the determination of textures by X-rays and the representation of the results in *pole figures* [2.1]–[2.3], [12.11]. Fig. 12.15 shows a grain in a polycrystalline sheet at the centre of a unit sphere. The rolling, transverse and normal directions define a coordinate system relative to the specimen. We now look for the normals to the $\{100\}$ planes of all the grains in the sheet by irradiating it with monochromatic X-rays, wavelength λ, at varying angles α and measuring the diffracted intensity at angles $(\alpha + 2\theta_{\{100\}})$ where $\sin \theta_{\{100\}}$ is given by $\lambda/2d_{100}$, the Bragg condition for planes d_{100} apart. The $\langle 100 \rangle$ plane normals C thus obtained intersect the sphere and are projected stereographically on to the equatorial plane (C'). The result for a statistical distribution of orientations is shown in fig. 12.16(a). It should, however, be noted that: (1) It is not possible from this

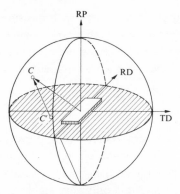

Fig. 12.15. (100) rolling texture in a sheet. RD = rolling direction, TD = transverse direction, RP = sheet plane normal. A $\{100\}$ plane with normal C reflects monochromatic X-radiation and is plotted stereographically in the pole figure as C'.

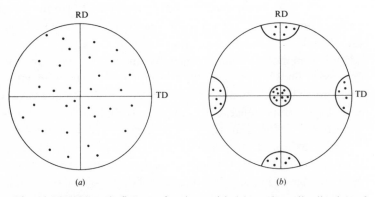

Fig. 12.16. (100) pole figures of a sheet with (*a*) random distribution of orientations, (*b*) cube texture.

representation to see which orientations belong to neighbouring grains. In this respect, therefore, the information is incomplete. (2) We could have plotted the $\{111\}$ plane normals instead, in which case the pole figures would have looked completely different although it contained the same information. Each pole figure must therefore be labelled with the appropriate $\{hkl\}$. (3) The representation of the tensile axis in the *stereographic standard triangle*, fig. 12.2, i.e. a (stereographically projected) *crystal related* coordinate system, is obviously the inverse of the *specimen related pole figure* although in principle it contains the same information. (4) The distribution of orientations is essentially independent of the shape of the grains.

In practice, with present-day automatic texture goniometers, the position of the specimen is moved relative to the primary beam rather than vice versa. In principle this provides a quantitative frequency distribution of the grain orientations [2.1].

More information on the texture is obtained from the three-dimensional *orientation distribution function* (ODF) $f(\alpha, \beta, \gamma)$ which gives the volume fraction of grains with angles α, β, γ relative to the specimen axes [12.11a]. The ODF follows from X-ray or electron diffraction or by inversion and superposition of a series of pole figures P_{hkl} for different h, k, l which are two-dimensional projections of f according to

$$P_{hkl}(\alpha, \beta) = \frac{1}{2\pi} \int f(\alpha, \beta, \gamma) \, \mathrm{d}\gamma.$$

If the sheet or wire has a texture, a pole figure such as that in fig. 12.16(b) is obtained instead of that in fig. 12.16(a). This example shows a 'cube texture' $(001)\langle 100 \rangle$ (which is highly desirable in Fe–3.5% Si, conventional transformer sheet, because it avoids magnetization losses). We see that the orientation distribution represents a spread about an ideal orientation. We have deduced this orientation above from the orientation change for infinite tensile or compressive deformation. Fig. 12.17 shows the $\{111\}$ pole figure for α-brass which has been rolled 95%. The $(011)\langle 21\bar{1} \rangle$ rolling texture mentioned above can be recognized in the most densely covered areas. A significant complication compared with the model in section 12.5.2 and which is probably responsible for some of the spread in the texture is the inhomogeneous nature of the microstructure. On deformation, the grains are split into differently oriented subgrains by the formation of various types of 'deformation bands'. These consist of low-angle grain boundaries [12.11]. True multiple slip as described in section 12.5.2 is no longer observed in each of these subgrains; instead only single or double slip occurs. This is more economical from the point of view of dislocation interactions and does not violate Taylor's condition for polycrystalline deformation on a macroscopic scale. A further complication in the orientation distribution of deformed metals with a low stacking fault energy is the occurrence of mechanical twinning, chapter 13, [12.11]. A deformation texture renders the material mechanically anisotropic because of its microstructure as well as because of the crystal structure of each of its grains. This has undesirable consequences in the deep drawing of sheet for example when different reductions in thickness cause 'earing', see fig. 12.18 and [12.11].

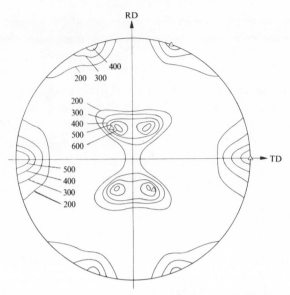

Fig. 12.17. (111) pole figure of 95% rolled Cu–30% Zn. Lines of equal X-ray intensity in arbitrary units. After P. A. Beck and co-workers, see [2.1].

Fig. 12.18. Cups deep-drawn from Al sheet with different textures (left: four ears at 45° to the rolling direction; middle: no earing; right four ears at 0 and 90° to the rolling direction [12.11]).

12.6   Grain boundary sliding and superplasticity

12.6.1   *Homogeneous grain boundary sliding*

At high temperatures $(T > 2/3T_m)$ a polycrystal can deform by (high-angle) grain boundary sliding, a phenomenon which can have deleterious effects, for example on the stability of electric light bulb filaments. While small grains mean high strength at low temperatures (see section 12.5.1) the opposite is true at high T. On the other hand the grain boundaries follow a zig-zag path through the specimen, as shown for example in fig. 12.19. Unless the stress amplitude is very small, stress

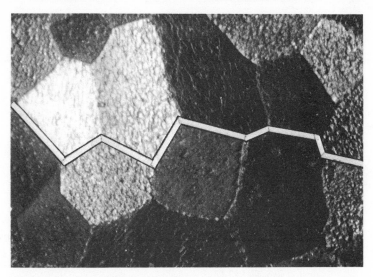

Fig. 12.19. A (horizontal) shear stress causes a polycrystal to slide on a non-planar set of grain boundaries.

concentrations are produced at the corners of the grains during grain boundary sliding. Displacements u greater than 10^{-6} cm are possible only if these stress concentrations can be relieved by transport of material, that is diffusion. R. Raj and M. Ashby [12.12] have developed a good model to describe this process. The grain boundary configuration is represented by a two-dimensional hexagonal network, fig. 12.20. The profile of a displacement surface (e.g. mode 1) is expanded into a Fourier series. The first Fourier component is described by its wavelength λ and amplitude h. If the grains on this surface are displaced by a shear stress τ then normal stresses shown in fig. 12.21 are induced,

$$\sigma_n = -2\frac{\tau\lambda}{\pi h}\sin\frac{2\pi y}{\lambda}, \qquad (12\text{-}20)$$

resulting in an additional chemical potential $\Delta\mu = \sigma_n\Omega$ for the vacancies. The vacancies migrate down this gradient from the zones of dilatation to the zones of compression at the boundary, and it is this process which determines the displacement velocity $\dot{u}$ at the grain boundary [12.12]

$$\dot{u} = \frac{8}{\pi}\frac{\tau\Omega}{kT}\frac{\lambda}{h^2}D_L\left\{1 + \frac{\pi\delta}{\lambda}\frac{D_G}{D_L}\right\}. \qquad (12\text{-}21)$$

Both lattice diffusion (D_L) and grain boundary diffusion (D_G) (grain boundary thickness δ, see section 8.4.1) have been taken into account. Both

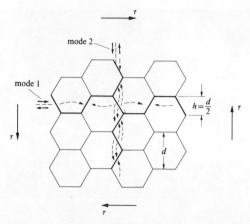

Fig. 12.20. Idealized model of a polycrystal as a hexagonal network of grain boundaries permitting grain boundary sliding according to two orthogonal modes. The accompanying vacancy flux is shown by the dotted arrows [12.12].

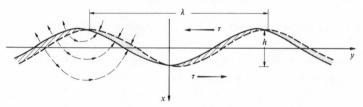

Fig. 12.21. Sliding on a wavy grain boundary leads to the formation of a cavity (cross-hatched) which is transported as a vacancy flux into the overlap zones (dotted) [12.12].

processes have been described by F. Nabarro, C. Herring and R. Coble individually in a completely different way but with almost identical results. These authors calculated the change in shape of a grain by a flow of vacancies due to an external stress, see fig. 12.20, i.e. 'diffusion creep' by lattice or grain boundary diffusion. $\dot{u}$ depends on the waviness of the grain boundaries, i.e. the shape of the grain, and is proportional to λ/h^2 for lattice diffusion and to δ/h^2 for grain boundary diffusion. The latter predominates at small λ and T. The larger the 'grain shape factor' (h/λ) the smaller is the contribution made by grain boundary sliding to the creep rate $\dot{a}_G = \dot{u}/h$. Tungsten light bulb filaments retain their shape if the grains are drawn out into fibres. Relatively large precipitates also increase the waviness of the grain boundary and decrease $\dot{a}_G$. Unlike the grain boundary viscosity,

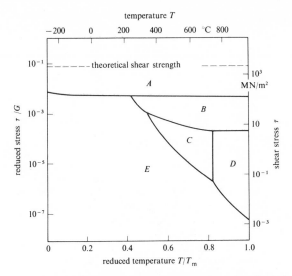

Fig. 12.22. Deformation mechanisms for silver (grain size 32 μm, $\dot{\varepsilon} = 10^{-8}$ sec^{-1}). A: dislocation glide; B: dislocation climb; C: grain boundary sliding determined by grain boundary diffusion; D: by bulk diffusion; E: elastic deformation [12.13].

which describes the reaction of the grain boundary to short time loading, $\dot{a}_{\mathrm{G}}$ is not an inherent property of the boundary.

A comparison of $\dot{a}_{\mathrm{G}}$ (12-21) with the strain rate by dislocation climb $\dot{a}_{\mathrm{S}}$ (12-17) or slip (for the special case of the Peierls mechanism, (12-7) or dislocation intersection, (12-11)) shows which process dominates in the particular range of stress τ and temperature T (with the grain size d and $\dot{a}$ as parameters). From this information Ashby has constructed 'deformation mechanism maps', fig. 12.22, which can vary from metal to metal not qualitatively, but in a quantitative and informative fashion [12.13], [12.14a], [12.14b].

12.6.2 *Superplasticity* [12.14]

Very fine grained metals and alloys (grain size several μm), in particular two-phase eutectic or eutectoid microstructures, often exhibit unusually large, homogeneous strains up to fracture (100–1000%) at elevated temperatures if deformed in a characteristic strain rate range $\dot{\varepsilon}_{\mathrm{SP}}$, (see section 2.6.1). This *superplastic* behaviour permits metals to be blown like glass, i.e. easily formed into complicated shapes. Naturally grain boundary sliding must be involved but according to section 12.6.1, $\dot{\varepsilon}_{\mathrm{G}}$ is an order of magnitude smaller than $\dot{\varepsilon}_{\mathrm{SP}}$. $\dot{\varepsilon}_{\mathrm{G}}$, however, depends on the stress in

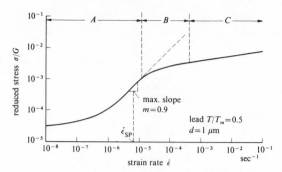

Fig. 12.23. Creep rate in fine-grained lead at half the melting point: mainly grain boundary sliding at low stresses (A), dislocation climb at high stresses (C). Superposition of both mechanisms in region B, superplasticity is observed at the maximum slope (d ln σ/d ln ἐ) [12.15].

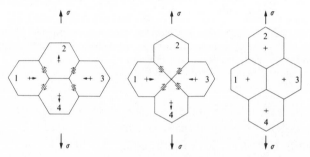

Fig. 12.24. Ashby–Verrall process of grain boundary sliding and relative displacement of four grains under tension [12.15].

the 'correct' manner which, as shown in (2-26), permits stable deformation without work hardening in a tensile test: $m'_{SP} \equiv (\mathrm{d} \ln \sigma/\mathrm{d} \ln \dot{\varepsilon}) \approx 1$. Fig. 12.23 shows a typical curve (ln σ) against (ln $\dot{\varepsilon}$) for these materials and conditions. At $\dot{\varepsilon}_{SP}$ superplastic behaviour is observed. No elongation of the grains is visible in the tensile direction as occurs in the tensile deformation of polycrystals at low temperatures (section 12.5.2) but rather a mutual displacement of the grains, which *acquire new neighbours* [12.15]. Fig. 12.24 shows the corresponding Ashby model. In this model material is again transported by diffusion but the diffusing volume is much smaller and the diffusion paths much shorter than in homogeneous grain boundary sliding. An estimate of $\dot{\varepsilon}_{SP}$ for Ashby's inhomogeneous flow gives $\dot{\varepsilon}_{SP} \approx 7\dot{\varepsilon}_G$, showing the same dependence on T, σ and grain size d [12.15] but at much larger strains. At high τ and $\dot{\varepsilon}$, dislocation climb occurs and homogeneous extension to fracture once more decreases (fig. 12.23).

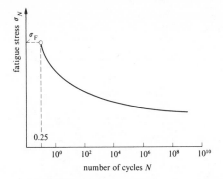

Fig. 12.25. Typical SN-curve gives the stress necessary for fracture after N cycles.

12.7 Cyclic deformation and fatigue

The ductile fracture following necking, which concludes the tensile test described above, is far less important technically than the *fatigue fracture* observed after prolonged cyclic loading, for example of rotating or vibrating machine parts, at $\sigma_N = \sigma_F/4$ to $\sigma_F/2$, where σ_F is the tensile stress for ductile fracture. Fatigue behaviour is often studied in symmetrical push–pull, tension–compression cycling with a constant stress amplitude σ. The 'SN-curve' (Wöhler curve), fig. 12.25, shows the stress amplitude σ_N plotted against $\log N$, where N is the number of cycles to fracture. For $N = \frac{1}{4}$, i.e. at the peak of the *tensile loading part* of the cycle, the stress for fracture is σ_F, that for the tensile test; on the plateau between $N = 10^4$ and 10^8, typical fatigue processes are responsible for failure. Between $N = \frac{1}{4}$ and 10^4 lies the transition region of low cycle fatigue. The SN-curve is not strongly dependent on temperature. Understanding the physical background to the typical fatigue failure demands considerable effort and many additional observations [12.16]–[12.18]. It is, however, significant that *plastic* deformation occurs during the N cycles and furthermore that fracture is nearly always initiated at the *surface*. Hence the condition of the surface and its chemical environment are very important in the control of fatigue in practice.

The starting point for an interpretation of fatigue in terms of physical metallurgy is the shape of the stress–strain curve after a reversal of the stress. Fig. 12.26 shows that the reverse deformation by shear on the basal plane of zinc starts at much lower stresses than that at which the forward deformation ended. This *Bauschinger effect* is also observed in fcc metals at all points along the stress–strain curve. It proves that the dislocation arrangement responsible for tensile work hardening is not very stable to

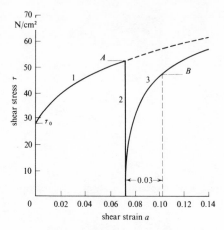

Fig. 12.26. Cyclic shear deformation of a zinc single crystal at 77 K on the basal plane: (1) loading, (2) unloading, (3) reverse loading. $(A - B)/A$ defines the Bauschinger effect.

stress reversal, as would be expected from the Seeger work hardening model (section 12.3), (but not from the forest model). After many cycles a new dislocation arrangement is formed, which consists essentially of dislocation dipoles and multipoles in which the positive and negative dislocations lock each other until the stress reaches the order of the passing stress τ_p, section 11.2.1. This arrangement generates less internal stress per unit strain than monotonic deformation. The fatigue hardening is relatively slight and leads to a limiting value of the flow stress in tests conducted at constant *plastic* strain amplitude ε_{pl}. A plot of the limiting stress vs the plastic strain amplitude ε_{pl} is called the *cyclic stress–strain curve*. It often shows a plateau during which a more favourable dislocation arrangement forms by cross-slip or climb and this state spreads along the specimen. In so-called *persistent slip bands* (psb) sharp dislocation walls form; further deformation occurs by dislocation movement within the cells between the walls. The psb's are softer than the surrounding matrix and finally fill the whole specimen [12.18a], [12.18b]. Cyclic deformation at constant ε_{pl} or the equivalent slow increase in applied stress avoids large, sudden plastic strain amplitudes in the first few cycles. At high strain rates it is difficult for the dislocations to form minimum energy configurations such as dipoles and multipoles. The quasi-equilibrium nature of the dislocation arrangement produced by a slow build-up of stress in the fatigue test is illustrated by the fact that monotonically deformed metals *soften* to the same stress as previously undeformed specimens work harden (see work softening in

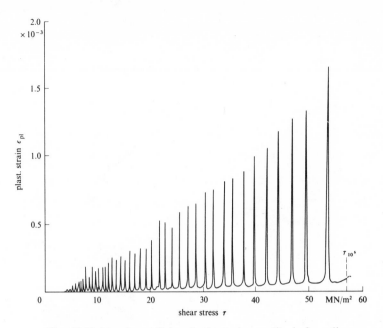

Fig. 12.27. Envelope of the plastic strain amplitude for a linear increase in stress in the course of 23 170 cycles to $\tau_{10^6} = 57\,\mathrm{N/mm^2}$ for an Al single crystal [12.19].

section 12.4). During the process of hardening (or softening) the dislocation configuration repeatedly becomes unstable. Dipoles break up and new ones are formed with a passing stress corresponding to the new, increased applied stress. The plastic strain increases dramatically for a short period. Fig. 12.27 shows the envelope of the plastic cyclic strain for a linear increase in stress amplitude during 23 170 cycles to the value of the fracture stress at 10^6 cycles of $\tau_{10^6} = 57\,\mathrm{N/mm^2}$ for an aluminium single crystal [12.19]. If the stress amplitude is built up more rapidly, the dislocation configuration does not stabilize between each strain burst but many dislocations leave the crystal at the surface in *coarse slip steps* (height several 100 nm). These seem to provide the decisive stress concentrations necessary for the formation of fatigue cracks. If dislocations with **b** parallel to the surface leave the crystal or if the slip lines are regularly polished off, no fatigue cracks are produced. Further deformation is concentrated in these coarse slip planes, in which atomic defects also accumulate, with the result that 'persistent slip bands' cannot be removed completely by chemical polishing. Occasionally lamellar extrusions will form there.

P. Neumann [12.9] describes the formation of a crack on a coarse slip

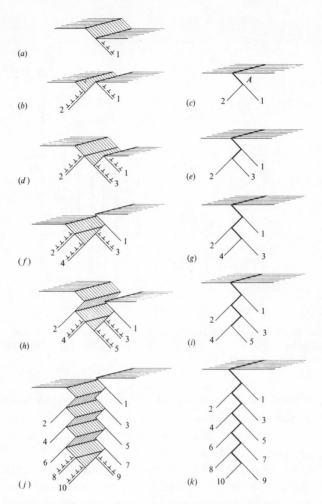

Fig. 12.28. P. Neumann's model [12.19] for the formation of a crack at coarse surface steps under symmetrical push–pull loading. Left: tensile stages, right: unloading stages.

surface as shown in fig. 12.28. The slip plane (1) is work hardened below the slip step. In the next tensile phase, a slip plane (2) experiences a high shear stress at a sharp surface edge and becomes operative (fig. 12.28(b)). On relaxing (fig. 12.28(c)), planes (1) and (2) slip back, but not necessarily in the reverse order. In addition to other irreversible effects the surface A may have become contaminated with oxygen preventing complete welding of the surface at A. In the next tensile phase the situation shown in (d) is obtained, etc. The Neumann model explains how an asymmetric

Fig. 12.29. The side face of a fatigued copper single crystal shows traces of two sets of slip lines round a crack which has propagated from a rounded notch to the undeformed (black) triangle below (P. Neumann). 17×.

propagation of the crack is possible under symmetrical push–pull loading. (In pulsed compression testing no fatigue failure is observed.) In ultra-high vacuum, crack formation is delayed because contact welding of the surface at A will sometimes occur. The necessary conditions for this mechanism to operate are coarse slip, i.e. high slip steps, and two sets of slip planes. Fine slip blunts the crack front and the stress concentration disappears. Both conditions appear to be characteristic of fatigue fracture. No fatigue cracks are observed in zinc, when it deforms by basal slip at low temperatures. Effects which coarsen slip in monotonic deformation such as cross-slip (i.e. high stacking fault energies), neutron irradiation, formation of solid solutions, precipitation of coherent particles (see chapter 14), etc., facilitate the formation of fatigue cracks. An oxide layer on aluminium and an unfavourable orientation lead to finer slip thus increasing the fatigue stress σ_N. The Neumann model is supported by a large number of metallographic observations [12.19]. For example the crack front in copper single crystals is parallel to the line of intersection of two $\{111\}$ slip planes (fig. 12.29).

The further growth of a crack is determined by the macroscopic stress distribution around the crack and takes place in two stages [12.18]. In stage I (crack length $a \ll$ specimen diameter), the crack grows very slowly and approximately in the plane of the largest shear stress, i.e. at 45° to the push–pull axis. In stage II the crack is at right angles to this axis. Propagation occurs by a mechanism similar to that in fig. 12.28 in which the two sides of the crack rest against one another during the compression half-cycle, whereas in the tensile half-cycle, slip alternating between two systems at the crack tip extends the crack of initial length l.

A crack subject to a stress σ is equivalent to a dislocation pile-up of length l and produces a stress concentration proportional to $\sigma \sqrt{l}$ at its ends, see section 11.2.3. It is understandable therefore that the crack propagation rate dl/dN (in stage II about 1 μm/cycle) can be described empirically as a function of $\sigma \sqrt{l}$ (proportional to the fourth power) [12.18]. In the range $N < 10^4$, Coffin's law is obeyed, $\varepsilon_{pl}^2 N = $ const for constant strain amplitude experiments.

12.8   **Fracture at small tensile strains – 'brittle fracture'**
[12.20], [12.22]
    In addition to the metals, mainly fcc, which undergo considerable plastic deformation and necking prior to fracture, there are others, for example bcc and cph, which break after very little tensile deformation at low temperatures or high strain rates. This is termed brittle fracture although, certainly in the case of metals, fracture is preceded by some plastic deformation. This is demonstrated by the fact that at 77 K and dependent on the grain size (12-18) a coarse-grained carbon steel fractures in tension at the stress at which it begins to deform plastically in compression, fig. 12.30. Various methods are used to observe the transition from brittle to ductile behaviour at a particular transition temperature T_{trans} (or transition strain rate, or transition grain size). Firstly, examination of a *fracture surface* ('fractography') produced at low temperatures reveals a smooth, 'crystalline' appearance; the brittle fracture occurs in the plane of greatest shear stress running straight through the grains. At higher temperatures the fracture surface is found in the necked region at right angles to the tensile axis and has a fibrous structure. Secondly, the energy A absorbed by a standard notched specimen on fracturing is measured in an impact test by the height to which a pendulum rises after fracturing the specimen. The specimen absorbs energy only during the plastic deformation preceding fracture. $A(T)$ shows a sharp drop to low temperatures at the transition from tough to brittle fracture, fig. 12.31. Thirdly, the upper yield point curve $\sigma_{uy}(T)$ intersects the fracture stress–temperature curve at the transition

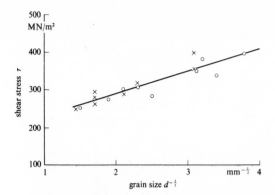

Fig. 12.30. Fracture stress in tension (×) and yield stress in compression (○) for carbon steel at 77 K (J. R. Low, Schenectady).

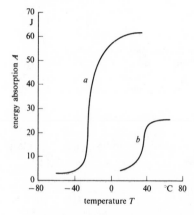

Fig. 12.31. Energy absorbed at fracture as a function of temperature for carbon steel (*a*). After irradiation by 1.9×10^{19} n/cm^2 the transition to brittle fracture (reduced absorption) occurs at a higher temperature (A. H. Cottrell, AERE Harwell) (*b*).

temperature (not necessarily the same value as in the first two examples) and the strain to fracture falls to zero. (In fcc metals the flow stress is less dependent on temperature than in bcc metals and the ductility is greater.) Transition temperatures $T_{\text{trans}} = (-60 \text{ to } +40)\,^\circ\text{C}$ are typical for steels and dangerously close to room temperature. The question of whether initiation or propagation of cracks is the determining stage remained unanswered for some time. Obviously crack nuclei are often formed by plastic deformation, for example by one of the dislocation mechanisms depicted in fig. 12.32. In example (*b*) the dislocations actually attract one another since $(a/2)[111] +$

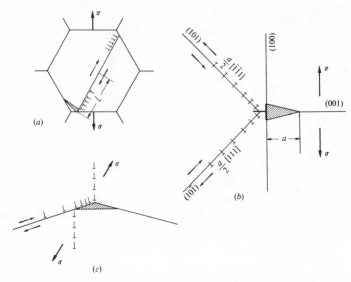

Fig. 12.32. Mechanisms for crack initiation from groups of dislocations. (a) Pile-up at a grain boundary (C. Zener); (b) attractive dislocation reaction (A. H. Cottrell); (c) shear in a low-angle grain boundary (A. N. Stroh). Specimen thickness t perpendicular to the plane of the drawing.

$(a/2)[111] = a[001]$, corresponding to the observed (001) fracture plane of ferrite. In all examples the elastic distortion energy is converted to surface energy when the crack opens. Microcracks incapable of growth are often observed in worked steel plates but do not impair the performance of the bow of a ship, a pipeline or a reactor pressure vessel! The dislocation configurations are produced solely by the action of shear stresses although investigations by P. W. Bridgman have shown that hydrostatic pressure strongly suppresses the ductile–brittle transformation, i.e. promotes ductility. This led A. H. Cottrell [12.21] to the currently accepted conclusion that brittle fracture is controlled by the ability of the cracks to grow. As early as 1920 A. Griffith estimated the saving in elastic distortion energy in a rod of thickness t resulting from the growth of a lens-shaped crack of diameter $2a$ in the plane at right angles to an applied stress σ. The surface energy of the crack is $2\tilde{E}_s \pi a^2$. The distortions ahead of the crack tip are of the order $\sigma/\hat{E}$ and extend over distances of the order of a. Thus the opening of the crack in the distorted material produces an energy change

$$\Delta E = 2\tilde{E}_s \pi a^2 - \frac{\sigma^2}{2\hat{E}} \tfrac{4}{3}\pi a^3. \tag{12-22a}$$

At fracture

$$\frac{d\Delta E}{da} = 0 \quad \text{and} \quad \sigma_F = \sqrt{\left(2\frac{\tilde{E}_S \hat{E}}{a_F}\right)}. \tag{12-22b}$$

Using experimental values for a steel ($\sigma_F = 700\,\text{N/mm}^2$, $\tilde{E}_S = 1.2\,\text{J/m}^2$) a value of $a_F \approx 1\,\mu\text{m}$ is obtained. Even larger crack nuclei are necessary if *plastic work* is expended at the crack tip in which case $\tilde{E}_S$ increases by a factor 100 to 1000. The importance of the parameter $K_c = \sigma_F\sqrt{a}$, the *fracture toughness*, in determining the ability of cracks to propagate (cf. end of section 12.7) can be seen from (12-22b). K_c is used in fracture mechanics to judge the fracture strength of various constructions [12.23a], [12.23b]. The Griffith fracture stress in particular does not depend sufficiently on the temperature and microstructure to explain the observed variation of the brittle–ductile transition in metals. The difficulty is not resolved by an atomistic treatment of the crack tip which avoids the singularity of the elastic solution (11-14) for the distortion at the crack tip and yields a type of Peierls force for the crack propagation.

The energy balance for the formation of the Cottrell crack nucleus (in a plate of thickness t) in fig. 12.32(b) has the form

$$\Delta E = \frac{G(nb)^2 t}{4\pi(1-v)}\ln\frac{t}{a} + 2\tilde{E}_S at$$

$$-\frac{\sigma^2(1-v)}{8G}\pi a^2 t - \frac{\sigma(nb)at}{2}. \tag{12-23}$$

The first term is the energy of the superdislocation which is being formed, the second is the surface energy, the third is the distortion energy saved on formation of the crack and the fourth is the work done in opening the crack. n is the number of dislocations which can pile up in a grain of diameter d under a stress σ, (12-18). For equilibrium it is again necessary that $d\Delta E/da = 0$. Equation (12-23) is thus transformed to a quadratic equation in a. The smaller solution a_1 corresponds to a stable crack, the larger a_2 to an unstable crack. $a_1 = a_2$ is the critical condition for the fracture stress. Together with (12-18), this condition means that

$$2\tilde{E}_S = \sigma_F bn = \sigma_F \frac{(\sigma_F - \sigma_1)d}{2G}. \tag{12-24}$$

The solution is plotted in fig. 12.33 together with the yield stress σ_0 given in (12-18) (but not resolved on to the slip plane). There is in fact one point of intersection for finite d, i.e. a transition grain size d_{trans} above which $\sigma_F < \sigma_0$,

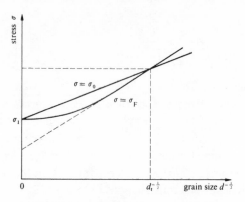

Fig. 12.33. Fracture stress σ_F and yield stress σ_0 as a function of the grain size d according to equations (12-24) and (12-18).

i.e. fracture occurs before yield. With decreasing temperature the friction stress for dislocation movement in the individual grains σ_1 (section 12.5) and the slope k_y in (12-18) become larger, i.e. the range of brittle fracture is extended to smaller grain size. An increase in k_y means (12-18) an increase in the stress σ_l to activate the sources. This can result from the pinning of dislocations by precipitated carbon or nitrogen or from neutron irradiation (fig. 12.31) as well as from a reduction in the temperature. A low stacking fault energy also leads to a stronger concentration of slip, higher slip steps, a larger n in (12-24) and hence to a lower σ_F and easier fracture. In metallurgical practice, emphasis is placed on reducing the grain size (grain refinement) and lowering the flow stress as a means of reducing the tendency to brittle fracture. The Cottrell theory of the ductile–brittle transition is confirmed by practice but it is still quantitatively unsatisfactory [12.20].

The limiting velocity of crack propagation is the speed of sound. If crack propagation is accompanied by plastic deformation, $\tilde{E}_{S,eff}$ is greater than $\tilde{E}_S$ and the dislocation velocity in the neighbourhood of the crack tip (section 11.4.1) competes with the crack velocity which then assumes much lower values (of the order of m/sec) [12.21].

13
Martensitic transformations

13.1 Mechanical twinning [13.1], [9.6]

Instead of deforming inhomogeneously by *slip* on relatively few slip planes (chapter 11) a crystal subjected to an externally applied stress can deform homogeneously in shear by the formation of twins. *Twinning* reproduces the initial structure of a crystal but changes its orientation. The *martensitic transformation*, however, a shear process very similar to twinning and the main subject of this chapter, transforms the crystal into a crystallographically different structure. It is usually initiated by a change in temperature, as has already been mentioned in section 6.1.2 when the Fe–C system was discussed. The relationship between the two shear processes is best illustrated by considering the fcc structure in which the stacking order of {111} planes is *ABCABC* (fig. 6.11). If a stacking fault (section 6.2.1) is introduced successively into each plane, or in other words if a Shockley partial dislocation $(a/6)\langle 112\rangle$ traverses each {111} plane (11.3.3.1) the sequence

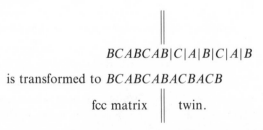

$$BCABCAB|C|A|B|C|A|B$$

is transformed to $BCABCABACBACB$

fcc matrix ‖ twin.

A mirror image of the stacking sequence of {111} planes has been produced relative to the first unsheared plane (*B*). This mirror symmetry is characteristic of twinning, fig. 13.1. We have already met the coherent twin boundary (the mirror plane) as a low energy, high-angle grain boundary in fig. 3.9. An incoherent twin boundary consists of all the Shockley partial dislocations which produce the twinning shear. If these dislocations are allowed to move only over every second {111} plane of the fcc lattice, a

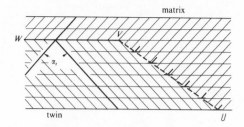

Fig. 13.1. Coherent (WV) and incoherent (VU) twin boundaries, the latter containing partial dislocations. The incoherent twin boundary (VU) being composed of like partials $(a/6)[112]$ produces a long-range stress. A. Sleeswyk (*Acta Met.*, **10** (1962), 705) has proposed (specifically for bcc twins) that glide dislocations should be incorporated into the boundary to relieve these stresses. In our fcc case the addition of two complete dislocations $(a/2)[0\bar{1}1]+(a/2)[\bar{1}0\bar{1}]$ on every third plane in VU relieves the stress of three $(a/6)[112]$ partials. (See also S. Mahajan and D. F. Williams, *Int. Met. Rev.*, **18** (1973), 43.)

special martensitic transformation fcc → cph takes place such as is observed in cobalt when it is cooled below 420 °C.

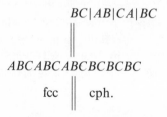

How can such coordinated dislocation movement in neighbouring {111} planes be possible? Certainly Shockley partial dislocations are formed by the splitting of perfect dislocations (as described in section 11.3.3.1) but not on (almost) every {111} plane. It is more likely that a relatively small number of partial dislocations move through the neighbouring {111} planes one after the other. A screw dislocation at right angles to the partial dislocation acts as a 'spiral staircase' by means of which the partial dislocation can move onto the next or next but one {111} plane. For this the screw dislocation must possess a component $\mathbf{b}_\perp = (a/3)\langle 111\rangle$ or $(2a/3)\langle 111\rangle$ at right angles to the twinning or martensite plane and also be a perfect dislocation of the matrix and of the twin [13.1*a*]. This is the principle of the 'pole mechanism', fig. 13.2. This mechanism for twin or martensite formation does not, however, operate as simply as this because after one half-cycle the arm of the partial dislocations running up the

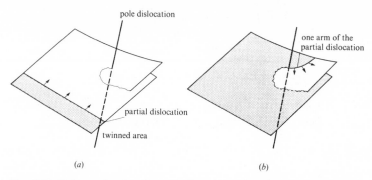

Fig. 13.2. Pole mechanism for twinning. A partial dislocation spirals up a pole (screw) dislocation converting one plane after another into the twin orientation.

staircase and the other arm running down are only a *few atoms apart* and moving in opposite directions. The passing stress is of the order of $G/20$ ($\approx 2500\,\mathrm{N/mm^2}$ for Cu), section 11.2.3, and thus much higher than the twinning stress τ_t (for Cu $150\,\mathrm{N/mm^2}$). Naturally, since twinning is generally observed to be preceded by slip, piled-up dislocations can provide a stress concentration. A flow stress of $150\,\mathrm{N/mm^2}$ is attained in copper at 4 K only after heavy deformation. The smaller the stacking fault energy, the easier is the nucleation of fcc twins by the dissociation of perfect dislocations [13.2]. After nucleation, the twin propagates across macroscopic distances under a decreasing load at almost the speed of sound. The participating partial dislocations possess such a large kinetic energy (11-21) that they are able to pass each other dynamically at closer range than would be possible under quasi-static conditions. The same applies incidentally to the cobalt transformation [13.3]. In zinc-rich α-brass, twin lamellae form very easily on account of the wide dislocation splitting, but they hardly grow. This is a result of their interaction with the forest dislocations which are also widely dissociated, see section 12.2.

Twinning plays an essential role also in the bcc and cph lattices, e.g. in the low temperature deformation of α-iron and in zinc subjected to a mechanical stress which cannot be relieved by basal slip. The twinning elements are given in table 13.1.

The shears associated with twinning are of the order of magnitude unity, yet the volume fraction of twins present is generally small. It is the resulting orientation change therefore which is important, rather than their contribution to shear. According to G. Wassermann twinning is of significance in the formation of fcc deformation textures. If the shear stress

Table 13.1. *Twinning elements*

Structure	Twinning plane	Twinning direction	Twinning shear α_t
fcc	$\{111\}$	$\langle 112 \rangle$	$1/\sqrt{2}$
cph	$\{10\bar{1}2\}$	$\langle 10\bar{1}1 \rangle$	$((c/a)^2 - 3)/(c/a)\sqrt{3}$
bcc	$\{112\}$	$\langle 111 \rangle$	$1/\sqrt{2}$

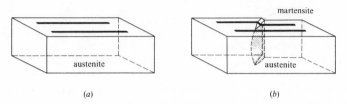

(a)   (b)

Fig. 13.3. Scratches on the surface (*a*) are sheared in the martensitic transformation (*b*) resulting in surface relief.

on the basal plane in cph metals is low, twinning is favoured. This brings the basal slip plane into a more favourable orientation [12.1].

13.2   Characterization of martensitic transformations [13.4], [9.6]

As a generalization of the above discussion on twinning and the cobalt fcc–cph transformation, the martensitic transformation can be characterized by the following features (in contrast to the diffusion controlled transformation, chapter 9).

(*a*) In addition to a volume change, the lattice deformation produces a *shape change* in the transforming region. This produces a relief or tilt on a polished surface, figs. 13.3 and 13.4, and distortion of the surrounding matrix (which constrains the martensite to form as plates or needles).

(*b*) A martensite needle or plate generally possesses an *internal structure*, which is the result of slip or twinning (see fig. 13.4). These two 'lattice invariant' shears are an integral part of the transformation. They largely compensate the distortion of the matrix associated with the lattice deformation, see fig. 13.5.

(*c*) The *total* deformation concomitant with the martensite transformation corresponds essentially to a shear parallel to an undistorted plane (in Co $\{111\}_{fcc} = \{0001\}_{cph}$), common to both the matrix and martensite phases, the so-called *habit plane*. (The definition of the habit

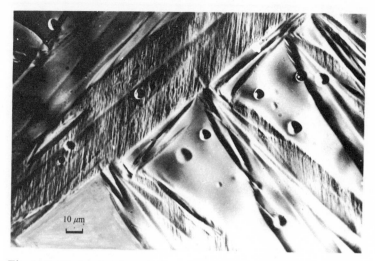

Fig. 13.4. Martensite plates in Fe–33.2% Ni with internal twinning (C. M. Wayman, Univ. Illinois).

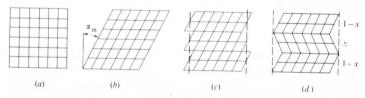

Fig. 13.5. Martensitic transformation (a) to (d) of a crystal region. Its external shape can be restored approximately by slip (c) or twinning (d).

plane for martensite regions with non-planar boundaries is given later.) In the case of Fe–C martensite fcc(γ) → b.c.tetr.(α'), which gave its name to this whole class of transformations, the habit plane is only macroscopically undistorted and generally irrational. It is a semi-coherent, highly mobile phase boundary.

(d) There are generally precise *orientation relationships* between the matrix and martensite lattices. A close packed plane in the matrix is parallel to a similar plane in the martensite; likewise a close packed direction. Since there are usually several such elements in the matrix a whole range of differently oriented martensite crystals can arise from a single matrix crystal.

(e) The distances between the atoms do not change significantly in the martensite transformation. In other words the relative movement of the

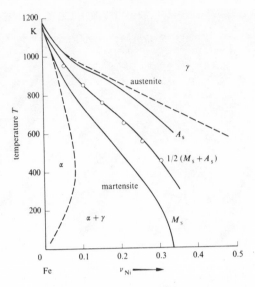

Fig. 13.6. Temperatures for the onset of the martensite transformation (M_s) and the reverse transformation (A_s) in the Fe–Ni phase diagram (dashed curves).

atoms is very small and less than the atomic separation (in contrast to the diffusion controlled transformation!). The nearest neighbours remain the same and short-range order in the matrix is retained in the martensite.

Some of the above criteria also apply to diffusion controlled transformations (see chapter 9, e.g. orientation relationships, habit plane), but not (a) and (e). The diffusion controlled transformation has a more 'civilian' character (individual atomic movement) in contrast to the 'military' nature of the martensite transformation (cooperative atomic movements). The *kinetics* of the two types of transformation are usually considerably different. The diffusional transformation is thermally activated and proceeds isothermally according to well-defined time laws (chapter 9). The martensite transformation occurs 'instantaneously' (i.e. almost with the speed of sound c_t) when a change in temperature causes a change in driving force, see fig. 9.23. In many cases, however, thermal activation can assist nucleation or movement of the martensite boundary with the result that the martensite reaction can also proceed isothermally (with $v \ll c_t$).

Fig. 13.6 shows the temperature M_s for the onset of spontaneous martensite formation on cooling Fe–Ni alloys, together with A_s, the temperature at which austenite forms spontaneously on reheating. The hysteresis illustrates the large expenditure of distortional energy required in the two processes. If the material is plastically deformed by external forces during cooling or heating the transformations start at the temperatures M_d, A_d. These lie closer together than M_s, A_s and also closer to the estimated equilibrium temperature for $\Delta F_{\alpha'\gamma} = 0$. This is indicated in fig. 13.6 by the approximation $T_{eq} \approx (M_s + A_s)/2$. In Fe–C alloys reheating causes the martensite to decompose into the equilibrium phases $\alpha + Fe_3C$ instead of transforming back into austenite, fig. 6.4. The fraction transformed increases with the degree of undercooling (or overheating) until the reaction ceases at M_f, A_f, see the TTT diagram shown in fig. 9.23.

There are alloys such as AuCd, Fe_3Pt, NiAl which show very little hysteresis in the martensite transformation ('thermoelastic martensite'). These alloys are characterized by long-range order and a high critical shear stress for plastic deformation by dislocation movement. Twinning within the martensite plate, a 'lattice invariant' shear process, serves to compensate in part for the lattice distortion produced by the martensitic transformation. The combination of thermoelastic transformation behaviour, long-range order and internal twinning appears to be responsible for a curious effect, which has considerable technological possibilities, the *shape memory effect* (called 'Marmem' because *mar*tensite re*mem*bers the shape of the specimen) [13.9]. If a specimen which has undergone the transformation to martensite is deformed while still at the low temperature it reassumes its original shape when it is reheated to temperatures above that for the reverse transformation. The explanation for this is that the low temperature deformation takes place by the movement of twin and habit planes in such a way that the orientation variant with the largest resulting strain is preferred. As a limiting case a single martensite crystal is produced which on heating above A_s reverts to the original austenite crystal. According to the lower symmetry of the martensite relative to the austenite structure there will be no manifold orientations as a result of reversion [13.14]. The Marmem alloys also display rubberlike 'ferroelastic' properties if lightly stressed in the martensite region, which also stem from the reversible movement of twin and phase boundaries. On deformation above M_s the material transforms, again showing large reversible strains.

13.3   **Crystallography of martensitic transformations** [13.4], [13.5]

In iron alloys an X-ray investigation reveals the following orientation relationships for the $\left\{ \begin{array}{c} \gamma \to \alpha' \\ A \to M \end{array} \right\}$ martensitic transformation (section 13.2(d))

$$\text{Fe–1.4\% C:} \quad (111)_\gamma \| (110)_{\alpha'} ; \quad [1\bar{1}0]_\gamma \| [1\bar{1}1]_{\alpha'} ,$$

after Kurdjumov and Sachs,

$$\text{Fe–30\% Ni:} \quad (111)_\gamma \| (110)_{\alpha'} ; \quad [\bar{2}11]_\gamma \| [1\bar{1}0]_{\alpha'} ,$$

after Nishiyama and Wassermann.

The habit plane, which in non-planar martensite plates is defined as the centre plane (midrib), is $\{225\}_\gamma$ in the 1.4% C alloy and $\{259\}_\gamma$ in the 30% Ni alloy. Greninger and Troiano investigated a Fe–22% Ni–0.8% C alloy in great detail and found an orientation relationship between the above extremes and a habit plane close to $\{3, 10, 15\}_\gamma$. It is obvious that the simple concepts of the cobalt transformation are insufficient to explain such highly and indeed usually irrationally indexed habit planes.

In the case of Fe–C martensite, E. C. Bain has specified a distortion matrix for the γ–α' transformation

$$\mathcal{B} = \begin{pmatrix} \eta_1 & 0 & 0 \\ 0 & \eta_2 & 0 \\ 0 & 0 & \eta_3 \end{pmatrix},$$

which translates one lattice into the other with a minimum of atomic displacements. The fcc lattice is first described as b.c.tetr. with

$$c_\gamma / a_\gamma = \sqrt{2},$$

fig. 13.7(a). It is then compressed in the γ-cube direction by about $\eta_2 = -0.83$ and extended at right angles by about $\eta_1 = \eta_3 = 1.12$, fig. 13.7(b) ('Bain-cell'). The volume difference of the two cells is only $\eta_1^2 \cdot |\eta_2| = 1.03$ to 1.05 whereas the axial ratio $c_{\alpha'}/a_{\alpha'}$ varies linearly with the carbon concentration from 1.00 to 1.08 (at 1.8 wt.% C). It is interesting to trace the position of the C atoms during the transformation (one is shown in fig. 13.7). As they are in the octahedral sites in the γ-phase (see section 6.1.2) they find themselves automatically on the tetragonal c axis in the α'. They are in fact responsible for the distortion of the c axis from the cubic cell size a. In this respect the Bain model is very satisfactory. The orientation relationships, however, can be explained only very crudely by fig. 13.7. Bain's model does not produce an invariant plane, the existence of which is linked with the condition that one principal strain η_1 is unity, another is greater and another less.

Fig. 13.8 shows that when a sphere is distorted to an ellipsoid, certain

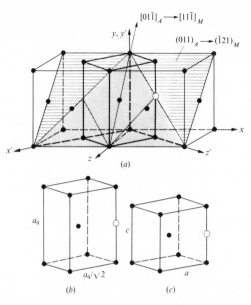

Fig. 13.7. Bain distortion of the body centred tetragonal cell in the fcc lattice, (*a*) and (*b*), to the martensite cell (*c*). A lattice invariant slip system is indicated in austenite and martensite. A carbon atom (○) at the centre of the fcc cell is situated on the *c*-axis of the martensite cell.

vectors (e.g. $\overline{AB}$) retain their original length, even though they are rotated (to $\overline{A'B'}$). Due to the requirement of an invariant habit plane therefore, pure rotations, described by matrices $\mathscr{R}$, must be allowed in addition to $\mathscr{B}$:

$$\mathscr{R} = \begin{pmatrix} 1 & 0 & 0 \\ 0 & \cos\varphi & \sin\varphi \\ 0 & -\sin\varphi & \cos\varphi \end{pmatrix}.$$

Finally the distortion in the transforming material must be compensated at least on average, i.e. macroscopically, by lattice invariant shears as was indicated in fig. 13.5. Consider the shape change in fig. 13.9 experienced by a hemisphere on the twinning plane as a result of a twinning shear $\mathscr{S}$. Naturally the twinning plane K_1 remains undistorted as also does the plane K_2, which transforms on twinning to K'_2. Since the angle of shear α_t is fixed and in general different from the martensite shear α_M, the volume fraction twinned x must be used as a free parameter to ensure that the two shears ($\mathscr{R}\mathscr{B}$ and $\mathscr{S}$) cancel out macroscopically. In the cobalt transformation, twinning of 50% of the fcc lattice is sufficient to compensate

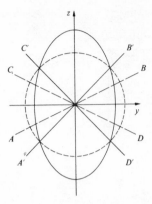

Fig. 13.8. If a sphere is distorted to an ellipsoid the directions AB, CD remain undistorted but are rotated.

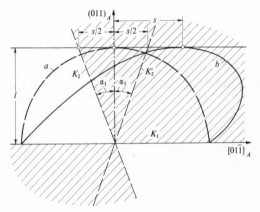

Fig. 13.9. A twinning shear through the angle $2\alpha_t$ and a displacement s transform the circle (a) into the curve (b); all vectors in planes K_1 and K_2 remain unchanged, those inside the cross-hatched area are lengthened and those outside are shortened.

macroscopically for the shear effect of the fcc → cph transformation, see section 13.1. Figs. 13.5(c) and (d) show that a plane can likewise be rotated back through an angle α_m by slip or twinning of a volume fraction x. The plane is then effectively uninfluenced by the martensite shear, i.e. becomes a habit plane.

The shape change is expressed by the matrix product $\mathscr{S}\mathscr{R}\mathscr{B}$, whereby the lattice invariant shear $\mathscr{S}_1$ and rotation $\mathscr{R}_1$ are found in a fraction x of the volume and, in the general case, another shear $\mathscr{S}_2$ and rotation $\mathscr{R}_2$ in the fraction $(1 - x)$. A position vector **r** in the crystal thus transforms to

$$\mathbf{r}' = [x\mathscr{S}_1\mathscr{R}_1\mathscr{B} + (1 - x)\mathscr{S}_2\mathscr{R}_2\mathscr{B}]\mathbf{r} = \mathscr{E}\mathbf{r}.$$

The habit plane is determined by the eigenvalue equation $\mathscr{E}\mathbf{r} = \mathbf{r}$. Naturally no physical significance is ascribed to the order in which the operations occur. The main object is to combine the requirement of an invariant plane with the assumed known operations $\mathscr{B}$ and $\mathscr{S}$.

The problem can be illustrated graphically using the stereographic projection.

A lattice invariant shear $\mathscr{S}$ of $2\alpha_t$ on K_1 can be represented for a cubic crystal as in fig. 13.10(a). K_2 becomes K'_2 without any distortion. All vectors in the dotted area are lengthened and the others shortened (see section 13.9). The Bain distortion $\mathscr{B}$ leaves all vectors on the cone of intersection of an ellipsoid with a sphere undistorted. This cone (section $B'D'$ in fig. 13.8) and the one defining its position before the Bain transformation (BD in fig. 13.8) are shown stereographically in fig. 13.10(b). In order that the combined effect of $\mathscr{B}$ and $\mathscr{S}$ produces an undistorted plane, the two figures 13.10(a) and 13.10(b) must be superposed as in fig. 13.10(c). The four vectors in which the cone and the K_1, K_2 planes intersect thereby remain undistorted. Any two of them define a plane. The intermediate vectors in the plane (e.g. **a** and **d**) are lengthened by $\mathscr{S}$ and shortened by $\mathscr{B}$. By varying the angle α_t these two length changes can be made to compensate exactly, i.e. these intermediate vectors also remain invariant. This is possible if and only if the angle enclosed by **a**, **d** is equal to that enclosed by **A**′, **D**′ (after the transformation). α_t (x in matrix terminology) can be obtained graphically. Finally, in order to obtain a macroscopically undistorted and unrotated habit plane, **A**′ must be rotated into **a** and **D**′ into **d**. The magnitude of this rotation ultimately determines the orientation relationship between the lattices (γ, α'). In this way all the parameters of the martensitic transformation can be calculated, in good agreement with the measurements of Greninger and Troiano, as shown in table 13.2. The theory outlined above was developed by M. Wechsler, D. Liebermann and T. A. Read [13.4]. Only the Bain distortion and the slip or twinning system of the lattice invariant shear were assumed.

13.4   The martensitic phase boundary

The lattice transformations described above take place at the austenite–martensite phase boundary. The inhomogeneous lattice invariant shear requires the interface to contain an array of dislocations which will now be discussed. Since the interface is obviously very mobile,

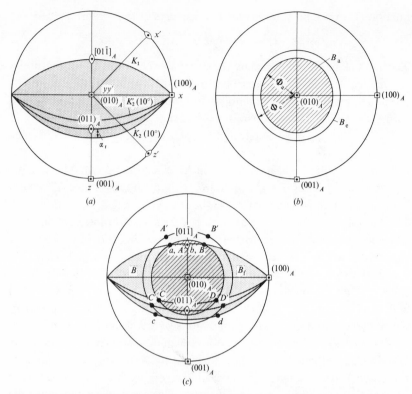

Fig. 13.10. Stereographic projections of a cubic lattice. (*a*) A homogeneous shear by $2\alpha_t = 20°$ leaves planes K_1, K_2 undistorted (fig. 13.9). The vectors inside the dotted area are lengthened and those outside are shortened. The coordinate systems xyz in austenite and $x'y'z'$ in martensite are as in fig. 13.7. In (*c*), *d* transforms to *D* and *c* to *C*.

(*b*) The Bain distortion transforms the cone B_a to B_e without the vectors in the cone surfaces suffering any change in length. Vectors in the cross-hatched area are, however, shortened and those outside lengthened. In (*c*), *D* transforms to *D'* and *C* to *C'*.

(*c*) Superposition of shear (*a*) and Bain distortion (*b*) results in opposing length changes in the cross-hatched/dotted area and four vectors *a*, *b*, *c*, *d* whose lengths remain unaltered.

the Burgers vectors of the dislocations cannot lie *in* the interface unless they are screw dislocations. The interface or habit plane is not usually the same as the plane of the martensitic and therefore of the lattice invariant shear. (It is, however, the same in the case of the cobalt transformation.) F. C. Frank [13.7] has proposed a detailed model for the $\{225\}_\gamma$ habit plane in low carbon steels. The Kurdjumov–Sachs orientation relationship in which the

Table 13.2. *Comparison of the experimental and theoretical crystallographic data for martensite* [13.4]

	Experiment	Theory	Discrepancy
Normal to the habit plane	Fe–22% Ni–0.8% C fcc → bct (twinned) ($c/a > 1$) $\begin{pmatrix} 10 \\ 3 \\ 15 \end{pmatrix} = \begin{pmatrix} 0.5472 \\ 0.1642 \\ 0.8208 \end{pmatrix}$	$\begin{pmatrix} 0.5691 \\ 0.1783 \\ 0.8027 \end{pmatrix}$	$< 2°$
Orientation relationship	$(111)_A \| (101)_M$ to within $1°$ $[1\bar{1}0]_A$ $2\frac{1}{2}°$ from $[111]_M$	$(111)_A$, $15'$ from $(101)_{M1,M2}$ $[1\bar{1}0]_A$ $3°$ from $[111]_{M1,M2}$	~ 0 $\sim \frac{1}{2}°$
Shear direction	$\begin{pmatrix} -0.7315 \\ -0.3828 \\ 0.5642 \end{pmatrix}$	$\begin{pmatrix} -0.7660 \\ -0.2400 \\ 0.5964 \end{pmatrix}$	$\sim 8°$
Shear angle	$10.66°$	$10.71°$	~ 0

$(111)_\gamma$ plane is parallel to $(110)_{\alpha'}$ applies. According to what has been said earlier, however, it is not parallel to the habit plane, with the result that the two planes, $(111)_\gamma, (110)_{\alpha'}$, intersect along a close packed direction $[1\bar{1}0]_\gamma \| [1\bar{1}1]_{\alpha'}$. The slight difference in repeat distance of the two planes can, according to Frank, be matched by mutually tilting the two sets of planes by $\frac{1}{2}°$. More serious are the difference of 6–7% in the distance between close packed rows in $\{225\}_\gamma$ and the 1% difference in the distance between the atoms in these rows in α' and in γ which give rise to considerable elastic distortion. (Thus $\{225\}_\gamma$ is no longer an undistorted habit plane in the strictest sense.) The $\{259\}_\gamma$ habit plane appears to be less distorted than the $\{225\}_\gamma$. The extent to which heavy distortion can be tolerated and which habit plane is selected in the various iron alloys is thus dependent on the interatomic binding forces [13.6]. The crystallographic theory of Bowles and Mackenzie also predicts a distorted habit plane [13.8].

After making these corrections it is apparent that the two structures can be matched easily by means of an arrangement of parallel screw dislocations $\| [1\bar{1}0]_\gamma$ in each sixth $(111)_\gamma$ or $(110)_{\alpha'}$ plane and in the $\{225\}_\gamma$ interface between the two lattices, fig. 13.11. A mechanism for the martensite transformation thus becomes apparent consisting of the

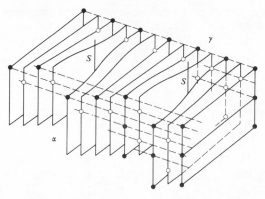

Fig. 13.11. Model of an austenite–martensite phase boundary in which the close packed planes $(111)_A$ and $(110)_M$ meet parallel to $[1\bar{1}0]_A$. A screw dislocation S lies in every sixth plane [13.7].

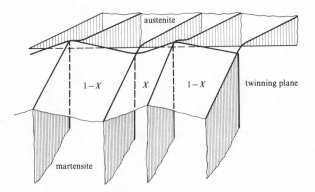

Fig. 13.12. Habit plane between austenite and twinned martensite (twinning shear direction in the interface).

formation and movement of these screw dislocations. If the lattice invariant shear occurs by twinning the matching is similar, fig. 13.12, but the dislocation structure of this boundary is still unknown. The twin lamellae have been shown by transmission electron microscopy to be often exceedingly narrow (1 nm–10 nm) [13.6].

Using this special model of the phase boundary, the total interfacial energy of a martensite plate in the shape of an oblate ellipsoid of rotation can be estimated, fig. 13.13. In addition, Dehlinger and Knapp have calculated the elastic distortional energy of the plate as a function of the thickness ratio c/r and the radius r. The distortional energy per mole of martensite for a plate of $r = 50$ nm and $c = 2$ nm is about 1200 J/mol and even larger for smaller plates.

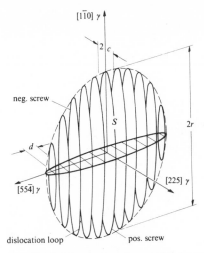

Fig. 13.13. Martensite plate which has a screw dislocation in every sixth plane of the interface (see fig. 13.11). The screws in the front and back surfaces are linked to dislocation loops (U. Dehlinger and H. Knapp, Stuttgart).

13.5   Nucleation of martensite [9.6]

The saving in free energy $\Delta F_{\gamma\alpha'}$ is also about 1200 J/mol at the M_s temperature for the spontaneous martensite transformation in steels. It is therefore most improbable that a martensite nucleus of atomic size could be obtained by homogeneous nucleation, section 9.1.1. The free energy of a critical nucleus ΔF^* at M_s is several thousand eV. It is therefore assumed that 'strain embryos' are present, distorted atomic arrangements in γ with an α'-like structure, such as are found in the neighbourhood of a group of screw dislocations in γ. (Another concept of nucleation is given at the end of this section.) The first question is how large must these nuclei be if they are to grow spontaneously [13.10], [13.10a]. The growth does not start immediately at the saddle point of the activation profile $\Delta F(c, r)$, fig. 13.14, because there are still finer energy troughs in its path originating from the discrete dislocation loop structure of the phase boundary. A new dislocation loop must be formed each time at the periphery of the martensite plate (*radial growth*). The activation energy for this is of the order of $\frac{1}{2}$ eV at the temperature M_s with the result that thermal activation is able to create dislocation loops in finite times. M. Cohen obtained agreement with the isothermal martensite growth kinetics for Fe–29% Ni– 0.2% Mn as a function of the undercooling for an embryo of $r = 59$ nm, $c = 2.4$ nm. In addition martensite forms in the absence of thermal

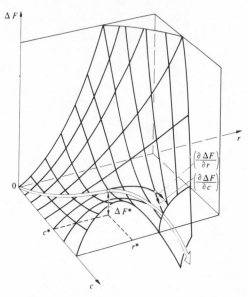

Fig. 13.14. Energy of formation of a martensite plate $\Delta F(r, c)$ as function of its dimensions (fig. 13.13). Dimensions of a critical nucleus (r^*, c^*). After M. Cohen [13.10].

activation ('athermally') with increasing undercooling. The reason for this is the increasing slope $-\partial\Delta F/\partial r$ which swamps the local undulations in the large activation barrier ΔF^* in fig. 13.14.

In addition to the isothermal radial growth there is also an *increase in the plate thickness* which forces the dislocations surrounding the martensite plate into the austenite. It begins when the gradient $(-\partial\Delta F/\partial c)$, which can be interpreted as a force on the dislocations created by the tendency to transform, becomes larger than the slip resistance of the surrounding austenite. This is the situation after ten new dislocation loops have been produced in radial growth, i.e. above $r = 78$ nm [13.10]. The growth of the martensite plate is halted when it meets other martensite regions and by interaction with dislocations produced by deformation at the reaction front. An external stress applied during the deformation can undoubtedly reduce the activation barrier to nucleation and hence raise the transformation temperature from M_s to M_d. The heavy lattice distortion around a martensite plate has an autocatalytic effect in the sense of promoting the formation of further plates. The origin of the first 10^6 nuclei per cm^3 in the undercooled austenite still remains to be explained. Cech and Turnbull [13.11] found a density of embryos of this order in their investigation of the transformation of Fe–30% Ni powder with a grain size of 10–$100\,\mu$m. (The crystals which have undergone the martensitic transformation are ferromagnetic and can be separated magnetically.) The smallest grains show the lowest M_s temperatures and therefore require the largest undercooling. This indicates that nucleation occurs at dislocations rather than on the surface (or grain boundaries) because the likelihood of finding a dislocation becomes vanishingly small as the size of the particle decreases. Small coherent iron precipitates in copper (20–130 nm in diameter) transform to martensite only after deformation, i.e. when there are dislocations in the γ-iron, see chapter 14 [13.12]. On the assumption that the $\{225\}_\gamma$ habit plane consists of screw dislocations it seems likely that random dislocation configurations in the sense of fig. 13.13 act as strain embryos.

It is difficult, however, to provide the necessary concentration of distortional energy in this manner. On the basis of electron microscope observations, W. Pitsch has proposed (unpublished), that a thin coherent martensite layer is first formed. Internal twinning ensures a match with the matrix. This layer possesses such a low interfacial and distortional energy that it ought to be able to form by homogeneous nucleation. The matrix then provides dislocations which are incorporated into the γ–α' interface permitting further growth of the martensite plate by the F. C. Frank and M. Cohen models from the 'midrib' which would thus be the nucleus described above. The nucleation of martensite is facilitated in many structures by the fact that they exhibit a low resistance to certain shears and that the amplitudes of the corresponding lattice vibrations are very large (see the shear instability of the bcc lattice described in section 6.2.2). Such dynamic Bain distortions have actually been observed recently in the electron microscope at temperatures a little above M_s [13.12a].

13.6   Hardening of steel [1.3], [13.13]

As indicated in section 9.6 the decomposition of austenite can be shifted to lower temperatures by rapid cooling so that fine pearlite, bainite or martensite form. This improves the mechanical properties. The strength of iron alloys resulting from the formation of martensite is mainly responsible for their outstanding technological importance. The high flow stress of the transformed alloy is explained partly by that of the martensite needles themselves and partly by the finely divided nature of the microstructure and the elastic distortion in the microstructure as a whole including the residual austenite. The strength of the martensite crystal itself relies mainly on *solid solution hardening* by carbon as demonstrated by fig. 13.15. The interaction of dislocations with dissolved solute atoms is

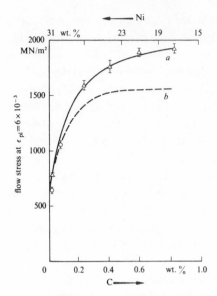

Fig. 13.15. Flow stress of Fe–Ni alloys in which the Fe:Ni ratio is varied so that $M_s = -35\,°C$ regardless of the carbon content. Curve (b) was measured immediately after the martensitic transformation and curve (a) after a 3 h anneal at $0\,°C$. After M. Cohen [1.3].

discussed in chapter 14. The second reason for the hardening effect of martensite, namely the shortening of the free paths of the dislocations due to the subdivision of the microstructure by martensite plates, twins, etc. was treated in section 12.5.1.

Dislocation densities of 10^{11}–10^{12} cm^{-2} are observed in the residual austenite as a result of the martensite shear, and these also contribute to the high flow stress of the microstructure. In practice the martensitic microstructure is too brittle for many purposes. The toughness is recovered by 'tempering' (annealing) during which some of the carbon precipitates as carbide ('heat treatment'). Depending on annealing temperature, time and composition metastable carbides, cementite or alloy carbides are formed. This reaction proceeds at room temperature or a little above thanks to the rapid diffusion of carbon (see chapter 8). In particular, the carbon segregates on to dislocations, for example on those in the $\gamma\alpha'$ phase boundary, and impedes their movement. This is the reason for the *stabilization* of the residual austenite when the quenching treatment is interrupted at room temperature (or above). The following examples show how desirable properties of steels can be obtained by alloying (in wt.%) and heat treatment.

(*a*)   *Carbon steels* (not alloyed except for desoxidants like Mn or Si)
        For many applications hardening by carbon suffices: with more
than 0.35% C it is martensite that hardens but beyond 0.7% C M_f falls below
300 K and the residual austenite softens the steel again. A disadvantage
with these steels is the small 'through hardening' of a massive specimen:
only near its surface can a high quenching rate be obtained from the γ-range
where carbon is brought in solution. In the interior there is decomposition
$\gamma \rightarrow \alpha + Fe_3C$, see section 6.1.2. Also martensite decomposes above 300 °C
and the strength decreases with annealing. Carbon steels are more brittle
than alloy steels; quenching stresses produce cracks.

(*b*)   *Alloyed tool steels*
        For the use in wear-resistant tools of high-strength carbide formers
like Cr ($\leqslant 13\%$), Mo, V, W, Mn ($\leqslant 1\%$) are alloyed in addition to about 1%
C. These additions lower the necessary cooling rates so that quenching can
be done in oil or air rather than in water. Carbon precipitates more slowly
now and this is favourable for deeper 'through hardening' of larger
specimens.

(*c*)   *High speed steels*
        Drilling tools should not soften at temperatures up to 600 °C. The
softening effect of the decomposition of martensite must then be
compensated for by the secondary (precipitation) hardening of alloy
carbides. A typical steel contains 0.8% C, 4% Cr, 1% V and 18% W (or 9%
Mo). It is (multiply) quenched from the γ-range and aged at 560 °C.

(*d*)   *Thermo-mechanically hardened steels*
        High-strength steels are obtained by *ausforming* (= deformation in
the austenitic state) as is shown in fig. 13.16 for a typical heat-resisting steel.
In the TTT diagram of the carbon steels *bainite* appears at the lower edge of
the pearlite field. Here the martensitic transformation concurs with a fine
precipitation of the carbon. In alloy steels different carbides are formed in
the bainite and pearlite fields, so that the two fields are separated in the TTT
diagram, fig. 13.16. Deformation is now performed in the temperature range
between the pearlite and bainite 'noses', but below the recrystallization
temperature and before cooling to form martensite. This 'ausforming'
introduces dislocations and subdivides the microstructure so finely that the
subsequently formed martensite produces high strength at good ductility.

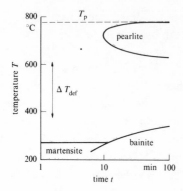

Fig. 13.16. TTT diagram of a steel with 0.4 wt.% C, 5 wt.% Cr, 1.3 wt.% Mo, 1.0 wt.% Si, 0.5 wt.% V deformed in the austenitic range ΔT_{def}.

(*e*)   *Maraging steels*

Low carbon steels (0.02% C) with additions of typically 18% Ni, 7.5% Co, 5% Mo, 0.4% Ti form a relatively soft Fe–Ni-martensite (see fig. 13.15). On tempering at 500 °C finely dispersed particles of Ni_3Ti, Ni_3Mo and Fe_2Mo precipitate and produce good high-temperature strength and ductility.

(*f*)   *Austenitic steels*

By larger additions of Ni, Mn and other γ-openers (section 6.1.2) austenite can be stabilized at low temperatures. Its mechanical properties are much better than those of the bcc ferrite. A small stacking fault energy of the austenitic steels is responsible for a planar dislocation arrangement and high work hardening rate (chapter 12). The *Hadfield steel* (with 12% Mn, 1.2% C) forms martensite locally on cold working in the temperature range between M_d and M_s which increases work hardening – at good ductility and wear resistance. Austenitic steels are often called *stainless steels* when their chromium content makes them resistant to corrosion. A typical alloy contains 18% Cr and 8% Ni. Austenitic steels can be used at higher temperatures than ferritic ones if their carbon content is small enough or stabilized in Nb- or Ta-carbides, so that no $Cr_{23}C_6$-carbides can form in grain boundaries. Further hardening can be produced by coherent, ordered γ'-precipitates of the type (Fe, Ni)$_3$(Al, Ti), chapter 14. Also austenitic steels are like γ-Fe often *non-ferromagnetic* but the three above mentioned useful attributes do not necessarily belong to the same austenitic steel. 90% of the steels used contain, for economic reasons, only carbon.

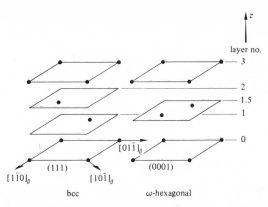

Fig. 13.17. {111} planes of the bcc structure (left) collapse into the hexagonal ω-structure (right).

13.7   The displacive ω-transformation [13.15]

Alloys with elements of the IVth group (Ti, Zr, Hf) which have the β-(bcc) or α-(cph) structure transform on quenching athermally into the ω-structure (AlB$_2$). This is hexagonal with an axial ratio $c/a = 0.61$ and three atoms per unit cell. It forms by 'collapsing' pairs of bcc {111} planes into one (0001) plane of the cph structure (fig. 13.17). This defines the orientation relation of the two structures and the four possible orientation variants of ω. If the collapse is not complete, the resulting atom layer at $z = 1.5$ remains wavy and a trigonal ω-variant results as is observed with concentrated alloys. A similar 'displacive' or pairing mechanism also relates the (0001)$_α$- and (11$\bar{2}$0)$_ω$-planes. One can describe it as a 'charge density wave' and recognize it already above the transformation temperature $T_{βω}$ by 'soft phonons', in the case of the longitudinal 2/3⟨111⟩ type. In pure group IV metals the ω-phase is stable only under hydrostatic pressure. For dilute alloys with increased number of d-electrons ω appears as a metastable phase as shown in fig. 13.18. Here ΔN_d^0 typically is 0.13,...,0.15 which corresponds to the addition of 14% V, 6.5% Cr or 5% Mn. The horizontal axis could read also hydrostatic pressure instead of increase in d-electron number as pressure transfers electrons from the s- to the d-band (ω is the most densely packed of the phases!). In fig. 13.18 one recognizes the ΔN_d-intervals in which the free energy decreases by the $α \rightarrow ω$ or $β \rightarrow ω$ transformations. The ω-transformation is therefore (d)-electronically driven, similarly to those in the Hume-Rothery-alloys (on the basis of s- and p-electrons), (see section 6.3.2). The ω-transformation also occurs on isothermal ageing of concentrated β-alloys. Then concentration

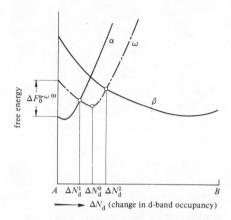

Fig. 13.18. Free energies of the α-, β- and ω-phases depending on the occupation of the d-band.

modulations occur leading to more dilute volume elements in which $\beta \to \omega$ may transform athermally. In equilibrium at atmospheric pressure an $(\alpha + \beta)$ phase mixture is stable, see fig. 13.18. The ω-phase is technologically undesirable as it embrittles Ti-alloys. But the displacive transformation in these alloys is (as the pure shear transformation of cobalt) physically more transparent than the martensitic one in steels.

14
Alloy hardening

Since it is primarily on account of their mechanical properties that metals are of such technological interest, this chapter is concerned with one of the vital themes of this book. Only by alloying do metals achieve sufficient strength, i.e. a sufficiently high yield stress, to be able to withstand the stresses applied in practice. Unlike the hardening due to deformation discussed in chapter 12 or that produced by the martensitic transformation (section 13.6) the hardening we are concerned with here is produced by the addition of alloying elements. As described in chapter 11, plastic deformation is brought about by dislocations. Hardening thus implies restriction of dislocation movement due to interaction of dislocations with the solute. The solute can be present in a variety of forms: as dissolved atoms, as precipitate particles, as ordered regions in a disordered matrix, as particles of a second phase, or as a constituent of a phase mixture. In all these examples hardening is achieved by *making the material heterogeneous.* We shall discuss these various alloy-heterogeneities one after the other.

14.1 Solid solution hardening [14.1], [14.2], [12.5]

In a solid solution the heterogeneity is on an atomic scale. We assume a very dilute solution of B in A in which the dislocation interacts with individual solute atoms. This interaction involves various mechanisms, which are the first problem to be discussed in the following. In reality the dislocation interacts simultaneously with many solute atoms, regarded here (at low temperatures) as immobile whereas the dislocation is mobile. The line tension prevents the dislocation from assuming the position of minimum (repulsive) interaction relative to all the neighbouring solute atoms. For this, it would have to bend too sharply and increase its length. The actual equilibrium position is determined by the combined minimum of the interaction and line energies. The calculation of the force necessary to move the dislocation from this equilibrium position is the second problem in understanding solid solution hardening (section 14.1.2).

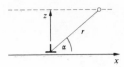

Fig. 14.1. Solute atom at a distance r from an edge dislocation.

At higher temperatures the solute atoms become mobile and diffuse to positions of minimum energy relative to the dislocations, which, among other things, causes pronounced yield points (section 12.2). These will be discussed in section 14.2.

14.1.1 *Interaction of a straight dislocation with a solute atom*

We consider a substitutional solute atom, which on introduction into the lattice changes the volume of the latter by ΔV. The distortion around the solute atom is assumed to be spherically symmetrical. An *edge dislocation* produces a compressional field above the slip plane ($0 < \alpha < \pi$) and a dilatational field below it ($\pi < \alpha < 2\pi$), which at a distance r from the dislocation is given by the hydrostatic stress component (see section 11.2.1 and fig. 11.10(b))

$$p = \tfrac{1}{3}(\sigma_{xx} + \sigma_{yy} + \sigma_{zz}) = -\frac{Gb}{3\pi r}\sin\alpha\,\frac{1+v}{1-v}. \tag{14-1}$$

The interaction energy between a dislocation at the origin and the solute atom at (r, α) is then $\Delta E^{\mathrm{p}} = p\Delta V\{3(1-v)/(1+v)\}$. The factor in the curly brackets takes into account the interaction energy included in ΔV [12.10]. This yields an interaction force on the dislocation in the slip direction (x, see fig. 14.1) of

$$K^{\mathrm{p}} = -\left.\frac{\partial \Delta E^{\mathrm{p}}}{\partial x}\right|_z = -\frac{Gb\,\Delta V}{\pi z^2}\,\frac{2(x/z)}{(1+(x/z)^2)^2}. \tag{14-2}$$

Fig. 14.2 shows $K^{\mathrm{p}}(x)$ for a constant distance z from the slip plane. A solute atom near the slip plane hinders the dislocation most, i.e. $K^{\mathrm{p}}_{\max}$ increases with decreasing z. As $z \to 0$, however, linear elasticity theory becomes inadequate. If the volume change $\Delta V = 3\Omega\delta$ is expressed in terms of the relative change δ in the lattice parameter a for the solute atom concentration v_B, $\delta = \mathrm{d}\ln a/\mathrm{d}v_B$, the *maximum parelastic interaction* force $K^{\mathrm{p}}_{\max} \approx Gb^2|\delta|$ is obtained (if the minimum value of z_0 is put equal to $b/\sqrt{6}$, i.e. one half the separation of the $\{111\}$ planes in the fcc lattice).

According to linear elasticity theory, a *screw dislocation* does not possess a hydrostatic stress field, see section 11.2.1, i.e. to the first order it has no

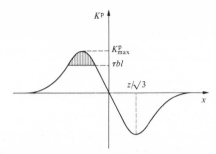

Fig. 14.2. Parelastic interaction force K^p due to a solute atom experienced by an edge dislocation on a slip plane at a distance z.

parelastic interaction with a substitutional solute atom. It does, however, interact with the tetragonal distortion field

$$\varepsilon^{SA} = \begin{pmatrix} \varepsilon_{11} & 0 & 0 \\ 0 & \varepsilon_{22} & 0 \\ 0 & 0 & \varepsilon_{33} \end{pmatrix}$$

around an interstitial atom (carbon, see section 6.1.2) in the bcc lattice. The interaction [14.3] is

$$\Delta E^p = \sum_{i,k} \varepsilon_{ik}^{SA} \sigma_{ik}^{screw} \Omega \tag{14-3}$$

where σ_{ik}^{screw} is given by equation (11-4). The resultant force exerted by the solute atom on the dislocation is similar in form to that given by (14-2) but with δ replaced by $(\varepsilon_{11} - \varepsilon_{22})/3$ [14.3]. Thus a solute atom producing an isotropic distortion $(\varepsilon_{11} = \varepsilon_{22})$ does not exert any force on the screw dislocation. Carbon in α-iron, with $\varepsilon_{11} = 0.38$, $\varepsilon_{22} = -0.03$ is a stronger obstacle to all dislocations than a normal substitutional atom. This is obvious for the following reasons. For substitutional atoms to have a finite solubility, Hume-Rothery rules require $\delta \leqslant 0.14$ (section 6.3.1). Yet interstitial atoms with $(\varepsilon_{11} - \varepsilon_{22}) \approx 1$ are still soluble in bcc metals. This is because metals can accommodate greater uniaxial distortion than isotropic distortion by solute atoms since the electron energy depends essentially on the specific volume.

In addition to the *parelastic interaction* which is caused by the permanent 'elastic moment' of the solute atom (more exactly a combination of 'double forces' each described by a $\pm$ pair of forces acting along the line joining them, see [11.3]) there is a *dielastic interaction*. In this case the distortion field of the dislocation induces an elastic moment in the neighbourhood of the solute atom if this possesses different elastic properties from those of the matrix. Since according to section 11.2.2 the energy density e of the distortion field of the dislocation is proportional to the shear modulus G, the dielastic interaction energy with the screw dislocation is

$$\Delta E^d = \eta e\Omega = \frac{Gb^2\Omega\eta}{8\pi^2 r^2} \tag{14-4}$$

where the 'shear modulus defect' $\eta \equiv d \ln G/dv_B$, see [14.5a]. Comparing this with (14-1) it can be seen that the dielastic interaction is of the second order and therefore decreases with $1/r^2$ instead of with $1/r$ as is the case with parelastic interaction. Correspondingly the interaction force K^d is $(3b/8\pi z)$ times smaller than K^p, (14-2), but otherwise they have the same functional form. On the other hand, however, η is often 20 times larger than δ so that both interactions make a significant contribution to solid solution hardening. The dielastic interaction is symmetrical in z, the parelastic, on the other hand, has a different sign above and below the slip plane for an edge dislocation (for which (14-4) likewise applies, apart from a factor $(1-v)^{-2}$). Thus the effects of solute atoms on different sides of the slip plane cannot simply be added together, see section 14.1.2 [14.4].

The elasticity calculation of the interaction is very unsatisfactory if the solute atom is very close to the dislocation. As was shown in section 6.3.3, if the valency of solute and matrix atoms differ, the solute atoms are surrounded by an oscillating electronic screen. The same applies to the dislocation core because the ionic charge density is different from that of the matrix. When the two screens overlap, a very short range *electrostatic interaction* occurs, about which little is yet known.

Finally, if the dislocation is dissociated into partial dislocations with an included stacking fault, the differences between the interactions of screw and edge dislocations with solute atoms become less pronounced. A new interaction, first proposed by H. Suzuki, is now possible between *stacking fault* and solute atom. A stacking fault in the fcc lattice can be imagined as a cph layer, section 6.2.1. The cph phase becomes more stable than the fcc phase as the solute concentration increases, i.e. with increasing electron concentration in the case of the noble metal alloys of interest here, section 6.3.2. As a consequence of the relationship between phase stability and solute concentration, mobile solute atoms are expected to segregate in the stacking fault [11.3]. This causes a decrease in the stacking fault energy (section 6.2.1), the fault becomes wider, and the energy of the whole dislocation decreases. An additional force is then necessary to tear the dislocation away from its 'Suzuki atmosphere', i.e. a yield drop is observed

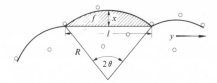

Fig. 14.3. A dislocation under a stress τ bows out between two obstacles separated by a distance l to a radius of curvature R until, having traversed an area f, it touches another obstacle.

(see section 12.2). The Suzuki interaction is unimportant if the solute atoms are immobile, i.e. for solid solution hardening, because the number of solute atoms in the stacking fault is to a first approximation independent of its position.

14.1.2   *The critical shear stress of a solid solution*

We must now sum statistically the interactions of the many solute atoms on both sides of the slip plane with a dislocation of finite line tension. The critical shear stress is the minimum force necessary to move the dislocation over the slip plane in the solid solution. Two plausible theories exist for $T = 0$ [14.1], [14.4]. For very dilute alloys (range of the interaction small compared with distance between solute atoms parallel to the slip plane) R. Fleischer defines a mean separation l between solute atoms in contact with a dislocation under a stress τ. This is the Friedel distance, see fig. 14.3, which is derived as follows: The cross-hatched area f in fig. 14.3 is

$$f = R^2\theta - R^2 \cos\theta \sin\theta \approx \tfrac{2}{3}R^2\theta^3 \approx \frac{l^3}{12R} = \frac{l^3\tau b}{12E_L} \qquad (14\text{-}5a)$$

(cf. (11-8)). The dislocation can be expected to touch another atom when f becomes equal to one half the area per solute atom, i.e. $f = 1/2c_F$ where c_F is the areal concentration of solute atoms. Thus

$$l = \sqrt[3]{\left(\frac{6E_L}{\tau b c_F}\right)}. \qquad (14\text{-}5b)$$

As its curvature increases with increasing stress, the dislocation comes into contact with more obstacles and l becomes smaller. Equilibrium of forces at the critical shear stress τ_c implies

$$\tau_c b l(\tau_c) = K_{max},$$

or

$$\tau_c b = K_{max}^{3/2} c_F^{1/2} / \sqrt{(6E_L)} \qquad (14\text{-}6)$$

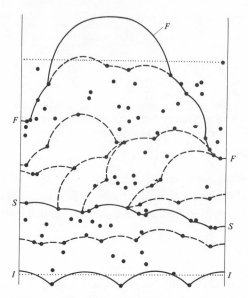

Fig. 14.4. Sequence of positions $(I \rightarrow S \rightarrow F)$ assumed by a dislocation in a field of point obstacles under the action of an upwards directed force (J. Dorn, Univ. California).

where K_{max} is the maximum obstacle strength. Computer simulations have confirmed this relationship for moderately strong point obstacles. In these models a dislocation at the bottom of a statistical distribution of point obstacles is subjected to a force τb per unit length, fig. 14.4. The total vertical force is calculated at each obstacle with which it comes into contact. If this is larger than K_{max}, the dislocation does not sense the obstacle and continues to move upwards until $K < K_{max}$ for all the obstacles with which it is in contact. τb is then increased until at $\tau_c b$ the dislocation finally overcomes all the obstacles and emerges at the top of the slip plane. The obstacle strength K_{max} can also be varied. The dislocation bows out between strong obstacles in the Orowan process, sections 11.2.2 and 14.3.

If the solid solution is no longer dilute in the sense defined above the dislocation will come into contact with obstacles of all interaction strengths K and not only those with $K = K_{max}$ or zero as assumed by Fleischer. R. Labusch has defined a distribution function $\tilde{\rho}(K) \, dK$ as the number of solute atoms touched by unit dislocation length with interaction forces between K and $K + dK$. If the force profile is known, fig. 14.2, it is possible to redefine $\tilde{\rho}$ as a distribution function $\rho(x)$ of the distances x from the solute

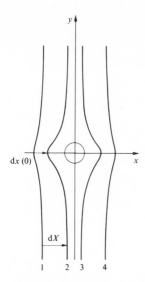

Fig. 14.5. Dislocation in the positions 1 to 4 at a repulsive obstacle at $y = 0$.

atoms. Equilibrium of forces then yields

$$\tau b = \int \tilde{\rho}(K)K \, dK = \int \rho(x)K(x) \, dx. \tag{14-7}$$

When a dislocation line meets a (repulsive) obstacle at $y = 0$, the segment closest to the obstacle at $x(0)$ lags behind relative to the average position X of the dislocation line, fig. 14.5, according to

$$dX - dx(0) = G(0) \, dK = G(0) \frac{dK}{dx} \, dx(0). \tag{14-8a}$$

$G(x)$ is the elastic Green's function which describes the shape of the dislocation (position 2 in fig. 14.5) under unit stress applied at $y = 0$ in the $(-x)$ direction. We then obtain [14.4]

$$G(0) = \frac{1}{2\sqrt{(E_L \alpha)}} \quad \text{with} \quad \alpha = \int \rho(x) \frac{dK}{dx} \, dx. \tag{14-8b}$$

α is the mean curvature of the solute atom/dislocation interaction potential. In the stationary state of overcoming the obstacles at the critical shear stress τ_c, the ratio of the density $\rho_c(0)$ of dislocation segments waiting in front of the obstacle to the mean density $\bar{\rho}_c = c_F$ is the inverse of the ratio of

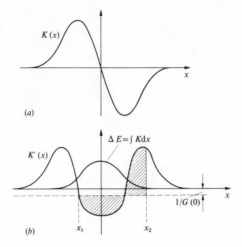

Fig. 14.6. Force and energy profiles, $K(x)$ and $\Delta E(x)$, of an interaction between a solute atom and a dislocation (a, b). Density of the dislocation elements in front of the solute atom

$$\rho_c(x)/c_F\,G(0)=[1/G(0)+K'(x)]\,;$$

$\rho_c(x)=0$ in the shaded region $x_1<x<x_2$ [14.4].

the velocities at these points

i.e. $\rho_c(0)/\bar{\rho}_c=\rho_c(0)/c_F=\bar{v}/v(0).$

Differentiation of (14-8a) with respect to time yields

$$\bar{v}=\frac{\mathrm{d}X}{\mathrm{d}t}, \qquad v(0)=\frac{\mathrm{d}x(0)}{\mathrm{d}t}$$

and hence the desired distribution

$$\rho_c(x)=c_F(1+K'(x)G(0)). \tag{14-9}$$

Fig. 14.6 shows the distribution function, which, however, according to the definition, cannot become negative. In the shaded region between x_1 and x_2 it is thus put equal to zero so that the total number $\int\rho_c(x)\,\mathrm{d}x$ remains unchanged. The form of the distribution $\rho_c(x)$ is easily interpreted. Dislocations pile up on the left-hand flank of the potential barrier, whereas there is no equilibrium position for the dislocation at the peak. α can now be eliminated from (14-7) and (14-8b) which together with (14-9) and neglecting the first term results in

$$\tau_c b = K_{\max}^{4/3}c_F^{2/3}z^{1/3}\,\mathrm{const}/E_L^{1/3}, \tag{14-10}$$

where the constant is a numerical factor of the order of magnitude 1, which contains the dimensionless definite integral

$$\int_0^1 \frac{\partial K/K_{max}}{\partial(x/z)}\, dK/K_{max}.$$

Compared with the Fleischer relation, (14-6), which is valid for extremely dilute solutions, Labusch's expression, (14-10), does not only contain different exponents, it also contains the range of the obstacle, measured according to fig. 14.2 by its distance z from the slip plane. The transition from equation (14-6) to equation (14-10) takes place when the dimensionless range of interaction reaches unity

$$\eta_0 = \frac{z}{b}\sqrt{(Gc_F/F_{max})}.\qquad\qquad(14\text{-}10a)$$

In addition the Labusch theory offers various advantages in its application:

(*a*) If there are several types of obstacle present differing in their distance z from the slip plane or if some of the obstacles are on the compressive and others are on the dilatational side of the dislocation, the theory can easily be formulated with several distribution functions ρ_i and evaluated. In practice only the solute atoms next to the slip plane make a significant contribution to solid solution hardening so that z in (14-10) has a value of atomic dimensions. Furthermore the parelastic and dielastic interactions of solute atoms on both sides of the slip plane are superposed to yield $K_{max}^{eff} \approx Gb^2\sqrt{(\delta^2 + \eta^2\beta^2)}$ where $\beta \approx 1/20$ for the fcc lattice.

(*b*) Even at moderate concentrations v_B it is necessary to take into account the groups of solute atoms produced by statistical fluctuations. The Labusch theory again provides a result of the form of equation (14-10), [14.4].

(*c*) Up to now, thermal activation has been neglected. Dislocations can, however, at finite temperatures T and dislocation velocities v, overcome part of the potential barrier, fig. 14.6, proportional to the Arrhenius factor $v_0 \exp(-E/kT)$ with the result that the distribution function $\rho_c(x)$ in front of the obstacle decreases. Each force–distance profile $F^p(x)$, including that of fig. 14.2, can be approximated near the maximum by a parabola. If the applied force on the dislocation in front of the obstacle is $F = \tau bL$ (with L from equation (14-5b)) and F is close to F_{max}^p the integration over the

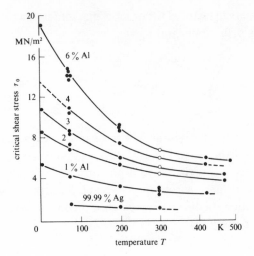

Fig. 14.7. Critical shear stress of *Ag*–Al single crystals as function of the deformation temperature and the Al content (A. Hendrickson and M. Fine, Northwestern Univ.)

hatched area in fig. 14.2 gives

$$E = -\int F^p \, dx \approx ZF^p_{max} \cdot (1 - F/F^p_{max})^{3/2}$$

$$= ZF^p_{max} \left\{ 1 - \left(\frac{\tau}{\tau_{c0}} \right)^{2/3} \right\}^{3/2} \tag{14-11a}$$

with τ_{c0} from equation (14-6) and Z measuring the obstacle width. For constant dislocation velocity v and strain rate, E must be a constant fraction $(= \ln v_0/v)$ of kT (with $v_0 = $ const). Then the temperature dependent part of the critical shear stress of a solid solution is

$$\Delta\tau_c(T, v) = \tau_{c0} \left\{ 1 - \left(\frac{T}{T_0} \right)^{2/3} \right\}^{3/2} \tag{14-11b}$$

with $kT_0 = ZF^p_{max}/\ln(v_0/v)$. τ_{c0} is the critical shear stress at $T = 0$. $\Delta\tau_c$ characterizes the decrease of the critical shear stress with increasing T towards a plateau which is shown in fig. 14.7 for *Ag*Al. The plateau stress is also proportional to τ_{c0} [14.5]. It is thought to be caused by dislocations overcoming groups of solute atoms without the help of thermal activation. At low temperatures and $\eta_0 < 1$ (equation 14-10a) τ_c has been found recently to decrease again. This is most likely a dynamical effect of the inertia of the dislocation which lets kinetic energy help in overcoming solute obstacles if

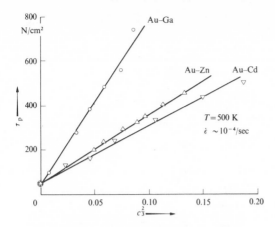

Fig. 14.8. Critical shear stress of gold solid solutions at 500 K according to equation (14-10).

the electron drag of the dislocation becomes small, section 11.4.2 [11.11], [11.11a], [14.4a].

14.1.3   *Experimental results for the solid solution hardening of single crystals* [12.5], [14.2]

The material parameter governing solid solution hardening is thus τ_{c0}, the critical shear stress of an alloy crystal at low temperatures, or because of the scaling law (14-11) also the shear stress τ_p in the plateau region. The concentration dependence of τ_p has been studied in particular for alloy crystals of Cu, Ag and Au. The experimental results are well described by a $v_B^{2/3}$ dependence, fig. 14.8. The slopes are satisfactorily explained by the Labusch combination of the solute parameters δ and η in K_{max}^{eff} (14-10), fig. 14.9. The hardening of Au by Cd and Zn is due predominantly to the difference in atomic size δ of $+5\%$ and -5% respectively, whereas in alloys with Ga and Ge the modulus defect $(-\eta)$ of over 100% is responsible. In the case of *Au*–Al both effects contribute to the hardening; δ and η have the same sign with the result that a small solute atom is a 'soft spot' in the crystal. (According to C. Zener the nearest neighbours of a small solute atom should lie in the high curvature region of the potential well, in other words in the region of high modulus, see fig. 7.17.) Thus, as predicted in section 14.1.2(a), a *quadratic combination* of δ and η in K_{max} is necessary to describe the experimental results. Similar results are obtained for alloys based on lead and cph magnesium. Fleischer has found that the hardening of polycrystalline α-iron by carbon in the

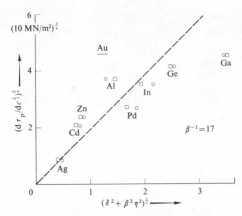

Fig. 14.9. Specific hardening of gold solid solutions (slope of the lines in fig. 14.8) plotted against the solute atom size (δ) and modulus parameter (η) according to equation (14-10). (Circles or squares are obtained for a different choice of second-order parelastic interaction between solute atom and screw dislocation [14.2].)

concentration range $v_C \leqslant 10^{-4}$ can be explained by (14-6). Solid solution hardening by carbon is largely responsible for the hardness of α-iron and martensite and thus for the strengthening of steels (section 13.6).

14.2 Dislocation locking and unlocking

At temperatures slightly above room temperature the solute atoms in the alloys discussed above become sufficiently mobile to diffuse to positions of large negative interaction energy $U(r, \alpha)$ near the dislocation. The time law for this decomposition process was discussed in section 9.2. Oversize substitutional solute atoms collect in the dilatation zone of an edge dislocation and undersize atoms in the compressive zone. As in the barometric formula the entropy at finite temperatures opposes a 'condensation' of solute atoms in the dislocation core. A Boltzmann distribution of solute atoms is established around the dislocation with a concentration

$$c(r, \alpha) = c_0 \exp\left(-\frac{\Delta E^P(r, \alpha)}{kT}\right). \tag{14-12}$$

This is called a Cottrell atmosphere [12.10]. At low temperatures a Fermi distribution should be used because each lattice site in the dislocation core can be occupied only once. Recently atom probing (section 2.4.2) has made visible the carbon distribution around screw dislocations in α-iron. The carbon atoms all sit within a 2 nm radius from the dislocation and are inhomogeneously distributed along the dislocation (Oxford, unpublished). For topological reasons (definition of the dislocation), the atmosphere is not completely able to compensate the long-range stress field of the dislocations which attract the solute atoms.

The arrival of solute atoms at the dislocation has a particularly strong influence on the *anelastic properties* (see section 2.7); the anelastic contribution to the strain ε_A of a dislocation bowing out between two pinning points separated by l (fig. 14.3) is proportional to the area covered by the dislocation $f = 2lx/3$. The amplitude x is obtained from the equation of motion for forced vibration (see section 11.4)

$$\rho_m b^2 \ddot{x} + B\dot{x} + E_L \frac{d^2 x}{dy^2} = \tau_0 b \exp[i\omega t]. \tag{14-13}$$

For small curvatures

$$\frac{d^2 x}{dy^2} = \frac{1}{R} \approx \frac{8x}{l^2}.$$

Thus with $E_L \approx Gb^2/2$ the characteristic frequency of the dislocation segment is

$$\omega_0 = \sqrt{(8E_L/\rho_m b^2 l^2)} \approx \left(\frac{2}{l}\right)\sqrt{\frac{G}{\rho_m}} \approx b\omega_{\text{Debye}}/l.$$

The inertia term can be neglected for frequencies of the applied stress $\omega \ll \omega_0$. The amplitude of the dislocation is then

$$x = \frac{\tau_0 b l^2}{8E_L} \exp[i(\omega t - \varphi)]; \qquad \varphi = \frac{B\omega l^2}{8E_L}. \tag{14-14}$$

Hence for M dislocation segments/cm^3, N dislocations per cm^2:

$$\varepsilon_A = \frac{2Mlxb}{3} = \frac{2Nbx}{3}.$$

The real part of the anelastic strain gives the modulus defect (see section 2.7) for $\varphi \ll 1$

$$\frac{\text{Re}(\varepsilon_A \exp[-i\omega t])}{\tau_0/G} = \frac{\Delta G}{G} = \frac{Nl^2}{6} \tag{14-15}$$

and the imaginary part determines the energy loss per cycle ($\varphi \ll 1$)

$$\frac{\Delta u}{u} = \frac{\text{Im}(\varepsilon_A \tau_0 \exp[-i\omega t])}{\tau_0^2/2G} = \frac{Nl^2 \varphi}{3} = \frac{Nl^4}{24} \frac{B\omega}{E_L}. \tag{14-16}$$

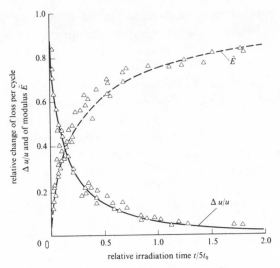

Fig. 14.10. Changes in the elastic modulus and the internal friction of a copper single crystal on irradiation with neutrons. (The characteristic time t_0 corresponds to a radiation dose of 2×10^{11} n/cm^2; $\omega = 12$ kHz [14.9].

It is thus apparent that a reduction of the free segment length by subdivision will have an extraordinarily marked effect on the anelastic contribution by the dislocations. This is observed experimentally, particularly clearly in the case of radiation-induced point defects (see chapter 10), which then diffuse to the dislocations [14.9], fig. 14.10. On the other hand the proportionality of $\Delta u/u$ to ω required by (14-16) is questionable much below the resonance frequency $\omega = \omega_0$. This lies in the mega to gigahertz range for the usual segment lengths $l \approx 10^4 b$ and is strongly overdamped [14.9]. At larger alternating stress amplitudes τ_0, the dislocation breaks away from the solute atoms that pin it and a hysteresis loop is observed, see A. Granato and K. Lücke [14.10].

The Cottrell atmosphere was formulated specifically for carbon in α-iron. In fcc solid solutions the Suzuki effect of solute atom segregation to stacking faults is more likely to be important. The dislocation is locked in its rest position by the atmosphere. An additional force must be applied to free the dislocation unless it has been rendered completely immobile. In any event a *pronounced yield point* is to be expected, as is observed in α-iron containing carbon, fig. 14.11. It differs from the yield drop caused by multiplication of dislocations (fig. 12.6) in its sharpness, which is responsible for technically undesirable deformation instabilities. The deformation

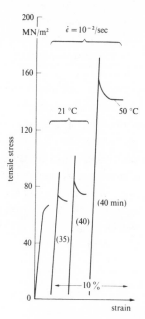

Fig. 14.11. Pronounced yield points in Fe–5 × 10⁻³ wt.% C single crystals at various deformation temperatures after ageing at 21 °C for the times (in min) given in brackets.

starts locally at a stress concentration (e.g. on the specimen surface) and propagates over the specimen in the form of a Lüders band. The localized deformation continues until the specimen fractures or work hardens. In the latter case it recommences in the region ahead of the Lüders band. The localized strain rate is thus much higher than the nominal strain rate. E. Hart [14.6] has described Lüders band deformation phenomenologically.

The process of unlocking a dislocation from a dilute or condensed Cottrell atmosphere has been investigated extensively particularly for Fe–C [12.10], [11.3]. It is difficult to find the right model for this process because various phenomena suggest that the carbon segregation in the dislocation core leads to localized formation of carbide particles: the yield drop can be 'overaged' (see sections 9.1.2 and 14.3); it is reasonably independent of the deformation temperature; it depends on the sign of the dislocation curvature resulting from ageing under tension/compression. All these observations are incompatible with the Cottrell atmosphere model, but are typical of discrete precipitates on the dislocation line.

At even higher temperatures the solute atom segregation on the dislocation is in any case small. A pronounced yield point can still be obtained in the Fe–C system if the carbon atoms along the cube edges of the α-iron which cause tetragonal distortion align themselves in the neighbourhood of the dislocation such that they match its distortional field at every point (cf. the remarks concerning the Snoek effect in section 6.1.2 and [14.7]). Both this 'Snoek atmosphere' and the 'Cottrell atmosphere' discussed above can move with the dislocation at high temperatures and low dislocation velocities thus reducing their hardening effect. This causes characteristic instabilities which manifest themselves in the stress–strain curve as the *Portevin–Le Chatelier effect*: repeated yield drops produce a saw tooth stress–strain curve. If the dislocation moves more slowly more solute atoms collect on it and increase the resistance to its movement; if it moves more quickly it can escape and move freely [12.10]. A situation of stationary drag is possible if the dislocation velocity v_D equals the drift velocity v_{SA} of the solute atom (carbon). According to Einstein (see section 8.5)

$$v_D = v_{SA} = \frac{D_C}{kT} K_D = \frac{D_c}{kT} \frac{\tau b}{n}. \tag{14-17}$$

n is the number of solute atoms dragged along per unit length of dislocation, subjected to a force τb, and thus from (14-12)

$$n = c_0 \iint \exp\left(\frac{-\Delta E^p(r, \alpha)}{kT}\right) r \, dr \, d\alpha. \tag{14-18}$$

Finally the strain rate $\dot{\varepsilon}_c$ for this microcreep can be expressed in terms of the variables (c_0, T, τ) with the aid of (12-2), (11-9), (14-3) and (8-14). The result has been confirmed by experiment [14.2], [14.8].

14.3   Precipitation hardening [14.11], [14.12]

The reactions discussed in chapter 9 result in the precipitation of a second phase, which differs from the matrix in its composition or degree of order. These precipitates represent effective obstacles to dislocation motion. Precipitation hardening is of great technological importance, e.g. in duralumin (section 9.1.2) and steels (section 13.6). The relevant parameters are the strength, volume fraction, spacing, shape and distribution of the particles. Several elementary interactions are the same as in solid solution hardening, section 14.1.1. The superposition of the interactions of a statistical particle distribution with the dislocation generally fulfils the requirements of the Fleischer theory, section 14.1.2. A new aspect is the possibility of overageing of the alloy with a consequent decrease in its strength, determined by the Orowan process, section 11.2.2.

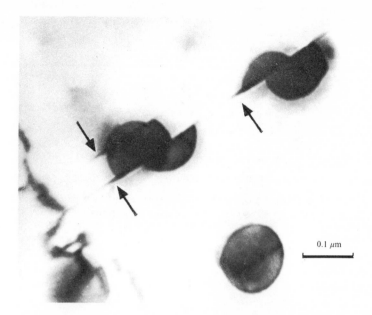

Fig. 14.12. Transmission electron micrograph of sheared Ni_3Al particles in a Ni–19% Cr–6% Al alloy after 2% deformation in tension (H. Gleiter and E. Hornbogen, Göttingen).

14.3.1   *Interactions between precipitate and dislocation*

Mechanisms (a) and (b)   As in solid solution hardening, a precipitate can interact with a passing dislocation parelastically by virtue of its distortion, (2-6), and dielastically (because it has a different shear modulus from that of the matrix). From section 14.1.1 the maximum interaction forces for a spherical particle of radius r_0 are $K_{max}^p \approx Gbr_0|\delta|$ and $K_{max}^d \approx Gb^2|\eta|/20$, respectively. δ and η are the differences in lattice parameter and shear modulus between matrix and particle.

Mechanism (c)   If the particle is coherent with the matrix, the dislocation can pass through the particle cutting it. It is necessary to provide surface energy for this because the particle is sheared on the slip plane by an amount equal to the Burgers vector, fig. 14.12. If $\tilde{E}_{\alpha\beta}$ is the specific particle–matrix interfacial energy, then the interfacial energy which must be supplied at the circumference of the particle corresponds to a repulsive force on the dislocation $K_{max}^{\alpha\beta} \approx \tilde{E}_{\alpha\beta}b$.

Mechanism (d)   If the particle is ordered, an antiphase boundary of energy $\tilde{E}_{APB}$ is produced when it is cut, fig. 14.13. This corresponds to a repulsive force $K_{max}^{APB} \approx \tilde{E}_{APB}r_0$. A second dislocation, however, restores the

350   *Alloy hardening*

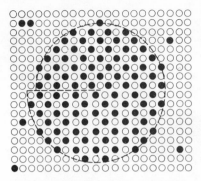

Fig. 14.13. An ordered Ni₃Al particle is cut by a perfect dislocation in the Ni (○) Al (●) matrix producing an antiphase boundary.

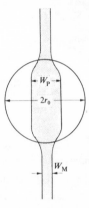

Fig. 14.14. The dissociation width W_M of a dislocation changes as it passes through a precipitate (W_P) with a smaller stacking fault energy.

order and is thus attracted to the particle. (Extending this concept, there is a general tendency in ordered alloys for super-dislocations to be formed from two perfect dislocations of the unordered structure with an antiphase boundary between them, see section 7.3.1.) If more than r_0/b dislocations cut the particle on the same plane, cutting becomes increasingly favourable, and this explains why slip is concentrated on relatively few slip planes in precipitation hardened alloys, see fig. 14.12.

Mechanism (e) If the precipitate has a stacking fault energy γ_P different from that of the matrix (γ_M) referred to the same crystal structure this also results in an interaction with the dislocation, see fig. 14.14. The maximum interaction force is $K_{max}^{SF} \approx 2r_0(\gamma_M - \gamma_P)$ (for $W_M < r_0 \approx W_P$). If these K_{max} values are substituted in the Fleischer equation (14-6) and c_F is replaced by the volume (or areal) fraction v_P of particles then for most of the above mechanisms (except b and c)

$$\tau_c b \approx \tilde{E}^{3/2} v_P^{1/2} r_0^{1/2} / \sqrt{(6E_L)}. \tag{14-19}$$

$\tilde{E}$ represents the interfacial energies in the mechanisms (d) and (e) or $Gb|\delta|$ in mechanism (a). In actual fact the situation in case (b), dielastic interaction with small particles, is more complicated [14.18]. The theory results in a $\tau_c(r_0)$ which is not observed in CuFe [14.19]. For a given v_P, τ_c is observed to increase parabolically with r_0 at small r_0 on coarsening (ageing) (see section 9.3), e.g. for Cu–Co. In this case $\delta_{Co} = -1.29\%$ and $\gamma_{cphCo} < 0$ so that mechanism (a) is certainly expected to contribute and possibly also (e). The same applies for the ageing of Al–Zn (after room temperature ageing the precipitates consist of up to 72% Zn).

There is virtually no difference in the size of the atoms in Al–Ag, but since $\gamma_{Ag} \ll \gamma_{Al}$, mechanism ($e$) can be expected. Precipitates of Ni_3Al in Ni–18% Cr–6% Al are distortion-free but ordered and hence mechanism (d) dominates (see fig. 14.13). The above theory is not directly applicable to the plate-like GP I zones in Al–Cu (see section 9.1.2). Analysis [14.12] shows that mechanism (c) dominates in this system because of the large surface to volume ratio of the zones. The zones which have to be cut in this alloy are of atomic thickness. Consequently this is the only process of those discussed above which is assisted by thermal activation (section 12.3). τ_c is generally found to be proportional to $\sqrt{v_P}$ as predicted by (14-19).

14.3.2   *Orowan process and geometrically necessary dislocations*

In section 11.2.2 it was shown that under the Orowan stress a dislocation can pass between two obstacles separated by l. Precipitation hardening is thus limited for large particles of radius r_0. There is a 'critical dispersion' r_{0c} corresponding to maximum hardening which can be calculated by equating τ_c from (14-19) with τ_l

$$\alpha \frac{Gb}{r_{0c}} \sqrt{v_P} = \frac{\tilde{E}^{3/2} \sqrt{r_{0c}}}{b^2 \sqrt{(3G)}} \sqrt{v_P},$$

therefore

$$r_{0c} = \frac{Gb^2}{\tilde{E}} (\alpha\sqrt{3})^{2/3}. \tag{14-20}$$

The constant α is still slightly dependent on the stacking fault energy and r_0 because r_0 determines the annihilation of neighbouring Orowan loops at an obstacle (see section 11.2.2). The Orowan process is effective also at small

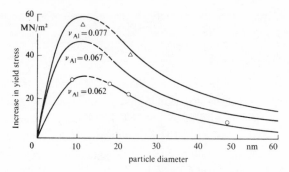

Fig. 14.15. Increase in the yield stress of Ni due to Ni₃Al precipitates of different sizes and of different volume fractions (H. Gleiter and E. Hornbogen, Göttingen).

radii, $r_0 \approx 10$ nm, if the specific obstacle strength $\tilde{E}$ $(= Gb/|\delta|$ for mechanism (a)) is large (since $r_{0c} \propto b/|\delta|$). The combination of τ_c from (14-19) and τ_l from (11-8) with $l = r_0/\sqrt{v_P}$ gives the dependence of the flow stress on the particle radius shown in fig. 14.15 for constant v_P. Further dislocations which traverse this slip plane and pass between the particles by the Orowan process form a system of concentric loops around the particles. Ashby, [14.13] and section 12.4.1, has called these loops around non-deformable particles in a deformable matrix 'geometrically necessary dislocations' in contrast to the 'statistically accumulating' dislocations in normal work hardening. The former are actually observed in transmission electron microscopy, especially at incoherent oxide particles (e.g. Al₂O₃ in Cu produced by internal oxidation of a Cu–Al matrix, see section 8.6), fig. 14.16. If cross-slip is possible the slip dislocation loops transform into prismatic loops. There is a marked tendency for the loops to enter the matrix–particle interface, thus destroying the coherence, or, in the form of slip dislocation loops to fracture the particle and thus to shear it. Apart from this, the high density of dislocations in these alloys explains the high work hardening, see fig. 14.17 [14.13].

14.4   Dispersion hardening and fibre-reinforcement [14.14], [12.7]

The internally oxidized alloys mentioned above are an example of dispersion hardening. Here, in contrast to precipitation hardening, incoherent, non-deformable particles are introduced into a soft matrix to increase its strength. Often the second, harder phase is in the form of fibres parallel to the loading direction. In this way composite materials with specific mechanical properties are tailormade according to physical principles which will be discussed below.

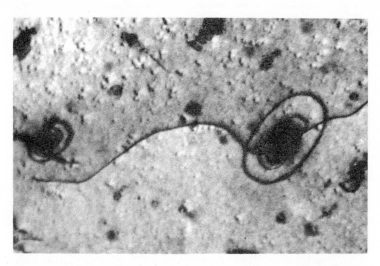

Fig. 14.16. Dislocation loops round Al_2O_3 particles in a Cu–30% Zn single crystal. The most closely spaced inner loops are about 100 nm in diameter. (P. B. Hirsch and F. J. Humphries, Oxford Univ.)

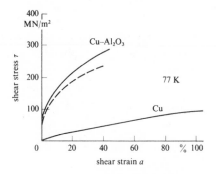

Fig. 14.17. Work hardening curves for copper single crystals with and without 1 vol.% Al_2O_3 particles. The dashed curve is the difference between the two full curves [14.13].

There are essentially three methods of preparing these composite materials: (1) the rapid directional solidification of a eutectic, see section 4.6; (2) the embedding of high strength fibres in a softer metallic matrix; (3) the 'in situ' preparation of fibre composite material by co-deformation of a powder mixture in a tube, e.g. by drawing. The strength of the material should be determined by the fibres. The function of the matrix is to bond the fibres, to transmit the load to the fibres and to protect their surfaces against

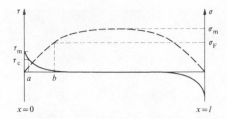

Fig. 14.18. Stress distribution in embedded fibres, length l, loaded in tension. Tensile stress σ, shear stress τ at the interface.

damage which could produce cracks. The propagation of transverse cracks is more difficult in a bundle of fibres than in a solid rod of the same material.

The strength of a sisal rope or a fibre glass rod serves as model in the preparation of a metallic composite, although here the physical processes are difficult to understand. It is not necessary to use infinitely long fibres as long as the matrix has a finite critical flow stress and the fibres a finite strength. This can be seen from fig. 14.18, which shows the distribution of the shear stress in the fibre–matrix interface and that of the tensile stress in a fibre of length l which is completely embedded in the matrix, both fibre nad matrix being subjected to an axial stress. The shear stress reaches a maximum τ_m at the end of the fibre, whereas the largest tensile stress σ_m is found at the centre. The ratio τ_m/σ_m depends on the elastic moduli of the two components and is about $1/10$ for metallic systems. Normally therefore the matrix deforms plastically in shear at $\tau_c < \tau_m$ at the fibre ends, $x < a$, before the fracture stress of the fibre σ_F is reached. For $x > a$ the tensile stress in the fibre increases until at $x = b$ it reaches the value σ_F. If $b < l/2$ the length of the fibre is no longer important. There is thus a critical length l_c (for a fibre radius $r \approx a$) [14.14]

$$l_c = r \frac{\sigma_F}{\tau_c}, \tag{14-21}$$

which the fibre must possess in order to be loaded up to the fracture stress. Otherwise the matrix just flows around the fibre. l_c/r is the critical 'aspect ratio'.

Fig. 14.19 shows stress–strain curves of copper tungsten composite specimens in which tungsten fibres with $r = 5$ and $10\,\mu m$ ($l_c/r = 10$) were embedded parallel to one another in copper in volume fractions up to $v_F = 40\%$ [14.15]. In stage I of the curve the fibre and matrix deform elastically. The modulus of the composite obeys the law of mixtures

$$\hat{E}_C = v_F \hat{E}_F + (1 - v_F)\hat{E}_M. \tag{14-22}$$

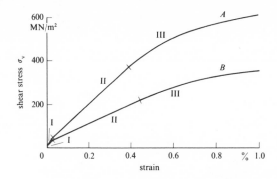

Fig. 14.19. Stress–strain curves for Cu–W composite rods (*A*) 20 μm diam., 20 vol.% tungsten; (*B*) 10 μm diam., 10 vol.% tungsten [14.15].

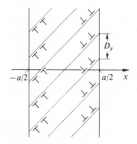

Fig. 14.20. Model [14.16] for dislocation pile-ups in the matrix at the fibre–matrix interface $(-a/2, a/2)$ to explain the work hardening.

In stage II, the matrix deforms plastically whereas the fibres still extend elastically. In this case $\hat{E}_M$ in (14-22) should be replaced by $(d\sigma/d\varepsilon)|_M$. The work hardening is 100 times larger than in pure copper. Above 0.5% extension the tungsten wires begin to deform plastically and fracture (stage III). The high work-hardening rate of the matrix which increases with v_F can be explained by dislocation pile-ups of the type shown in fig. 14.20. The free parameter of the theory [14.16] is obtained as $D_y = 10 \mu m$. The fracture stress σ_C of the composite follows the curve shown in fig. 14.21 as a function of v_F. σ_U is the fracture stress of the matrix and σ_M the stress in the matrix at which the fibres fracture. It is apparent that the volume fraction of fibres must exceed a critical value v_{crit} in order that strong fibres with low fracture strains can strengthen the matrix as it work hardens. The strength of the composite in tension depends strongly on the angle ϕ between tensile axis and fibre axis and is a maximum for $\phi = 0$.

Experimental composite materials have been investigated containing

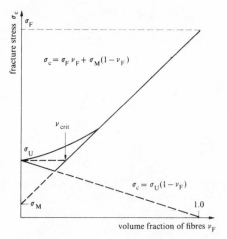

Fig. 14.21. Analysis of the fracture stress of a composite containing a volume fraction v_F of brittle fibres in a ductile matrix [14.14].

graphite and boron fibres or Al_2O_3 whiskers in Al and Ag. Whiskers grow from the vapour phase with diameters less than $10\,\mu m$ and since they are nearly perfect crystals they possess a high strength. G. Wassermann [14.17] has produced fibres of about $1\,\mu m$ diameter and an aspect ratio of 10^4 by 99% co-deformation of Fe–50% Ag powder mixtures. Such fibre materials are more than ten times stronger than the bulk metals, in other words well above the values expected from the law of mixtures, fig. 14.21. Other physical properties also exhibit strong anomalies when the constituent phases are so finely dispersed and so strongly work hardened.

15
Recrystallization [15.1], [1.3]

15.1 Definitions

In previous chapters we have discussed processes which introduce lattice defects into metals: deformation generates dislocations, irradiation can cause displacement cascades, martensitic and diffusion controlled transformations produce grain boundaries, etc. These processes change the microstructure creating a state of higher free energy. *Recrystallization* is the formation of a new microstructure with a lower free energy by a reaction in the solid state similar to the formation of the microstructure by the *crystallization* of a melt (chapter 4). In a typical recrystallization experiment, a heavily deformed metal is annealed at a temperature higher than half its melting point. This removes many of the lattice defects introduced by deformation and a *new arrangement of grain boundaries* is established. As was stated in chapter 3, the microstructure is, by definition, not in thermodynamic equilibrium. The grain boundaries remaining after recrystallization constitute a metastable arrangement separating grains with a certain orientation distribution. In other words the metal exhibits a characteristic *recrystallization texture*. Such textures have a vital influence on many of the physical properties of technological materials, e.g. magnetization losses in transformer sheet.

Recrystallization, the generation of a system of new grains, is fundamentally different from *recovery* which always precedes it and in which the lattice defects within a given arrangement of high-angle grain boundaries either anneal out or rearrange themselves. The recovery stages I to IV described in section 10.4 in which point defects created by irradiation, deformation or quenching anneal out are thus recovery in the true sense. Stage V is the recrystallization of deformed material. On closer examination the distinction is less precise because point defects often anneal out at dislocations, causing them to climb (section 11.1.3), and thereby enabling them to annihilate or arrange themselves in low-angle

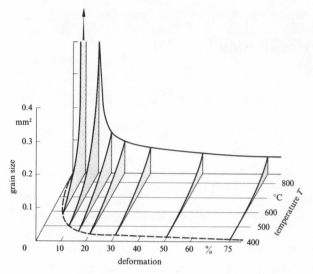

Fig. 15.1. Recrystallization diagram for electrolytic iron after annealing for 1 h (W. G. Burgers).

grain boundaries (section 3.2.1). The role of low-angle grain boundaries in the nucleation of recrystallization will, however, be discussed later.

Whereas recovery proceeds as long as there is a non-equilibrium concentration of point defects, the driving force for recrystallization is either the stored dislocation energy (primary recrystallization) or grain boundary energy (grain growth, secondary recrystallization). The expression 'secondary recrystallization' is normally employed to describe the renewed grain growth following primary recrystallization, i.e. after a new microstructure has been formed.

15.2　Primary recrystallization

15.2.1　*Phenomenology*

The grain size after recrystallization can be represented in a recrystallization diagram, fig. 15.1, as a function of the degree of previous deformation and annealing temperature (for a given, relatively long annealing time). This shows that a minimum ('critical') deformation is necessary to initiate recrystallization and that the larger the deformation the lower is the temperature at which recrystallization can begin. The region of light deformation and high annealing temperatures is exploited in the production of single crystals.

The second empirical aspect is the *recrystallization kinetics*. These can be described formally as in section 7.4(*b*) as Johnson–Mehl kinetics. The number N of recrystallization nuclei per unit volume can increase with time, e.g. $N = \dot{N}t$, where $\dot{N} = \text{const}$ so that according to (7-10) the recrystallized volume fraction $X(t)$ follows an S-shaped curve with an 'incubation period' or characteristic time for recrystallization of

$$t_i \approx (\dot{N}v^3)^{-1/4}. \tag{15-1}$$

v is the linear growth rate of a grain which is growing in three dimensions. In this period a mean grain size $d_i \approx vt_i = (v/\dot{N})^{1/4}$ is attained. It is apparent from (7-11) that v increases with the degree of deformation ε and, as a result of the Arrhenius factor, with the temperature. The recrystallization diagram $d_i(T, \varepsilon)$ together with (15-1) shows that $\dot{N}$ increases more rapidly with deformation than v, that it possesses a threshold value, and that its dependence on temperature is less than that of v.

The third empirical result obtained from a recrystallization experiment is the distribution of the orientations of the grains, i.e. the *recrystallization texture* which we shall discuss in section 15.4. The texture is determined by the orientation of the nuclei and by the dependence of their growth rate v on the orientation difference between them and the neighbouring grains, in other words on the nature of the grain boundary (sections 3.2.2 and 15.3).

15.2.2   *Stored energy* [15.2]

In order to describe primary recrystallization quantitatively we must use the gradient at the grain boundary of the energy e stored during deformation to represent the driving force for recrystallization rather than the 'degree of deformation' as in section 15.2.1. The energy e is stored in the material in the form of dislocations. It is released in stage V and can be measured calorimetrically, fig. 15.2. The stored energy in a copper single crystal deformed at 293 K to τ_{III} is $e \approx 5 \times 10^5$ J/m^3. (The area $\int \tau \, \mathrm{d}a$ under the work-hardening curve up to this stress (fig. 12.3) gives the work done by the machine as $6e$; apparently $\frac{5}{6}$ of the work is lost as heat.) As a rough approximation e can be expressed by the self-energy of the stored dislocations using (11-7) and (11-9)

$$e \approx E_L \cdot N \approx \frac{Gb^2}{2} \frac{\tau^2}{\alpha^2 G^2 b^2} \approx 25 \frac{\tau^2}{G}. \tag{15-2}$$

The increase in e with τ^2 is in fact observed experimentally. The energy of deformation actually stored is only 10^{-3} of the latent heat of fusion or 10^{-4} eV/atom. This small energy controls the process of recrystallization with all the associated vital microstructural changes.

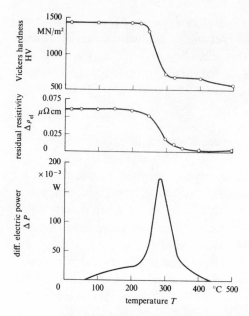

Fig. 15.2. Differential thermal analysis (section 2.5) of a copper rod plastically deformed in torsion. ΔP is proportional to the stored energy of deformation released at temperature T. The hardness and residual electrical resistance show a corresponding decrease [15.2].

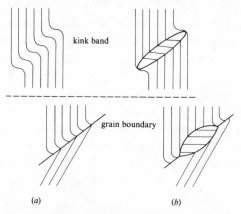

Fig. 15.3. Formation of subgrains (b) on annealing in regions of strong lattice curvature (deformation) (a).

15.2.3   *Polygonization* [15.3]

According to section 15.2.1 recrystallization begins in the most heavily deformed regions of the specimen. These are areas of high

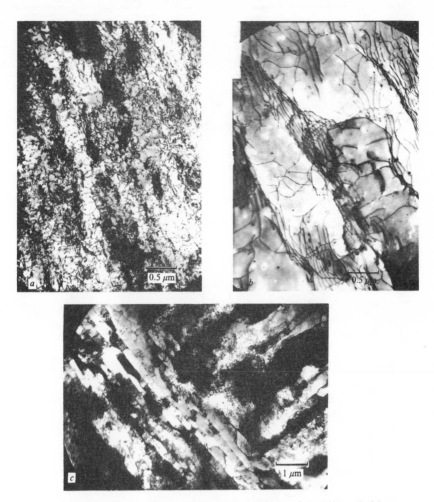

Fig. 15.4. Transmission electron micrographs of an 80% rolled iron
single crystal after annealing (*a*) for 20 min at 400 °C; (*b*) for 5 min at
600 °C; (*c*) as (*b*) but in the kink band where subgrains have already
formed (Hsun Hu in [15.2]).

dislocation density which form where the deformation is inhomogeneous.
For example, as was shown in section 12.5.1, dislocations can pile up at
grain boundaries. Regions in the neighbourhood of grain boundaries have
in fact been observed to act as nuclei for recrystallization. In reverse, a single
crystal homogeneously deformed up to 100% shear on *one* slip system does
not recrystallize at all, but recovers. If the orientation of the single crystal is
such that a second system operates, inhomogeneities arise: so-called
deformation bands (section 12.5.3) or specifically kink bands, fig. 15.3. It is

Fig. 15.5. Transmission electron micrograph of a recrystallized grain in iron after 80% deformation and a 125 min anneal at 600 °C. The new grain has arisen from the subgrains in a kink band. (Hsun Hu in [15.2].)

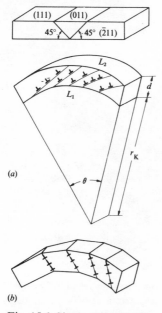

Fig. 15.6. Single crystal before and after plastic bending, and the definition of the experimental parameters for the density of the excess edge dislocations shown (*a*). Formation of polygon walls after an anneal (*b*).

found that the first 'subgrains' of the new microstructure form in such kink bands separated by low-angle grain boundaries, fig. 15.4. These low-angle grain boundaries can merge to form high-angle grain boundaries separating a new grain from the deformed matrix, fig. 15.5.

The creation of low-angle grain boundaries can be understood in terms of the model in fig. 15.6, which illustrates the process of *polygonization* in a plastically bent crystal. The bent crystal, in this case of Fe–3.25% Si, has an excess N_+ of dislocations of one sign on the slip planes, which is related to the radius of curvature r_K by the Nye relationship

$$N_+ = \frac{1}{r_K b} \tag{15-3}$$

(obtained by comparing the lengths of the upper and lower sides of the crystal, L_2 and L_1, which differ by the number n of extra half planes of the $N_+ = n/L_2 d$ edge dislocations which must be inserted, see fig. 15.6(a)). On annealing, the dislocations rearrange themselves into energetically more favourable configurations, namely low-angle grain boundaries, and the originally curved lattice planes of the crystal (e.g. edges L_2 and L_1) become the sides of a polygon (fig. 15.6(b)). X-ray reflections, which had become elongated as a result of the bending of the lattice planes, split up into a series of discrete spots. The rearrangement of dislocations from parallel to perpendicular to the slip plane in the specimen shown in fig. 15.6 can be observed directly (fig. 15.7) by etching the $(\bar{2}11)$ surface.

If the polygonization of kink bands provides the nuclei for recrystallization, one would expect the new grains to have orientations which could be derived from those of the old grains by rotations about axes perpendicular to the slip directions in the slip planes. It is still undecided whether oriented nucleation or selective growth depending on the growth rate of different grains determines the recrystallization texture (see section 15.4). The incubation time for recrystallization (15-1), which is determined largely by the nucleation rate $\dot{N}$, can thus be ascribed to the process of polygonization. The activation energy of $\dot{N}$ is then expected to be that for dislocation climb, i.e. according to section 12.4 that for self diffusion. If the specimen is heated up slowly it can have polygonized completely before recrystallization can even begin. In other words recovery has taken place and the driving force for primary recrystallization no longer exists.

15.3   Grain growth

15.3.1   *Experimental observations on grain boundary migration*

In principle, the growth rate v of a new grain in a deformed single crystal matrix can be measured as a function of orientation difference, position of grain boundaries, degree of deformation and temperature. There are, however, experimental difficulties in avoiding nucleation at other sites and recovery of the matrix during the anneal. This difficulty is circumvented if driving forces other than the stored energy of deformation

Fig. 15.7. Dislocation etch pits on a bent Fe–3.25% Si single crystal after a 1 h anneal at (*a*) 650 °C; (*b*) 700 °C; (*c*) 850 °C, 430×. After C. G. Dunn [15.3].

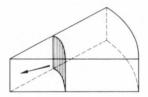

Fig. 15.8. Migration of a grain boundary due to its surface energy [15.5].

are used. Working on lead, Aust and Rutter [15.4] caused a grain boundary to move into a fine-grain, mechanically stable array of low-angle grain boundaries produced by solidification. A third method, fig. 15.8, utilizes an azimuthally oriented grain boundary in a single crystal with the shape of a circular sector. The boundary contracts towards the apex as a result of its self energy [15.5]. Further methods of measuring v are discussed in reference [15.6].

The most important result obtained from such measurements is the dependence of the grain boundary velocity on the orientation difference θ across the boundary, fig. 15.9. The activation energy $-\mathrm{d}\ln v/\mathrm{d}(1/kT) \equiv Q$ is also strongly dependent on θ, fig. 15.10. It thus turns out that the

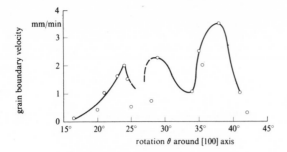

Fig. 15.9. Grain boundary velocity in lead at 200 °C. After K. Aust and W. Rutter [15.4].

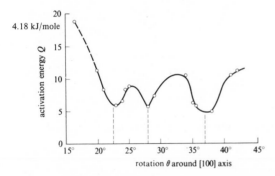

Fig. 15.10. Activation energy of the grain boundary velocity as a function of the orientation difference for lead [15.4].

empirically very mobile grain boundaries (corresponding to $\theta = 38°$ about $\langle 111 \rangle$ or $\theta = 23°$, $28°$, $37°$ about $\langle 100 \rangle$) are precisely the 'special grain boundaries' with a high density of coincidence sites described in section 3.2.2. In general v increases with θ for high-angle grain boundaries in the range $\theta = 15$ to $40°$ whereas for low-angle grain boundaries below $\theta = 10°$ the opposite is observed. In terms of the dislocation model for low-angle grain boundaries, section 3.2.1, it is plausible that the mobility of dislocations decreases with increasing dislocation density. In considering these results, the purity of the metal must be taken into account. Fig. 15.11 shows that the velocity of special boundaries (with a high density of coincidence sites) in lead is influenced much less by the addition of Sn than that of ordinary, i.e. randomly oriented boundaries. The same is true for the activation energy Q, fig. 15.12. The preferential growth of selected grains of certain orientations is strongly promoted if not actually caused by alloying additions even in the ppm range. This has a considerable effect on the

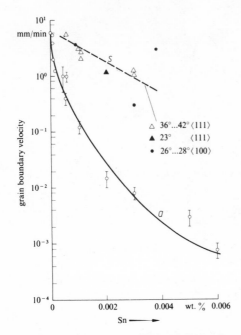

Fig. 15.11. Grain boundary velocity of special boundaries (*s*) and randomly oriented boundaries (*a*) at 300 °C in lead containing some tin [15.4].

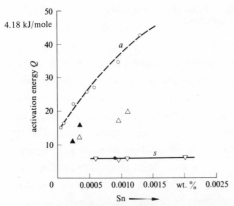

Fig. 15.12. Activation energy of the grain boundary velocity of special (*s*) and general (*a*) grain boundaries in lead as a function of the tin content [15.4].

recrystallization texture and also on the kinetics (15-1). A *recrystallization temperature* T_R is defined as the temperature at which a certain volume fraction X_R recrystallizes in a given time ($t_R \approx t_i$), (7-10) and (7-11). For many

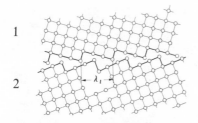

Fig. 15.13. Model of a high-angle grain boundary showing a periodic succession of crystal ledges.

metals of normal purity (≈ 100 ppm impurities) $T_R \approx 0.4 T_m$ (melting temperature).

Addition of an alloying element with an atomic size difference $\delta \approx 1\%$ raises T_R by about 10 K, for $\delta \approx 10\%$ by 100 K. There is thus a parelastic interaction between grain boundaries and solute atoms, which will be discussed in more detail in section 15.3.3.

15.3.2   *Theory of the velocity of 'pure' grain boundaries*

We start with the atomic arrangement in a high-angle grain boundary as proposed in the model of 'structural units' described in section 3.2.2. Fig. 15.13 shows that the most densely packed planes of grains 1 (deformed) and 2 (already recrystallized) end at the boundary in a periodic arrangement of *crystal ledges*. The period λ_l depends on the orientation difference between the grains as well as on the position of the boundary. If the boundary is not symmetrical with respect to both grains, the ledge periodicities $\lambda_{l1}, \lambda_{l2}$ are different. λ_l is smaller for 'special' than for normal boundaries (see chapter 3). Jogs can arise along a crystal ledge, as shown in fig. 15.14 (separation of jogs λ_j or $\lambda_{j1}, \lambda_{j2}$). According to H. Gleiter [15.7] the grain boundary migrates by atoms becoming dissociated from the jogs in the crystal ledges of (1), diffusing along the boundary to the jogs in the crystal ledges of (2) and reattaching themselves. The free energy required to dissociate an atom from a jog in the recrystallized crystal (2) is the activation energy ΔF_{j2}, and in the deformed crystal (1) the smaller energy $\Delta F_{j1} = \Delta F_{j2} - \Delta F_{12}$ where ΔF_{12} is the stored (free) energy of deformation. The diffusion coefficient for grain boundary diffusion (see section 8.4) is

$$D_G = v_0 b^4 c_V \exp(-\Delta F_M / kT), \tag{15-4}$$

where c_V is the areal density of vacancies in the grain boundary due essentially to the open structure of the boundary, ΔF_M is the migration

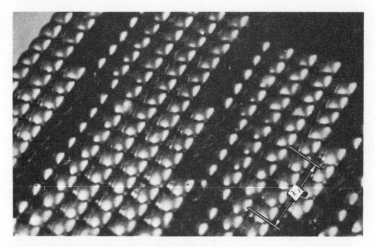

Fig. 15.14. Jogs in the crystal ledges of a high-angle grain boundary (looking down on the surface of a grain, [15.7]).

energy of an atom in the high-angle grain boundary (section 8.4.1), b the atomic separation, v_0 a characteristic jump frequency. The frequency with which the atoms change position as the boundary moves, i.e. the flux of particles per unit time and area in direction $1 \rightarrow 2$ is calculated in three stages:

(*a*) The transfer of an atom at a jog on the surface of crystal 1 to a position along the ledge of this crystal (and back) takes place at the net rate

$$J_{jl} = \frac{v_0}{\lambda_j \lambda_l} \exp[-\Delta F_{j1}/kT] - \frac{v_0 b}{\lambda_j \lambda_l} c_{l1}, \tag{15-5}$$

where c_{l1} is the linear density of atoms already separated from jogs along one ledge. It is assumed that these atoms are mobile along this ledge, i.e. do not require thermal activation.

(*b*) The atoms jump from the crystal ledge on to the crystal surface and thus into the grain boundary with a net flux

$$J_{lB} = \frac{1}{\lambda_l} \{ c_{l1} v_0 \exp[-\Delta F_l/kT] - b c_{B1} v_0 \exp[-\Delta F_M/kT] \}, \tag{15-6}$$

where ΔF_l is the energy required to dissociate an atom from the ledge of the crystal. c_{B1} is the areal concentration of dissociated atoms in the grain boundary ahead of a ledge. Equations analogous to (15-5) and (15-6) can be derived for crystal 2. In the stationary state of grain boundary migration

$$J_{jl}^{(1)} = J_{lB}^{(1)} = -J_{jl}^{(2)} = -J_{lB}^{(2)}.$$

(c) The difference in the densities c_{B1} and c_{B2} at the ledges in crystals 1 and 2 disappears over the distance λ_l as a result of grain boundary diffusion as described in section 8.1 at a rate of

$$J_{12} = \frac{D_G}{\delta} \frac{c_{B1} - c_{B2}}{\lambda_l}. \tag{15-7}$$

(δ is the 'thickness' of the grain boundary.) In the stationary state $J_{12} = J_{jl} = J_{lB}$ and the grain boundary velocity is given by

$$v = J_{12} b^3. \tag{15-8}$$

Combining equations (15-4) to (15-8) for both crystals, eliminating the parameters $c_{l1}, c_{l2}, c_{B1}, c_{B2}$ and assuming $\Delta F_{12} \ll kT$

$$v = \frac{b^3 v_0}{2\lambda_l} \frac{(\Delta F_{12}/kT) \exp(-(\Delta F_{j2} + \Delta F_l)/kT)}{\lambda_j \exp(-\Delta F_l/kT) + b + \delta/2c_v b^2}. \tag{15-9}$$

Gleiter gives a detailed description of the atomic jump processes involved in grain boundary migration in reference [15.7]. The following points emerge from an analysis of equation (15-9) in comparison with experiment (section 15.3.1):

(a) v is determined on the one hand by the process of dissociation of atoms from crystal ledges and their jogs and on the other by the vacancies present in the boundary which permit grain boundary diffusion. The structure of the grain boundary determines which term in (15-9) predominates. In general it may be assumed that $\Delta F_l \gg kT, c_v b^2 \ll 1$. In this case the third term in the denominator predominates and the grain boundary velocity v becomes proportional to the vacancy concentration c_v, i.e. to the 'porosity' of the grain boundary. It can be assumed that a grain boundary moving into deformed material contains more vacancies than it would in equilibrium (excess vacancy concentration Δc_v). It must continuously absorb the free volume of the dislocations and vacancies (agglomerates) in the deformed microstructure. There are experimental indications that irradiating cold-worked copper with neutrons increases the high-angle grain boundary velocity v on recrystallization by means of the resultant Δc_v. On the other hand v is noticeably smaller in thin Al wires (radius 10^{-1} mm) than in thick, and this can be attributed to the rapid removal of excess vacancies (Δc_v) along the grain boundary to the surface [15.8]. According to (15-9) the grain boundary velocity is strongly dependent on the grain boundary orientation via λ_l. In the case of a grain boundary lying parallel to a close-packed plane, i.e. $\lambda_l \to \infty$, v is very small in agreement, for example, with the well known low mobility of twin boundaries in recrystallization. Twist boundaries have no crystal ledges and are thus less mobile than tilt boundaries (section 3.2.1). (Ledges also form naturally on crystal surfaces where screw dislocations terminate in the grain boundary.)

(*b*) Equation (15-9) shows that to a first approximation v is proportional to $\Delta F_{12}/kT$, i.e. to the driving force for recrystallization (7-11) if $\Delta F_{12} \ll kT$. Even under these conditions, however, deviations from this relationship are observed, [15.5].

(*c*) The activation energy $Q = -\,\mathrm{d}\ln v/\mathrm{d}(1/kT)$ should lie between $(\Delta F_j + \Delta F_l)$ and $(\Delta F_j + \Delta F_l + \Delta F_{VF})$ where $\Delta F_{VF} = -\,\mathrm{d}\ln c_V/\mathrm{d}(1/kT)$ is the (free) energy of vacancy formation in the grain boundary. The experimental value for deformed Al single crystals is

$$Q = 96.2\ \mathrm{kJ/mol} = 23\ \mathrm{kcal/mol}$$

[15.8] which is comparable with the activation energy of grain boundary diffusion (section 8.4.2) and thus with the above limiting values. Frequently, however, Q values are found which are larger than the activation energy of bulk self diffusion. This cannot be understood in terms of the above theory and is attributed in the next section to an interaction between grain boundaries and solute atoms.

15.3.3   *Retardation of the grain boundary by solute atoms* [15.9], [15.10]

It can be expected that the solute atom concentration in the region of a grain boundary will differ from the mean concentration because the grain boundary is a more open structure than the crystal and can therefore provide sites with a lower potential energy for solute atoms of a different size ($\delta > 0$). A moving grain boundary, even in a high purity metal, will come into contact with many solute atoms and thus accumulate a solute atom atmosphere analogous to the Cottrell atmosphere around a dislocation (section 14.2). This solute atom atmosphere is dragged along by the moving boundary. During recrystallization it drifts by virtue of its interaction with the boundary. The grain boundary in its turn is held back by the solute atom atmosphere. At high velocities the grain boundary can break away from the atmosphere, a process analogous to the dynamic yielding in dislocation movement. This has been described by K. Lücke and co-workers [15.9] and will be outlined in the following. Fig. 15.15 shows the parelastic interaction potential $E_p(x)$ between solute atoms and a grain boundary and also the solute concentration $c(x)$ near a boundary moving to the right. The solute atoms have accumulated behind the boundary (relative to the mean concentration c_0). The velocity of the grain boundary

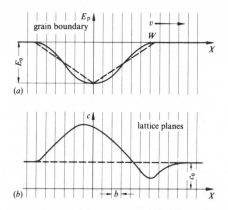

Fig. 15.15. Interaction potential $E_p(x)$ between a solute atom and a grain boundary (a). The solute concentration $c(x)$ near a grain boundary moving to the right (distribution coefficient > 1) (b) [15.9].

is given by (7-11) and (15-9)

$$v = \frac{Db^2}{kT}(P - P_F), \qquad P_F = \int_{-\infty}^{+\infty} c(x)\frac{dE_p}{dx}dx, \qquad (15\text{-}10)$$

where P is the difference in density of stored energy across the grain boundary, i.e. a pressure on the grain boundary, and P_F is the frictional drag due to the solute atoms. The solute atoms distribute themselves according to the solution of a diffusion equation including a drift term in a coordinate system moving at a velocity v (see chapter 8). The flux of solute atoms is in this case (constant solute diffusion coefficient D_{SA})

$$j_x = D_{SA}\left(\frac{dc}{dx} - \frac{c}{kT}\frac{dE_p}{dx}\right) - v(c - c_0). \qquad (15\text{-}11)$$

For the calculation, $E_p(x)$ is replaced by a saw tooth potential of width w and v is expressed in terms of the velocity $v_{free} = (Db^2/kT)P$ of freely mobile grain boundaries (without solute atom atmosphere).

The results are shown in figs. 15.16 and 15.17 in the form of reduced $v(c_0)$ and $v(P)$ curves. According to these, for medium solute atom concentrations and not too small values of P there are three solutions for v of which the middle one is unstable. The small velocity corresponds to the boundary with a solute atom atmosphere and the large one to the free boundary. The history of the specimen determines which v is found in practice. v is proportional to P for both solutions. The temperature dependence of v for the impure boundary, particularly if it breaks away from its solute atom atmosphere, is much greater than for the free

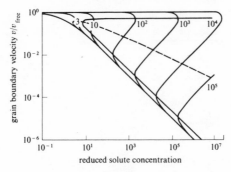

Fig. 15.16. Calculated velocity of a grain boundary with a solute atom atmosphere relative to that of a free boundary as a function of the normalized solute concentration $c_0(2K^2D/D_{SA})$ with

$$K^2 = \left[\frac{kT}{E_0}\exp\left(\frac{E_0}{kT}\right) - \exp\left(-\frac{E_0}{kT}\right) - 2\frac{E_0}{kT}\right],$$

see fig. 15.15. The parameter of the curves is the normalized driving force $P(KDb^3/D_{SA}E_0)$, $E_0 \approx 0.2\,\text{eV}$.

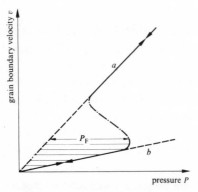

Fig. 15.17. Calculated grain boundary velocity as function of the driving force P for (a) free grain boundaries and (b) those with solute atom atmosphere. P_F is the frictional force due to the solute atoms [15.9].

boundary. The difference in Q is determined by E_0. The assumptions made in this model (particularly in the above presentation) are, however, quite restrictive. In particular (for $k_0 < 1$) strong segregations of solute atoms can form in the grain boundary, in which case the theory becomes invalid [15.6]. Fig. 15.18 shows the results of grain boundary velocity measurements in Al alloys which indicate something of the discontinuity predicted by the theory [15.10].

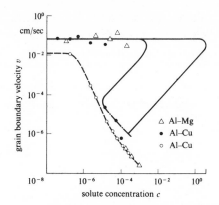

Fig. 15.18. Measured grain boundary velocities in Al single crystals as function of the solute atom concentration. After P. Gordon, and O. Dimitrov and co-workers, see [15.9].

15.3.4   *Recrystallization of a two-phase alloy* [15.11]

Many technically important alloys are two-phase, consisting often of a matrix with a small volume fraction $v_P < 10\%$ of a second phase (steels, aluminium alloys or even materials with included bubbles or voids). The particles of the second phase interact with the recrystallization front (the grain boundary between the old and new microstructures). This can lead to various different situations:

(*a*) The particles remain unchanged during the recrystallization anneal (for example SiO_2 particles in Cu, section 8.6, or so-called SAP, partly sintered and oxidized aluminium powder). They then retard recrystallization in the same way that they hinder deformation in precipitation hardening, see section 14.3. In comparing the recrystallization of a matrix and a heterogeneous alloy it is necessary to bear in mind the differences in the driving force for primary recrystallization due to the different dislocation densities (see 14.3.2).

(*b*) Small particles can be dragged along by the boundary.

(*c*) Large incoherent particles (radius $r_0 > 1\,\mu m$) can accelerate the nucleation of the new microstructure in their vicinity (e.g. oxides in iron).

(*d*) In the course of the recrystallizing anneal precipitates at the recrystallization front can coarsen by the growth of some particles at the expense of others (section 9.3). This increases the driving force for recrystallization but slows down the recrystallization front (e.g. Ni with Ni_3Al).

C. Zener has drawn an analogy between process (*a*), i.e. the slowing down of the recrystallization front by particles with a mean separation *d*, and the

Orowan mechanism of precipitation hardening, section 14.3.2. If a particle is lying in the grain boundary, it represents an energy saving of the order of $\Delta E = \tilde{E} \pi r_0^2$. This must be provided (over the distance r_0) if the grain boundary is to break away from the particle. From geometrical considerations the number n of particles which are intersected by unit area of a plane grain boundary is $n = 3v_P / 2\pi r_0^2$. The maximum binding force per unit area of grain boundary is therefore

$$P_{max} = n \frac{\Delta E}{r_0} = \frac{3v_P \tilde{E}}{2r_0}. \tag{15-12}$$

A driving force of this magnitude would bow out the grain boundary to a radius of curvature $R = 2\tilde{E} / P_{max}$ like the dislocation in the Orowan process, section 11.2.2. As in the case of the Orowan stress for dislocations, a critical pinning condition is reached for $R = d$ which leads to Zener's condition for critical particle separation

$$d_{crit} = \frac{4r_0}{3v_P}. \tag{15-13}$$

A dispersion of particles with a separation $d < d_{crit}$ thus anchors the boundary. In this case there can be no further discontinuous recrystallization, i.e. at a reaction front. Instead the material recrystallizes *in situ*, i.e. forms a new microstructure between the particles by homogeneous rearrangement of the excess dislocations to high-angle grain boundaries. It is interesting that in this case the deformation texture is often retained.

15.4   Recrystallization textures [12.11], [15.12]

As has already been discussed earlier in this chapter there are several reasons why recrystallized grains can have a preferred orientation: either the recrystallization grains have a preferred orientation or the growth rate of the growing grains depends on their orientation relative to the deformed matrix (see section 15.3.1). Both these theories, proposed by W. G. Burgers and P. A. Beck respectively, have found support in the literature. (Other nucleation mechanisms have also been proposed [15.13].) In the case of fcc metals, the concepts of oriented nucleation in a kink band described in section 15.2.3 would result in a rotation of the recrystallization texture about $\langle 112 \rangle$ with respect to the deformation texture. Observations in the high-voltage electron microscope (in which thick foils are transparent) show that heating of deformed copper and aluminium indeed produces such recrystallized nuclei [15.17]. During growth, multiple twinning occurs which dramatically changes the orientation of the growing

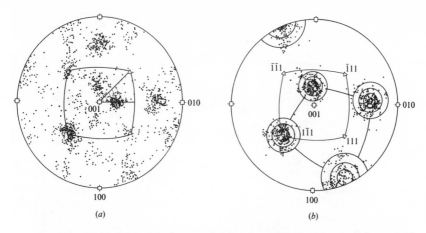

Fig. 15.19. (111) pole figures of the recrystallized grains in deformed Al single crystals with axes in the standard triangle (*a*). Transformation of this pole figure so that the initial and final orientations of the grains transform one into the other by a rotation about [1$\bar{1}$1] (*b*). B. Liebmann and K. Lücke, Göttingen, after [15.12].

grain. The reason for twinning seems to be a decrease of grain boundary energy (to that of a 'special' boundary) and an increase of mobility, particularly in impure aluminium. The final recrystallization texture of Al, fig. 15.19, then relates to that of the deformed material by rotations of 40° around ⟨111⟩. These correspond almost exactly to 'special' boundaries, i.e. the very mobile boundaries in impure fcc metals. The simple cube texture (001)⟨100⟩ is obtained on recrystallizing heavily rolled fcc metals. Nuclei with this orientation do not appear to be present in the rolling texture itself [15.14]. They are obtained from the rolling texture (110)⟨1$\bar{1}$2⟩ by a 55° rotation about a ⟨110⟩ direction. (On the other hand, recrystallization twins to the cube texture arise by a 15° rotation in the other direction.) It is interesting that transmission electron microscopy shows that there is a definite orientation relationship between the first recrystallized grains and the *neighbouring* grains, which suggests oriented nucleation. The cube texture oriented crystals subsequently predominate by virtue of their very mobile grain boundaries [15.14]. 90% of the recrystallized grains produced by annealing {111}⟨011⟩ oriented bcc Fe–3% Si single crystals arise from a 27° rotation of the matrix about ⟨110⟩, which again corresponds to highly mobile 'special' grain boundaries. The axis of rotation ⟨110⟩ in the bcc lattice is, however, at right angles to the ⟨1$\bar{1}$1⟩ slip direction in the {121} slip plane, as would be expected from oriented nucleation in kink bands.

A number of crucial experiments can be understood only on the basis of

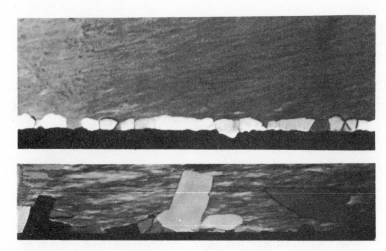

Fig. 15.20. Cross-section of an 80% rolled Al single crystal, rubbed on the underside with emery paper and annealed 2.5 min at 325 °C. A 10 μm thick layer of new grains has formed on the underside (*a*). After a further 12.5 min anneal, some of these grains have grown through the deformed matrix to the upper surface (*b*) (Paul Beck [15.12]).

selective growth [15.12]. An aluminium single crystal was rolled 80% to produce a pronounced deformation texture. The underside was rubbed lightly with emery and then numerous recrystallization nuclei were produced by annealing at 325 °C. These had virtually random orientations. On further annealing, recrystallized grains also appeared on the upper side of the crystal, fig. 15.20. They, however, exhibited a distinct recrystallization texture derived from the rolling texture by a 40° rotation about $\langle 111 \rangle$. From the complete range of nuclei available only grains of this orientation grew through the specimen. Since a very small volume fraction of the orientations present in the (never really sharp) deformation texture are able to determine the orientation of the recrystallization nuclei, it is difficult to arrive at a definite explanation of the mechanism of texture formation. For example, assuming the radius of a nucleus to be 1 μm and the grain size after recrystallization to be 10 μm, the recrystallization texture can be determined by the practically unobservable volume fraction of 10^{-3}.

15.5   Secondary recrystallization (grain growth) [15.15], [15.16]

After primary recrystallization is complete the structure is at best in metastable equilibrium. Even if the stored energy of deformation has been dissipated there is still an additional free energy $e \approx 10^4$ J/m^3 (grain size 50 μm) associated with the grain boundaries. As was shown in section

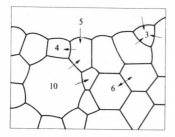

Fig. 15.21. Two-dimensional grain boundary arrangement in a sheet in which the number of corners determines the direction of growth of the grains. J. E. Burke after [1.3].

3.3 it is not possible to achieve mechanical equilibrium between the grain boundary tensions within a three-dimensional periodic arrangement of identical grains. A tendency to grain boundary movement still persists as shown in fig. 15.21 for the two-dimensional case. In this case, however, a mechanically stable grain boundary arrangement exists consisting of hexagonal grains with 120° angles between the sides. Grains with more sides grow, those with less shrink. The kinetics of grain growth are described by a driving force, proportional to the grain boundary area per grain volume, i.e. inversely proportional to the grain size. The same kinetics were described in (7-12) and (7-13) for the growth of domains. They also govern grain growth although there are some complications due to the interaction of the grain boundary with dissolved solute atoms and with particles. These can result in anomalous grain growth in which a very few grains grow extremely quickly, whereas the normal continuous growth of the majority of grains is retarded. The necessary condition for anomalous grain growth is a sufficiently high annealing temperature and a not too sharp primary recrystallization texture (in which case the orientation differences between primary grains, i.e. the mobility of their grain boundaries would be small). Even this grain growth often ceases when the grain size reaches twice the specimen thickness, because grain growth is then two-dimensional (with the possibility of 120° stabilization) and grain boundaries become anchored at surface grooves. If the surface energy is strongly anisotropic it can itself influence the recrystallization texture. In thin transformer sheet, Fe–3% Si, especially if it contains oxygen, ('tertiary') recrystallization produces a texture in which the crystal planes of low surface energy assume a position parallel to the sheet surface. The material thus possesses the $(001)\langle 100 \rangle$ cube texture which from the point of view of magnetization losses is extremely favourable. Thick sheets of this material (with additions of MnS) exhibit the magnetically soft 'Goss-texture' $(110)\langle 001 \rangle$ after secondary recrystallization.

Recrystallization is one of the most complicated and, despite the numerous investigations commensurate with its technical importance, one of the least understood phenomena in metallurgy. As we have seen, the study of recrystallization depends on a complete knowledge of the defect configuration following deformation, the thermally activated movement of the defects, the interactions with solute atoms in any form, and the physics and thermodynamics of interfaces, which together constitute the major part of physical metallurgy.

REFERENCES

1
1.1   Kittel, Ch.: *Introduction to Solid State Physics*. New York: J. Wiley, 4th Edn, 1971.
1.2   Cottrell, A. H.: *Theoretical Structural Metallurgy*. 2nd Edn. New York: St. Martin's Press 1955; new edn: *An Introduction to Metallurgy*. London: Arnold 1975.
1.3   Shewmon, P. G.: *Transformations in Metals*. New York: McGraw-Hill 1969.
1.4   Fast, J.-D.: *Interaction of Metals and Gases*. Eindhoven: Philips Techn. Library; vol. 1: 1965; vol. 2: 1971.
1.5   Fromm, E.; Gebhardt, E. (eds.): *Gase und Kohlenstoff in Metallen*. Berlin, Heidelberg, New York: Springer 1976.
1.6   Alefeld, G. (ed.): *Hydrogen in Metals*, vol. I, II. Berlin, Heidelberg, New York: Springer 1978.
1.7   Porter, D. A.; Easterling, K. E.: *Phase Transformations in Metals and Alloys*. New York: van Nostrand Reinhold 1981.

2
2.1   Barrett, C. S.; Massalski, T. B.: *Structure of Metals*. 3rd Edn. Oxford: Pergamon 1980.
2.2   Cullity, B. D.: *Elements of X-ray Diffraction*. Reading/Mass.: Addison-Wesley 1956.
2.3   Wassermann, G.: *Praktikum der Metallkunde und Werkstoffprüfung*. Berlin, Heidelberg, New York: Springer 1965.
2.4   Marton, L.: *Methods of Experimental Physics, vol. 6A/B: Solid State Physics*. New York: Academic Press 1959.
2.5   Chalmers, B.; Quarrell, A. G.: *The Physical Examination of Metals*. 2nd Edn. London: William Clowes 1960.
2.6a   Exner, H. E.: Ch. 10A in: Cahn, R. W.; Haasen, P. (eds.): *Physical Metallurgy*. 3rd Edn. Amsterdam: North Holland 1983.
2.6b   Brandon, D. G.: *Modern Techniques in Metallography*. London: Butterworths 1966.
2.6c   Freund, H. (ed..): *Handbuch der Mikroskopie in der Technik, Bd. II, Mikroskopie der metallischen Werkstoffe*. Frankfurt: Umschau-Verlag 1968, 1969.
2.6d   de Hoff, R. T.; Rhines, F. N. (ed.): *Quantitative Microscopy*. New York: McGraw-Hill 1968.
2.7   Hornbogen, E.: *Durchstrahlungs-Elektronenmikroskopie fester Stoffe*. Weinheim: Verlag Chemie 1971.
2.8   von Heimendahl, M.: *Einführung in die Elektronenmikroskopie*. Braunschweig: Vieweg 1970.
2.9   Schimmel, G.: *Elektronenmikroskopische Methodik*. Berlin, Heidelberg, New York: Springer 1969.
2.10   Alexander, H.: *Z. f. Metalldke.*, **51** (1960), 202.
2.11   Amelinckx, S.; Dekeyser, W.: *Solid State Physics*, **8** (1959), 327.

2.12   Gerold, V.: in *Erg. exakt. Naturw.*, **33** (1961), 105.
2.13   Guinier, A.: *Théorie et technique de la radiocristallographie*. 2e édition. Paris: Dunod 1956.
2.14   Müller, E. W.; Tsong, T. T.: *Field Ion Microscopy: Principles and Applications*. Amsterdam: Elsevier 1969.
2.15   Brenner, S. S.; McKinney, J. T.: *Surface Science*, **23** (1970), 88.
2.16   Predel, B.: Habilitationsschrift Münster, 1963.
2.17   Hart, E. W.: *Acta Met.*, **15** (1967), 351.
2.18   Cottrell, A. H.: *The Mechanical Properties of Matter*. New York: J. Wiley 1964.
2.19   Siebel, E.: *Handbuch d. Werkstoffprüfung, Bd. II*. Berlin: Springer 1939.
2.20   Zener, Cl.: *Elasticity and Anelasticity of Metals*. Chicago: Univ. Press 1948; see also: Nowick, A. S.; Berry, B. S.: *Anelastic Relaxation in Crystalline Solids*. New York: Academic Press 1972.
2.21   Gonser, U.: *Z. f. Metallkde.*, **57** (1966), 85; see also: Keune, W.; Trautwein, A.: *Metall*, **25** (1971), 27.
2.22   Frauenfelder, H.: *Mößbauer Effect*. New York: Benjamin 1962.
2.23   Wegener, H.: *Der Mößbauereffekt und seine Anwendungen*. Mannheim: Bibliogr. Inst. 1965.
2.24   Bethge, H.; Heidenreich, J. (Hrsg.: *Elektronenmikroskopie in der Festkörperphysik*. Berlin: Dt. Verlag der Wiss. 1982.
2.25   Spence, J. C. H.: *Experimental High-resolution Electron Microscopy*. Oxford: Clarendon Press 1981.
2.26   Cockayne, D. J. H.; Ray, J. L. F.; Whelan, M. J.: *Phil. Mag.*, **20** (1969), 1265.
2.27   Wagner, R.: *Crystals*, vol. 6. Berlin: Springer 1982.
2.28   Zhu, F.; Wendt, H.; Haasen, P.: *Scripta Met.*, **16** (1982), 1175.
2.29   Guinier, A.; Fournet, G.: *Small Angle Scattering of X-rays*. New York: J. Wiley 1955.
2.30   Kostorz, G.: *Small Angle Scattering in Neutron Scattering*. New York: Academic Press 1979, p. 227.
2.31   Hirsch, P. B.; Howie, A.; Nicholson, R. B.; Pashley, D. W.; Whelan, M. J.: *Electron Microscopy of Thin Crystals*. London: Butterworth 1965.

3
3.1   Hornbogen, E.: *Prakt. Metallogr.*, **5** (1968), 51.
3.2   American Soc. for Metals: *Metal Interfaces*. Cleveland 1952.
3.3   Gleiter, H.: Habil. Schrift Bochum 1971; see also: Weins, M. J. in: Chaudhari, P.; Matthews, J. W.: *Grain Boundaries and Interfaces*. Amsterdam: North Holland 1972, p. 138.
3.4   Brandon, D. G.: *Acta Met.*, **12** (1964), 813; **14** (1966), 1479.
3.5   Gleiter, H.: *Acta Met.*, **17** (1969), 565.
3.6   McLean, D.: *Grain Boundaries in Metals*. Oxford: Clarendon 1957.
3.7   Guy, A. G.: *Introduction to Materials Science*. New York: McGraw-Hill 1971.
3.8   Hermann, G.; Gleiter, H.; Bäro, G.: *Acta Met.*, **24** (1976), 353.

4
4.1   Chalmers, B.: *Principles of Solidification*. New York: J. Wiley 1964.
4.2   Tiller, W.: Ch. 9 in: Cahn, R. W.: *Physical Metallurgy*. 2nd Edn. Amsterdam: North Holland 1970.
4.3   Laudise, R. A.: *The Growth of Single Crystals*. Englewood Cliffs: Prentice Hall 1970.
4.4   Winegard, W. C.: *An Introduction to the Solidification of Metals*. London: Institute of Metals 1964.
4.5   Pfann, W. G.: *Zone Melting*. New York: J. Wiley 1958.
4.6   Chadwick, G. A.: *Progr. Mater. Science*, **12** (1964), 97; see also: *Metallography of Phase Transformations*. London: Butterworths 1972.
4.7   Kurz, W.; Sahm, P. R.: *Gerichtet erstarrte eutektische Werkstoffe*. Berlin, Heidelberg, New York: Springer 1975.

4.8 Jackson, K. A.; Hunt, T.: *Trans. AIME*, **236** (1966), 1129.
4.9 Polk, D.: *Acta Met.*, **20** (1972), 485.
4.10 Piller, J.; Haasen, P.: *Acta Met.*, **30** (1982), 1.
4.11 Wagner, R.; Gerling, R.; Schimansky, F. P.: *Scripta Met.*, **17** (1983), 203.
4.12 Luborsky, F. E.: *Amorphous Metallic Alloys*. London: Butterworth 1983.
4.13 Jones, H.: *Rapid Solidification of Metals and Alloys*. London: Inst. of Metallurgists 1982.
4.14 Güntherodt, H.-J.; Beck, H. (eds.): *Glassy Metals*. Berlin: Springer; vol. I: 1981; vol. II: 1982.
4.15 Häussler, P.: *Z. Phys.*, **B53** (1983), 15.

5
5.1 Gaskell, D. R.: Ch. 6 in: Cahn, R. W.; Haasen, P. (eds.): *Physical Metallurgy*. 3rd Edn. Amsterdam: North Holland 1983.
5.2 Darken, L. S.; Gurry, R. W.: *Physical Chemistry of Metals*. New York: McGraw-Hill 1953.
5.3 Becker, R.: *Theorie der Wärme*. Berlin, Göttingen, Heidelberg: Springer 1955.
5.4 Schmalzried, H.: *Festkörperthermodynamik*. Weinheim: Verlag Chemie 1974.
5.5 McLellan, R. B.: *J. Mater. Sci. Eng.*, **9** (1972), 122.
5.6 Wagner, C.: *Thermodynamics of Alloys*. London: Addison-Wesley 1952.
5.7 Prince, A.: *Alloy Phase Equilibria*. Amsterdam: Elsevier 1966.
5.8 Hansen, A.; Anderko, K.: *Constitution of Binary Alloys*. 2nd Edn. New York: McGraw-Hill 1958.
5.9 Elliot, R.: *Constitution of Binary Alloys*, First. Suppl. New York: McGraw-Hill 1965.
5.10 Shunk, F. A.: *Constitution of Binary Alloys*, Secd. Suppl. New York: McGraw-Hill 1969.
5.10a Moffat, W. G.: *Handbook of Binary Phase Diagrams*. Schenectady: General Electric Co. 1981 ff.
5.11 Petzow, G.; Lukas, H. L.: *Z. f. Metallkde.*, **61** (1970), 877.
5.12 Hoffman, D. W.: *Met. Trans.*, **3** (1972), 3231.
5.13 Bennett, L. H.; Massalski, T. B.; Giessen, B. C. (eds.): *Alloy Phase Diagrams*. New York: North Holland 1983.

6
6.1 Hornbogen, E.: Ch. 16 in: Cahn, R. W.; Haasen, P. (eds.): *Physical Metallurgy*. 3rd Edn. Amsterdam: North Holland 1983.
6.2 Zener, Cl.: Ch. 1.2 in: Rudman, P. S.; Stringer, J.; Jaffe, R. I.: *Phase Stability in Metals and Alloys*. New York: McGraw-Hill 1967.
6.3 Heine, V.; Weaire, D.: *Solid State Physics*, **24** (1970), 250.
6.4 Kaufman, L.: *Progr. Mater. Sci.*, **14** (1969), 57; see also: Kaufman, L.; Bernstein, H.: *Computer Calculation of Phase Diagrams*. New York: Academic Press 1970.
6.5 Kaufman, L.: Ch. 2.3 in: Rudman, P. S.; Stringer, J.; Jaffe, R. I.: *Phase Stability in Metals and Alloys*. New York: McGraw-Hill 1967.
6.6 Pettifor, D. G.: *J. Phys.*, **C3** (1970), 366.
6.7a Leibfried, G.: *Handb. Phys.*, VII, 1. Berlin, Göttingen, Heidelberg: Springer 1955, S. 104.
6.7b Friedel, J.: *J. de Phys.*, **35** (1974), L59.
6.8 Dehlinger, U.: *Theoretische Metallkunde*. Berlin, Göttingen, Heidelberg: Springer 1955.
6.9 Massalski, T. B.: Ch. 4 in: Cahn, R. W.; Haasen, P. (eds.): *Physical Metallurgy*. 3rd Edn. Amsterdam: North Holland 1983.
6.10a Stroud, D.; Ashkroft, N. W.: *J. Phys.*, **F1** (1971), 113.
6.10 Blandin, A.: Ch. 2.2 in: Rudman, P. S.; Stringer, J.; Jaffe, R. I.: *Phase Stability in Metals and Alloys*. New York: McGraw-Hill 1967.
6.11 King, H. W.: in: Massalski, T. B. (ed.): *Alloying Behaviour and Effects in Concentrated Solid Solutions*. New York: Gordon and Breach 1965, p. 85.

6.12   Girgis, K.: Ch. 5 in: Cahn, R. W.; Haasen, P. (eds.): *Physical Metallurgy*. 3rd
        Edn. Amsterdam: North Holland 1983.
6.13   Laves, F.: Ch. 8 in: Westbrook, J. H.: *Intermetallic Compounds*. New York: J.
        Wiley 1967.
6.14   Miedema, A. R.; de Boer, E. R.; de Chatel, P. F.: *J. Phys.*, **F3** (1973), 1558.
6.15   Pettifor, D. G.: Ch. 3 in: Cahn, R. W.; Haasen, P. (eds.): *Physical Metallurgy*.
        Amsterdam: North Holland 1983.
6.16   Shechtman, D.; Blech, I.; Gratias, D.; Cahn, J. W.: *Phys. Rev. Lett.*, **53** (1984),
        1951.

7
7.1   Sato, H.; Toth, R. S.: in: Massalski, T. B. (ed.): *Alloying Behaviour and Effects
        in Concentrated Solid Solutions*. New York: Gordon and Breach 1965, p. 295.
7.2   Cohen, J. B.: Ch. 13 in: *Phase Transformations*. Metals Park, Ohio: Amer. Soc.
        for Metals 1970.
7.3   Sato, H.: Ch. 10 in: Eyring, H.; Henderson, D.; Jost, W. (eds.): *Physical
        Chemistry*, vol. X. New York: Academic Press 1970.
7.4   Clapp, P. C.; Moss, S. C.: *Phys. Rev.*, **171** (1968), 764.
7.5   Inden, G.; Pitsch, W.: *Z. f. Metallkde.*, **62** (1971), 627; **63** (1972), 253.
7.6   Marcinkowski, M. J.: in: Thomas, G.; Washburn, J. (eds.): *Electron Microscopy
        and Strength of Crystals*. New York: Interscience 1963, p. 333.
7.7   Rudman, P. S.: Ch. 21 in: Westbrook, J. H.: *Intermetallic Compounds*. New
        York: J. Wiley 1967.
7.8   Sauthoff, G.: *Acta Met.*, **21** (1973), 273.
7.9   Marcinkowski, M. J.; Brown, N.: *J. Appl. Phys.*, **33** (1962), 537.
7.10   Clapp, P. C.: in Kear, B. H.; *et al.* (eds.): *Ordered Alloys*. Baton Rouge:
        Claytor's Publ. Div., p. 25; see also Rudman, P. S.: p. 37, l. c.
7.11   Pearson, W. B.: *A Handbook of Lattice Spacings and Structures of Metals and
        Alloys*. London: Pergamon Press; vol. 1: 1958; vol. 2: 1967.
7.12   Marcinkowski, M. J.; Brown, N.: *J. Appl. Phys.*, **32** (1961), 375.
7.13   de Fontaine, D.: *Acta Met.*, **23** (1975), 553.

8
8.1   Shewmon, P. G.: *Diffusion in Solids*. New York: McGraw-Hill 1963.
8.2   Crank, J.: *The Mathematics of Diffusion*. Oxford: Clarendon Press 1967.
8.3   Manning, J. R.: *Diffusion Kinetics for Atoms in Crystals*. Princeton: van
        Nostrand 1968.
8.4   Lazarus, D.: *Solid State Physics*, **10** (1960), 71.
8.5   Schmalzried, H.: *Festkörperreaktionen*. Weinheim: Verlag Chemie 1971.
8.6   Adda, Y.; Philibert, J.: *La diffusion dans les solides*. Paris: Presses Univ. de
        France 1966.
8.7   Queré, Y.: *Défauts ponctuels dans les métaux*. Paris: Masson 1967.
8.8   Balluffi, R. W.: *Phys. Stat. Sol.*, **42** (1970), 11.
8.9   Mullins, W. W.: *J. Appl. Phys.*, **30** (1959), 77.
8.10   Bonzel, H. P.; Gjostein, N. A.: *J. Appl. Phys.*, **39** (1968), 3480.
8.11   Kuczynski, G.: *Acta Met.*, **4** (1956), 58.
8.12   Thümmler, F.; Thomma, W.: *Met. Revs.*, **XII** (1967), 69.
8.13   Hehenkamp, Th.: in: Seeger, A.; *et al.* (eds.): *Vacancies and Interstitials in
        Metals*. Amsterdam: North Holland 1970, p. 91.
8.14   Hauffe, K.: *Reaktionen in und an festen Stoffen*. Berlin, Göttingen, Heidelberg:
        Springer 1955; see also: *Oxidation von Metallen und Metallegierungen*. Berlin,
        Göttingen, Heidelberg: Springer 1956.
8.15   Rice, S.: *Phys. Rev.*, **112** (1958), 804; see also: Vineyard, G. H.: *J. Phys. Chem.
        Sol.*, **3** (1957), 121.
8.16   Le Claire, A. D.: Ch. 5 in: Eyring, H.; *et al.* (eds.): *Physical Chemistry*, vol. X.
        New York: Academic Press 1970.
8.17   Wever, H.: *Elektro- und Thermotransport in Metallen*. Leipzig: J. Ambr. Barth
        1973.

8.18   Gerl, M.: in: *Atomic Transport in Solids and Liquids*. Tübingen: Verlag Z. f. Naturforschg. 1971, S. 9.
8.19   Ashby, M. F.: *Acta Met.*, **22** (1974), 275.
8.20   Cantor, B.; Cahn, R. W.: in [4.12].

9
9.1   Fine, M. E.: *Introduction to Phase Transformations in Condensed Systems*. New York: McMillan 1964.
9.2   Hilliard, J. E.: Ch. 12 in: *Phase Transformations*. Metals Park, Ohio: American Soc. for Metals 1970.
9.3   Kahlweit, M.: Ch. 11 in: Eyring, H.; *et al.* (eds.): *Physical Chemistry*, vol. X. New York: Academic Press 1970.
9.4   Servi, I. S.; Turnbull, D.: *Acta Met.*, **14** (1966), 161.
9.5   Hornbogen, E.: *Aluminium*, **43** (1967), 115.
9.6   Christian, J. W.: *The Theory of Transformations in Metals and Alloys*. Oxford: Pergamon Press 1965, 2nd Edn., part I, 1975.
9.7   Livingston, J. D.: *Trans. AIME*, **215** (1959), 566.
9.8   Cahn, J. W.: *Trans. AIME*, **242** (1968), 166.
9.9   Cahn, J. W.: *Acta Met.*, **7** (1959), 18.
9.10   Brown, L. M.; Cook, R. H.; Ham, R. K.; Purdy, G. R.: *Scripta Met.*, **7** (1973), 815.
9.11   Cook, H. E.: *Acta Met.*, **18** (1970), 297.
9.12   Lee, Y. W.; Aaronson, H. I.: *Acta Met.*, **28** (1980), 539.
9.13   Wendt, H.; Haasen, P.: *Acta Met.*, **31** (1983), 1649.
9.14   Langer, J. S.; Schwartz, A. J.: *Phys. Rev.*, **A21** (1980), 948.
9.15   Wendt, H.; Liu, Z.; Haasen, P.: *Proc. Intern. Conf. on Early Stages of Decompos. of Alloys*, Sonnenberg 1983. Oxford: Pergamon 1984; Kampmann, R.; Wagner, R.: *ibid*.
9.16   Langer, J. S.: in: Riste, T. (ed.): *Fluctuation Instabilities and Phase Transitions*. New York: Plenum Press 1975, p. 19.
9.17   Zhu, F.; Wendt, H.; Haasen, P.: [as 9.15].
9.18   Biehl, E.; Wagner, R.: in: Aaronson, H. I. (ed.): *Proc. Int. Meet. on Phase Trans*. Pennsylvania: AIME 1982, p. 185.
9.19   v. Alvensleben, L.; Wagner, R.: [as 9.15].
9.20   Massalski, T. B.: in: *Rapidly Quenched Metals IV*. Sendai: Japan Inst. Met. 1981, p. 203.
9.21   Wendt, H.; Haasen, P.: *Scripta Met.*, **19** (1985), 1053.

10
10.1   Seeger, A.: *J. Phys.*, **F3** (1973), 248.
10.2   Seeger, A.; *et al.* (eds.): *Vacancies and Interstitials in Metals*. Amsterdam: North Holland 1970.
10.2a   Hehenkamp, Th.: *Progr. Mater. Sci.*, in preparation.
10.3a   Wollenberger, H.: in: Cahn, R. W.; Haasen, P. (eds.): *Physical Metallurgy*. 3rd Edn. Amsterdam: North Holland 1983.
10.3   Damask, A. C.; Dienes, G. J.: *Point Defects in Metals*. New York: Gordon and Breach 1963.
10.4   Girifalco, L. A.; Hermann, H.: *Acta Met.*, **13** (1965), 583.
10.5   Berger, A. S.; Seidman, D. N.; Balluffi, R. W.: *Acta Met.*, **21** (1973), 123.
10.6   Leibfried, G.: *Bestrahlungseffekte in Festkörpern*. Stuttgart: Teubner 1965.
10.7   Thompson, M. W.: *Defects and Radiation Damage in Metals*. Cambridge: Univ. Press 1969.
10.8   Chadderton, L. T.: *Radiation Damage in Crystals*. London: Methuen 1965.
10.9   Diehl, J.: Ch. 5 in: Seeger, A. (ed.): *Moderne Probleme d. Metallphysik*, Bd. 1. Berlin, Heidelberg, New York: Springer 1965.
10.10   Eshelby, J. D.: *J. Appl. Phys.*, **25** (1954), 255.
10.11   Diehl, J.; Diepers, H.; Hertel, B.: *Canad. J. Phys.*, **46** (1968), 647.

10.12   Ehrhardt, P.; Schilling, W.: *Phys. Rev.*, **B8** (1973), 2604; see also: Ehrhardt, P.: *Report KFA Jülich 810 FF* (1971).
10.13   Bullough, R.; Lidiard, A. B.: *Comm. on Solid St. Phys.*, **4** (1972), 69; Seeger, A.: l. c., p. 79.
10.14   Corbett, J. W.; *et al.* (eds.): *Radiation-induced Voids in Metals*. Washington: US Atom. Energy Comm. 1972.

11
11.1   Weertman, J.; Weertman, J. R.: *Elementary Dislocation Theory*. New York: Macmillan 1967.
11.2   Friedel, J.: *Dislocations*. Oxford: Pergamon Press 1964.
11.3   Hirth, J. P.; Lothe, J.: *Theory of Dislocations*. New York: McGraw-Hill 1968; 2nd Edn. 1982.
11.4   Nabarro, F. R. N.: *Theory of Crystal Dislocations*. Oxford: Clarendon Press 1967.
11.5   Seeger, A.: in: *Handbuch d. Phys.*, Bd. VII, 1. Berlin, Göttingen, Heidelberg: Springer 1955, S. 1.
11.6   Seeger, A.; Haasen, P.: *Phil. Mag.*, **3** (1958), 470.
11.7   Alexander, H.; Haasen, P.: *Solid State Phys.*, **22** (1968), 28.
11.8   Kröner, E.: *Kontinuumstheorie der Versetzungen und Eigenspannungen*. Berlin, Göttingen, Heidelberg: Springer 1959.
11.9   Seeger, A.; Schiller, P.: Ch. 8 in: Mason, W. P.: *Physical Acoustics*, vol. IIIA. New York: Academic Press 1966.
11.10   American Soc. for Metals: *Proc. 2nd Intern. Conf. on Strength of Metals and Alloys*. Asilomar, Cleveland: ASM 1970.
11.10a   Seeger, A.; Wüthrich, Ch.: *Nuov. Cim.*, *33B* (1976), 38.
11.11   Granato, A. V.: *Phys. Rev.*, **B4** (1971), 2196.
11.11a   Schwarz, R. B.; Labusch, R.: *J. Appl. Phys.*, **49** (1978), 5174.
11.12   Haasen, P.: Ch. 2 in: Eyring, H.; *et al.* (eds.): *Physical Chemistry*, vol. X. New York: Academic Press 1970.
11.13   Neuhäuser, H.: in: Nabarro, F. R. N. (ed.): *Dislocations in Solids*, vol. 6. Amsterdam: North Holland 1983.

12
12.1   Schmid, E.; Boas, W.: *Kristallplastizität*. Berlin: Springer 1935.
12.2   Seeger, A.: *Handb. d. Physik*, Bd. VII, 2. Berlin, Göttingen, Heidelberg: Springer 1958, S. 1; see also: Seeger, A. (ed.): *Moderne Probleme der Metallphysik*, Bd. 1. Berlin, Göttingen, Heidelberg: Springer 1965.
12.3   Nabarro, F. R. N.; Basinski, Z. S.; Holt, D. B.: *Adv. in Phys.*, **13** (1964), 192.
12.4   Vreeland, T.: in: Rosenfield, A. R.; *et al.* (eds.): *Dislocation Dynamics*. New York: McGraw-Hill 1968, p. 529.
12.5   Haasen, P.: Ch. 21 in: Cahn, R. W.; Haasen, P. (eds.): *Physical Metallurgy*. 3rd Edn. Amsterdam: North Holland 1983.
12.6   Weertman, J.; Weertman, J. R.: Ch. 20 in: Cahn, R. W.; Haasen, P. (eds.): *Physical Metallurgy*. 3rd Edn. Amsterdam: North Holland 1983.
12.7   Kelly, A.: *Strong Solids*. Oxford: Clarendon Press 1966, 1973.
12.8   Eshelby, J. D.: *Proc. Roy. Soc.*, **A241** (1957), 376.
12.9   Kocks, U. F.: *Trans. AIME*, **1** (1970), 1121.
12.10   Cottrell, A. H.: *Dislocations and Plastic Flow in Crystals*. Oxford: Clarendon Press 1953.
12.11   Wassermann, G.; Grewen, J.: *Texturen metallischer Werkstoffe*. 2nd Edn. Berlin, Göttingen, Heidelberg: Springer 1962.
12.11a   Bunge, H.-J. (ed.): *Quantitative Texture Analysis*. Oberursel: DGM 1981.
12.12   Raj, R.; Ashby, M. F.: *Trans. AIME*, **2** (1971), 1113.
12.13   Ashby, M. F.: *Acta Met.*, **20** (1972), 887.
12.14   Stüwe, H. P.: *Z. f. Metallkde.*, **61** (1970), 704.
12.14a   Langdon, T. G.: *Trans. AIME*, **13A** (1982), 689.

12.14b   Arieli, A.; Mukherjee, A. K.: *Trans. AIME*, **13A** (1982), 717.

12.15   Ashby, M. F.; Verrall, R. A.: *Acta Met.*, **21** (1973), 149.

12.16   McLean, D.: *Mechanical Properties of Metals*, New York: J. Wiley 1962.

12.17   Grosskreutz, J. C.: *Phys. Stat. Sol.*, **47** (1971), 11, 359.

12.18   Munz, D.; Schwalbe, K.; Mayr, P.: *Dauerschwingverhalten metallischer Werkstoffe*. Braunschweig: Vieweg 1971.

12.18a   Mughrabi, H.: *Mat. Sci. Engg.*, **33** (1978), 207.

12.18b   Neumann, P.; Hunsche, A.: *Acta Met.*, **34** (1986), 207.

12.19   Neumann, P.: *Acta Met.*, **17** (1969), 1219.

12.20   Kochendörfer, A.: *Z. f. Metallkde.*, **62** (1971), 1, 71, 173, 255.

12.21   Cottrell, A. H.: *Trans. AIME*, **212** (1958), 192.

12.22   Thomson, R. M.: Ch. 23 in: Cahn, R. W.; Haasen, P. (eds.): *Physical Metallurgy*. 3rd Edn. Amsterdam: North Holland 1983.

12.23   Tetelman, A. S.; McEvily, A. J.: *Fracture of Structural Materials*. New York: J. Wiley 1967.

12.24   Ashby, M. F.; *Phil. Mag.*, **21** (1970), 399.

12.25   Thompson, A. W.: *Work Hardening in Tension and Fatigue*. New York: AIME 1977, p. 89.

12.26   Estrin, Y.; Mecking, H.: *Acta Met.*, **32** (1984), 57.

13

13.1   Reed-Hill, R. E.; *et al.* (eds.): *Deformation Twinning*. New York: Gordon and Breach 1964.

13.1a   Sleeswyk, A. W.: *Phil. Mag.*, **29** (1974), 407.

13.2   Peissker, E.: *Z. f. Metallkde.*, **56** (1965), 155.

13.3   Seeger, A.: *Z. f. Metallkde.*, **44** (1953), 247; **47** (1956), 653.

13.4   Liebermann, D. S.: Ch. 1 in: *Phase Transformations*. Metals Park, Ohio: American Soc. for Metals 1970.

13.5   Wayman, C. M.: *Introduction to the Crystallography of Martensitic Transformations*. New York: Macmillan 1964.

13.6   Pitsch, W.: *Archiv Eisenh. Wes.*, **30** (1959), 503.

13.7   Frank, F. C.: *Acta Met.*, **1** (1953), 15.

13.8   Bowles, J. S.; Mackenzie, J. D.: *Acta Met.*, **2** (1954), 129, 138, 224.

13.9   Wayman, C. M.; Shimizu, K.: *Metals Sci. J.*, **6** (1972), 175.

13.10   Cohen, M.: *Met. Trans.*, **3** (1972), 1095; see also: Raghavan, V.; Cohen, M.: *Acta Met.*, **20** (1972), 333.

13.10a   Olson, G. B.; Cohen, M.: in: Aaronson, H. J.; *et al.* (eds.): *Solid–Solid Phase Transformation*. New York: AIME 1982, p. 1145.

13.11   Cech, R. E.; Turnbull, D.: *Trans. AIME*, **194** (1952), 489.

13.12   Easterling, J.; Miekk-Oja, H. M.: *Acta Met.*, **15** (1967), 1133.

13.12a   Cornelis, I.; Oshima, R.; Tong, H. C.; Wayman, C. M.: *Scripta Met.*, **8** (1974), 133.

13.13   Hornbogen, E.: Ch. 16 in: Cahn, R. W.; Haasen, P. (eds.): *Physical Metallurgy*. 3rd Edn. Amsterdam: North Holland 1983.

13.14   Wayman, C. M.: Ch. 15 in: Cahn, R. W.; Haasen, P. (eds.): *Physical Metallurgy*. 3rd Edn. Amsterdam: North Holland 1983.

13.15   Sikka, S. K.; Vohra, Y. K.; Chidambaram, R.: *Progr. Mat. Sci.*, **27** (1982), 245.

14

14.1   Fleischer, R. L.; in: Peckner, D.: *The Strengthening of Metals*. London: Reinhold 1964, p. 93.

14.2   Haasen, P.: Ch. 15a in: Nabarro, F. R. N.: *Dislocations in Solids*. Amsterdam: North Holland 1979.

14.3   Cochhardt, A. W.; Schöck, G.; Wiedersich, H.: *Acta Met.*, **3** (1955), 533.

14.4   Labusch, R.: *Phys. Stat. Sol.*, **41** (1970), 659; *Acta Met.*, **20** (1971), 917; see also: Labusch, R.; Ahearn, J.; Grange, G.; Haasen, P. in: *Rate Processes in Plastic Deformation*. Cleveland: American Soc. for Metals 1974.

14.4*a*   Labusch, R.: *Cz. J. Phys.*, **B31** (1981), 165.
14.5   Basinski, Z. S.; Foxall, R. A.; Pascual, R.: *Scripta Met.*, **6** (1972), 807.
14.5*a*   Kröner, E.: *Phys. Kond. Mat.*, **2** (1964), 262.
14.6   Hart, E. W.: *Acta Met.*, **1** (1955), 146.
14.7   Schöck, G.; Seeger, A.: *Acta Met.*, **7** (1959), 469.
14.8   Brion, H. G.; Haasen, P.; Siethoff, H.: *Acta Met.*, **19** (1971), 283.
14.9   Thompson, D. O.; Paré, V. K.: Ch. 7 in: Mason, W. P.: *Physical Acoustics*, vol. II. New York: Academic Press 1966.
14.10   Lücke, K.; Granato, A. V.: in: Fisher, J. C.; *et al.* (eds.): *Dislocations and Mechanical Properties of Crystals.* New York: J. Wiley 1957, p. 425.
14.11   Haasen, P.: *Nachr. Gött. Akad. Wiss.*, **6** (1970); *Contemp. Phys.*, **18** (1977), 373.
14.12   Brown, L. M.; Ham, R. K.: Ch. 2 in: Kelly, A.; *et al.* (eds.): *Strengthening Methods in Crystals.* Amsterdam: Elsevier 1971.
14.13   Ashby, M. F.: Ch. 3 in: Kelly, A.; *et al.* (eds.): *Strengthening Methods in Crystals.* Amsterdam: Elsevier 1971.
14.14   Kelly, A.: Ch. 8 in: *Strengthening Methods in Crystals.* Amsterdam: Elsevier 1971.
14.15   Kelly, A.; Lilholt, H.: *Phil. Mag.*, **20** (1969), 311.
14.16   Neumann, P.; Haasen, P.: *Phil. Mag.*, **23** (1971), 285.
14.17   Wassermann, G.; Wahl, H. P.: *Z. f. Metallkde.*, **61** (1970), 326.
14.18   Melander, A.; Persson, P. A.: *Acta Met.*, **26** (1978), 267.
14.19   Wendt, H.; Wagner, R.: *Acta Met.*, **30** (1982), 1561.

15
15.1   Cahn, R. W.: Ch. 25 in: *Physical Metallurgy.* 3rd Edn. Amsterdam: North Holland 1983.
15.2   Clarebrough, L. M.; Hargreaves, M. E.; Loretto, M. H.: in: Himmel, L. (ed.): *Recovery and Recrystallization of metals.* New York: Interscience 1963, p. 63.
15.3   Hibbard, W. R.; Dunn, C. G.: in: *Creep and Recovery.* Cleveland: American Soc. for Metals 1957, p. 52.
15.4   Rutter, J. W.; Aust, K. T.: *Acta Met.*, **13** (1965), 181; see also: *Creep and Recovery.* Cleveland: American Soc. for Metals 1957, p. 131.
15.5   Rath, B. B.; Hu, H.: in: Hu, H.: *The Nature and Behaviour of Grain Boundaries.* New York: Plenum Press 1972, p. 405.
15.6   Gleiter, H.; Chalmers, B.: *Progr. Mater. Sci.*, **16** (1972).
15.7   Gleiter, H.: *Acta Met.*, **17** (1969), 853.
15.8   in der Schmitten, H.; Haasen, P.; Haeszner, F.: *Z. f. Metallkde.*, **51** (1960), 101.
15.9   Lücke, K.; Rixen, R.; Rosenbaum, F. W.: in: Hu, H.: *The Nature and Behaviour of Grain Boundaries.* New York: Plenum Press 1972, p. 245.
15.10   Gordon, P.; Vandermeer, R. A.: Ch. 6 in: *Recrystallization, Grain Growth and Texture.* Metals Park, Ohio: American Soc. for Metals 1966.
15.11   Hornbogen, E.; Kreye, H.: in: Grewen, J.; Wassermann, G. (eds.): *Texturen in Forschung und Praxis.* Berlin, Heidelberg, New York: Springer 1969, S. 274.
15.12   Beck, P. A.; Hu, H.: Ch. 9 in: *Recrystallization, Grain Growth and Texture.* Metals Park, Ohio: American Soc. for Metals 1966.
15.13   Burgers, W. G.: Ch. 22 in: Gilman, J. J. (ed.): *The Art and Science of Growing Crystals.* New York: J. Wiley 1963.
15.14   Hu, H.: in: Grewen, J.; Wassermann, G. (eds.): *Texturen in Forschung und Praxis.* Berlin, Heidelberg, New York: Springer 1969, p. 200.
15.15   Walter, J. L.: in: Grewen, J.; Wassermann, G. (eds.): *Texturen in Forschung und Praxis.* Berlin, Heidelberg, New York: Springer 1969, p. 227.
15.16   Hillert, M.: *Acta Met.*, **13** (1965), 227.
15.17   Wilbrandt, P.-J.; Haasen, P.: *Z. f. Metallkde.*, **71** (1980), 385; Berger, A.; Wilbrandt, P.-J.; Haasen, P.: *Acta Met.*, **31** (1983), 1433.

INDEX